EXPLORING MATHEMATICS

WITH YOUR

COMPUTER

by

Arthur Engel

University of Frankfurt

35

THE MATHEMATICAL ASSOCIATION OF AMERICA

NEW MATHEMATICAL LIBRARY

NEW MATHEMATICAL LIBRARY

published by

The Mathematical Association of America

The New Mathematical Library (NML) was begun in 1961 by the School Mathematics Study Group to make available to high school students short expository books on various topics not usually covered in the high school syllabus. In three decades the NML matured into a steadily growing series of some thirty titles of interest not only to the originally intended audience, but to college students and teachers at all levels. Previously published by Random House and L. W. Singer, the NML became a publication series of the Mathematical Association of America (MAA) in 1975. Under the auspices of the MAA the NML will continue to grow and will remain dedicated to its original and expanded purposes.

Typeset in T$_E$X by NYU's Academic Computing Facility

©1993 by
The Mathematical Association of America

Library of Congress Catalog Card Number 92-64177
Complete Set ISBN 0-88385-600-X
Vol. 35 ISBN 0-88385-639-5

Manufactured in the United States of America

Preface

A computer is a powerful instrument for performing numerical experiments and simulations on an unprecedented scale. Many young mathematicians now use it for research from the very beginning. The computer-generated data lead to clues, then to solutions and can even suggest elegant proofs. Good examples are provided by the short final Section 65, which solves a difficult problem and by the very tough Problem 17, due to Euler, in Exercises to Sections 17-18; the computer makes it accessible.

This is a mathematics-, not a programming book. It is intended for students, mathematics teachers and mathematicians who are just starting to explore mathematics on their own computers. In studying it, and especially in working through its exercises, readers will get to know many new, elementary topics and learn as much from the extensive exercises as from the examples.

We use Turbo Pascal. We need only a fragment of this dialect, easily picked up by readers as they work their way through the examples and exercises. There is little difference between dialects of Pascal, so you will need only slight modifications for a different Pascal implementation. Most of our programs are complete. Each serves as a reading exercise to help you learn Pascal; for this reason the programs are short and, for the most part, comprehensible without comment. Yet, many are not trivial!

The book contains 65 topics, loosely structured into seven chapters; these, and even the individual topics, are independent of each other. If you know the rudiments of Pascal you can tackle them in any order; if not, you are advised to cover them in sequence since new features of Pascal are introduced when needed.

The book originated in an inservice course for high school mathematics teachers, and a seminar *Mathematical Discoveries with a PC*. The idea was to give teachers interesting and challenging material for use in their classes; teachers had complained that they had to learn a language by programming trivial examples. Boredom had prevented effective learning!

I thank Nina Yuan, then a high school student, who caught many errors and made many useful comments. I thank Dr.Stephan Rosebrock for his crucial contribution and technical help in transmitting e-mail between Frankfurt and New York. He overcame every difficulty that arose. My special thanks go to NML-editors Anneli Lax and Peter Ungar. Peter's contribution is huge. Our correspondence has grown to nearly the size of this book; his detailed suggestions changed almost every page. In a final check, Eric Boesch spotted many subtle errors and made a number of substantial improvements.

Arthur Engel, 1993

Contents

Chapter 6   Numerical Algorithms

Chapter 7 Miscellaneous Problems

CHAPTER 1

Introductory Problems

1. The Factorial: First Encounter with Recursion

The factorial function is defined on the nonnegative integers by means of

$$0! = 1, \qquad n! = n(n-1)!$$

This can be translated into the Pascal program in Fig. 1.1.

```
program factorial;
var n:integer;

function fac(u:integer):integer;
begin
  if  u<2 then fac:=1
  else fac:=u*fac(u−1)
end;

begin
  write('n=');  readln(n);
  writeln(fac(n));  readln
end.
```
Fig. 1.1

Each program must have a name, which must begin with a letter and may continue with letters or digits to any length. We used **factorial** as program name. Then come the variable declarations. Each variable must be declared by giving its *type* (**integer**, **real**, **boolean**, or some range like 0..255). The notation 0..255 stands for the set $\{0, 1, \ldots, 255\}$. Now comes the function definition. A function is a special subprogram, and it must have a name which differs from the program name. We used **fac**. The rest of the function heading,

$$\text{fac}(u : \text{integer}) : \text{integer};$$

tells the computer that the independent variable u is an integer and the value of the function is also an integer. The function fac is called *recursive* because the name fac occurs in the body of the definition. The last four lines beginning and ending with the brackets **begin** ... **end.** are the main program. You can find more about brackets in the Short Summary of Pascal. The first statement write('n=') tells the computer to write on the screen n=. The next statement readln(n) tells the computer to wait for you to type in a number n. Suppose you type in the number 5. Then the computer evaluates fac(5) and writes the result, 120, with carriage return. To compute fac(5) the computer uses the definition of the function fac. The *function call* fac(5) generates a computational process as in Fig. 1.2 or 1.3. Each recursive computational process consists of an expansion followed by a contraction. The expansion occurs as the process builds up a chain of deferred operations, in our case multiplications. The contraction occurs as the multiplications are actually performed.

$$
\begin{array}{ll}
\text{fac}(5) \\
5*\text{fac}(4) \\
5*4*\text{fac}(3) \\
5*4*3*\text{fac}(2) \\
5*4*3*2*\text{fac}(1) \\
5*4*3*2*1 \\
5*4*3*2 \\
5*4*6 \\
5*24 \\
120
\end{array}
$$

Fig. 1.2   Fig. 1.3

The fac-programs in Figs 1.4 to 1.9 introduce almost all *control structures* of Pascal. For the beginner it is well worth to study them in detail.

```
program faciter1; {n> 0}
var n,p,c:integer;
begin
  write('n='); readln(n);
  c:=n; p:=1;
  repeat
    p:=p*c; c:=c-1
  until c=0;
  writeln(p); readln
end.
```

Fig. 1.4 $n!$ by counting upwards.

```
program faciter2; {n>0}
var n,p,c:integer;
begin
  write('n='); readln(n);
  p:=1; c:=0;
  repeat
    c:=c+1; p:=p*c
  until c=n;
  writeln(p); readln
end.
```

Fig. 1.5 $n!$ by counting downwards.

```
program faciter3;
var  n,p,c:integer;
begin
  write('n=');  readln(n);
  c:=0;  p:=1;
  while c<n do
  begin c:=c+1;  p:=p*c
  end;
  writeln(n,'!=',p);
  readln
end.
```

Fig. 1.6. $n!$ by counting upwards.

```
program faciter4;
var n,p,c:integer;
begin
  write('n=');  readln(n);
  c:=n;  p:=1;
  while c>0 do
  begin
    p:=p*c;  c:=c-1
  end;
  writeln(n,'!=',p);
readln end.
```

Fig. 1.7. $n!$ by counting downwards.

```
program faciter5;
var i,n,p:integer;
begin
  p:=1;
  write('n=');
  readln(n);
  for i:=1 to n do p:=p*i;
  writeln(n,'!=',p);
readln end.
```

Fig. 1.8
$n!$ with a for-loop upwards.

```
program faciter6;
var i,n:integer; p:real;
begin
  write('n=');  readln(n);
  p:=1;
  for i:=n downto 1 do
    p:=p*i;
  writeln(n,'!=',p:0:0);
readln end.
```

Fig. 1.9
$n!$ with a for-loop downwards and
declaration of **p** as a real number. The
notation **p:0:0** will be explained soon.

The **readln** before the final **end** makes Turbo Pascal keep the result
on the screen until you press "Return". You could also view the result
without the **readln** by pressing AltF5.

2. Fibonacci's Sequence: a Two-Term Recursion

For a good account on this section see Vorob'ev in the bibliography.
The Fibonacci sequence is defined by means of the recursion

$$\text{fib}(0) = 0, \quad \text{fib}(1) = 1, \quad \text{fib}(n) = \text{fib}(n-1) + \text{fib}(n-2), \quad n \geq 2.$$

This definition translates into the program in Fig. 2.1.

The function definition and the main program are between the brackets
begin...end. Two statements are always separated by a semicolon (;)
and each program ends with a period (.). Suppose we type in for n the
number 5. Then the computer replaces the *formal parameter u* by the
actual parameter 5 and develops the tree in Fig. 2.2. Finally the value
fib(5) is returned to the place where fib(5) was first called.

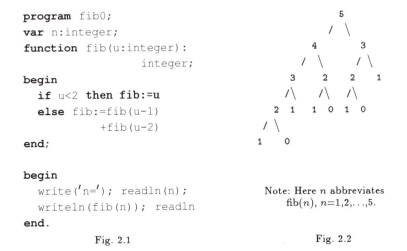

```
program fib0;
var n:integer;
function fib(u:integer):
                integer;
begin
  if u<2 then fib:=u
  else fib:=fib(u-1)
          +fib(u-2)
end;

begin
  write('n='); readln(n);
  writeln(fib(n)); readln
end.
```

Fig. 2.1

Note: Here n abbreviates
fib(n), n=1,2,...,5.

Fig. 2.2

Here we have a so-called *tree-recursive process*. The same function value is computed many times and the computation time grows exponentially with n. The required space is proportional to the depth of the tree. The program goes down the tree until it finds a set of arguments for which the function value is defined without reference to the value of the function elsewhere. Then it starts going up the tree. At each stage the parts of the tree branching out of nodes where the function value is already known can be discarded. Also, if the program has not yet started calculating the function value corresponding to a certain node, the part of the tree branching out of that node has not yet been constructed. Thus the data stored at any one time belong to nodes of a descending path of the tree and nodes adjacent to that path. So the computer usually does not run out of space, but the time requirements are enormous. Fig. 1.3 shows that in our first example we have a *linear recursion* with space and time requirements proportional to n.

We compute fib(22)=17711 and fib(23)=28657. Now fib(24) = fib(22) + fib(23) = 46368. But the computer gives fib(24) = -19168. Why? The range for integers runs from -2^{15} to $2^{15} - 1 = $ maxint. After we pass maxint we get $-32768, -32767, \ldots$. (This is explained in Section 39.) Finally we get fib(24) = -19168. To avoid overflow we use **real** variables and at the same time we will write a vastly more efficient program fib1. It uses for the first time a *procedure*. A procedure is a subprogram. A function is a special procedure which returns a value. We start with (fib(0),fib(1))=(0,1) and we use the assignment $(a, b) \leftarrow (b, a + b)$ repeatedly. After n steps we get (fib(n),fib($n + 1$)). $(a, b) \leftarrow (b, a + b)$ can be read: replace (a, b) by $(b, a + b)$.

```
program fib1;
procedure fib(a,b:real);
begin
  write(a:16:0);
  if a<1E+11 then fib(b,a+b)
end;
begin
  fib(0,1); readln
end.
```

```
program fibit;
var i,n,a,b,copy:integer;
begin
  write('n='); readln(n);
  a:=0;b:=1;i:=0;
  while i<>n do
  begin
    copy:=a;a:=b;
    b:=a+copy;i:=i+1
  end;
  writeln('fib(',i,')=',a);
  readln end.
```

Fig. 2.3   Fig. 2.4 Computes fib(n).

0	1	1	2	3
5	8	13	21	34
55	89	144	233	377
610	987	1597	2584	4181
6765	10946	17711	28657	46368
75025	121393	196418	317811	514229
832040	1346269	2178309	3524578	5702887
9227465	14930352	24157817	39088169	63245986
102334155	165580141	267914296	433494437	701408733
1134903170	1836311903	2971215073	4807526976	7778742049
12586269025	20365011074	32951280099	53316291173	86267571272
139583862440				

Fig. 2.5

The 4th line in Fig. 2.3 has the form write(a:m:n). Here m is the number of spaces on which a will be printed, with right alignment. If a needs more than m places, then this instruction is ignored. By using $m = 16$ we get exactly 5 columns of numbers since on one line of the screen there are 80 characters. The integer n gives the number of decimal places to which a is rounded. We use $n = 0$ since a is an integer.

To print the table in Fig. 2.5 one should use the statement write(lst,a:16:0). But then the output will not appear on the screen. That is, write(a) and write(lst,a) send the output to the screen and to the printer, respectively. Here lst stands for list. From Turbo 4.0 on you must also add uses printer after the program name. The iterative program in Fig. 2.4 writes fib(n) on the screen. The last line in Fig. 2.5 is rounded because *real* numbers in Turbo Pascal have an accuracy of 11 significant digits.

3. Another Linear Recursion. Bisection Method

We consider R. Perrin's sequence, defined by

$$v(0) = 3, \ v(1) = 0, \ v(2) = 2, \ v(n) = v(n-2) + v(n-3) \ \text{ for } \ n \geq 3.$$

We will write two fast recursive programs, which print the ratio $q(n) = v(n)/v(n-1)$. The first stops as soon as $v(n-1) > 10^8$ holds, the second prints 50 terms of the sequence $q(n)$, starting with $n = 3$. Study these programs until you understand them. We will return to the interesting sequence $v(n)$ again and again.

```
program lin_rec;

procedure v(a,b,c:real);
begin
  write(b/a:16:10);
  if a <= 1E+08 then
    v(b,c,a+b)
end;

begin
  v(2,3,2); readln
end.
```

```
program lin_rec1;

procedure v(a,b,c:real;
                 n:integer);
begin
  if n>0 then
  begin
    write(b/a:16:10);
    v(b,c,a+b,n-1)
  end
end;
begin
  v(2,3,2,50); readln
end.
```

Fig. 3.1   Fig. 3.2

1.5000000000	0.6666666667	2.5000000000	1.0000000000	1.4000000000
1.4285714286	1.2000000000	1.4166666667	1.2941176471	1.3181818182
1.3448275862	1.3076923077	1.3333333333	1.3235294118	1.3222222222
1.3277310924	1.3227848101	1.3253588517	1.3249097473	1.3242506812
1.3251028807	1.3247049867	1.3247362251	1.3247787611	1.3246492986
1.3247604639	1.3247049867	1.3247126437	1.3247288503	1.3247093499
1.3247218789	1.3247177382	1.3247164894	1.3247195193	1.3247170266
1.3247182160	1.3247180989	1.3247177043	1.3247181492	1.3247178740
1.3247179579	1.3247179924	1.3247179218	1.3247179776	1.3247179522
1.3247179536	1.3247179631	1.3247179530	1.3247179590	1.3247179573
1.3247179564	1.3247179580	1.3247179668	1.3247179573	1.3247179573
1.3247179571	1.3247179573	1.3247179572	1.3247179572	1.3247179573
1.3247179572	1.3247179573	1.3247179572	1.3247179572	1.3247179572

Fig. 3.3

If $q(n)$ converges, it is easy to see that the limit is a root of the equation

(1) $x^3 - x - 1 = 0.$

Indeed,

$$v(n) = v(n-2) + v(n-3) \Rightarrow \frac{v(n)}{v(n-1)} = \frac{v(n-2)}{v(n-1)} + \frac{v(n-3)}{v(n-2)} * \frac{v(n-2)}{v(n-1)}$$

Let x be the limit of $q(n)$. Then

$$x = 1/x + 1/x^2, \quad \text{or} \quad x^3 = x + 1.$$

So $q(n)$ does converge to a root of (1). For $f(x) = x^3 - x - 1$ we have $f(1) = -1$, $f(2) = 5$. Thus there is a root between 1 and 2, and there are no other real roots. We will find the real root recursively in the interval $[a, b]$, starting with $a = 1$, $b = 2$, using the bisection method in Fig. 3.4.

```
program bis;
var a,b:real;

function f(x:real):real;
begin f:=x*x*x-x-1
end;

function bi(a,b:real):real;
var m:real;
begin m:=(a+b)/2;
   if f(b)-f(a) < 1E-09 then bi:=m
   else if f(m)>0 then bi:=bi(a,m)
       else bi:=bi(m,b)
end;

begin
   write('a,b=');  readln(a,b);
   writeln('bis=',bi(a,b):12:10);
readln end.
```

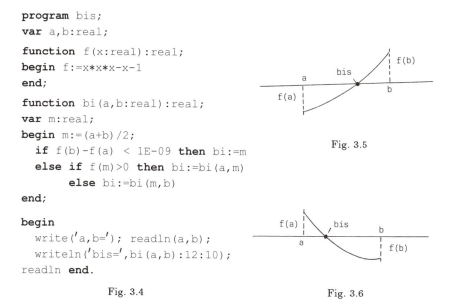

Fig. 3.5

Fig. 3.4   Fig. 3.6

Again we get **bis**=1.3247179572. The program **bis** is valid for functions increasing from a to b. In Exercise 2 we will rewrite **bis** so that it is also valid for functions decreasing from a to b.

In the program **bis** we use a new feature of Turbo Pascal:

$$\texttt{write('a, b ='); readln(a, b);}$$

The first statement causes $a, b =$ to appear on the screen. Now we type in a, then press the space bar, and finally type in b with carriage return.

4. The Bold Gamble. Recursion at its Best

Define the *bold gamble*: My current fortune is $x < 1$, my goal is 1. If $x = 1$ then I quit. If $0 < x \leq 0.5$ then I stake x, all I have. If I win, my fortune will become $2x$; a loss ruins me. If $0.5 < x < 1$, then I bet $1 - x$. If I win, then my fortune is $x + (1 - x) = 1$, and I quit. If I lose, I still have $x - (1 - x) = 2x - 1$ left.

Suppose in one round my chance of winning is p, my chance of losing q, and $p + q = 1$. Let $f(x)$ be the probability of eventual success under bold gamble starting with fortune x. Then we have

$$(2) \quad f(0) = 0, \quad f(1) = 1, \quad f(x) = \begin{cases} p\,f(2x) & \text{if } 0 < x \leq \frac{1}{2}; \\ p + q f(2x - 1) & \text{if } \frac{1}{2} < x < 1. \end{cases}$$

If $p \neq .5$, the function f is of a strange type, called *singular*. It is continuous and almost everywhere differentiable with $f'(x) = 0$, yet it is increasing. The recursion formula (2) expresses $f(x)$ in terms of $f(y)$ where y is the result of shifting the binary form of x to the left by one place and, if the resulting number is > 1, deleting the integer part.

If x has a k-digit binary expansion then k applications of (2) will express $f(x)$ in terms of $f(1) = 1$. Our program based on (2) will return a value even for values of x which are not finite binary fractions. This is because in most computers all numbers are rounded to finite binary fractions. If x is rational, its binary expansion is periodic or terminates. If the expansion is periodic, then repeated applications of (2) will eventually lead to the same argument y a second time. This gives a linear equation for $f(y)$

```
program bold;
var x,p:real;

function f(x,p:real):real;
begin
if (x=0) or (x=1) then
   f:=x
else
   if x<=0.5 then
      f:=p*f(2*x,p)
   else f:=p+(1-p)*f(2*x-1,p)
end;

begin
   write('x,p='); readln(x,p);
   writeln(f(x,p)); readln
end.
```

Fig. 4.1

```
program bold1;
var x,p:real;

function f(x,p:real):real;
begin
if (x=0.0) or (x=1.0) then
   f:=x
else if x<=0.5 then
      f:=p*f(2.0*x,p)
   else
      f:=p+(1.0-p)*f(2.0*x-1.0,p)
end;

begin
   write('x,p='); readln(x,p);
   writeln(f(x,p)); readln
end.
```

Fig. 4.2

from which we can compute $f(y)$, and then we can compute $f(x)$. A "closed" expression is out of the question. In spite of this the Pascal program **bold** quickly finds $f(x)$ for any x from $[0,1]$.

One can show that bold gamble is optimal for $p < 0.5$. For $p = 0.5$ the gambling strategy has no influence on $f(x)$. For $p > 0.5$ bold gamble is bad; you should play as timidly as the rules of the game permit.

In the program **bold** we find the constants 0, 1, 2 in0 several places. Now x is a **real**, whereas 0, 1, 2 are **integers**. This is legal, but the **integers** are converted to **reals** each time the statements are executed. We can "help" the computer by writing 0.0, 1.0, 2.0 instead of 0, 1, 2. This tells the computer that these constants are to be considered as **reals**. We get the program **bold1**, which is about 7% faster than **bold** on my AT. To detect this speedup I evaluated $f(0.6789, 0.3456)$ 1000 times; **bold** and **bold1** required 70 and 65 seconds, respectively. Usually we do not care about this speedup, except in simulations, where such a gain does matter.

5. The Josephus Problem

During the Jewish rebellion against Rome (A.D. 70) 40 Jews were caught in a cave. In order to avoid slavery they agreed upon a program of mutual destruction. They would stand in a circle and number themselves from 1 to 40. Then every seventh person was to be killed until only one was left who would commit suicide. We know the story from the only survivor, who did not carry out the last step. He later became the historian Flavius Josephus.

This story is the source of the

Josephus problem: *n persons are arranged in a circle and numbered from 1 to n. Then every k-th person is removed, with the circle closing up after each removal. What is the number $f(n)$ of the last survivor? What is the number of the s-th removed person?*

The problem is especially simple for $k = 2$. We try to express $f(2n)$ and $f(2n + 1)$ in terms of $f(n)$. In Fig. 5.1, with $2n$ persons around the circle, we eliminate numbers 2, 4,..., $2n$, and we are left with numbers 1, 3,..., $2n - 1$, which are renumbered from 1 to n. In Fig. 5.2, with $2n + 1$ persons, we eliminate numbers 2,4,...,$2n$, and we are left with numbers 3, 5,..., $2n + 1$ which are renumbered 1, 2,..., n. Since $f(n)$ denotes the last survivor on the inner circle, we see that his original number (on the outer circle) is

$$f(2n) = 2f(n) - 1 \quad \text{or} \quad f(2n + 1) = 2f(n) + 1.$$

In addition we have

$$f(1) = 1.$$

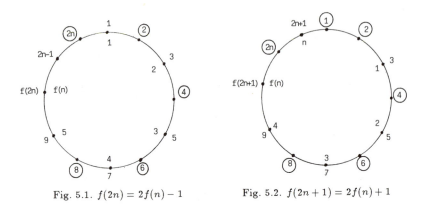

Fig. 5.1. $f(2n) = 2f(n) - 1$ Fig. 5.2. $f(2n+1) = 2f(n) + 1$

These three equations enable us to compute f recursively, see Fig. 5.4.

The relations for f translate into the recursive program jos. Here n div 2 is integer division by 2. It is defined for integers only and is the truncated value of $n/2$, i.e., any part after the decimal point is omitted.

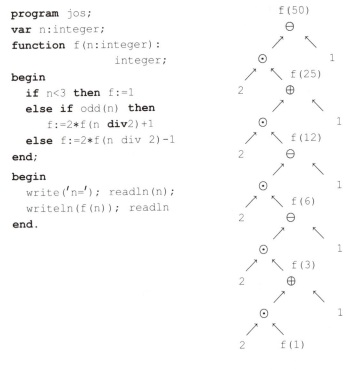

```
program jos;
var n:integer;
function f(n:integer):
            integer;
begin
   if n<3 then f:=1
   else if odd(n) then
      f:=2*f(n div2)+1
   else f:=2*f(n div 2)-1
end;

begin
   write('n=');  readln(n);
   writeln(f(n));  readln
end.
```

Fig. 5.3   Fig. 5.4

Fig. 5.4 shows that the recursion tree does not branch and its depth is proportional to $\log n$. In computer language it has time complexity $O(\log n)$. The O-notation is defined in the Summary of Notation.

As in section 2, we rewrite our program using variables of type **real** so that we can deal with numbers greater than maxint=32767. For $n = 10^9$ the program in Fig. 5.5 gives the correct value $f(n) = 926,258,177$ and for $n = 8 \cdot 10^9$ the value $f(n) = 7,410,065,409$, which is also correct.

```
program jos;
var n:real;

function f(n:real):real;
begin
  if n<3 then f:=1
  else if n/2=int(n/2)
    then f:=2*f(n/2)-1
    else f:=2*f((n-1)/2)+1
end;

begin
  write('n='); readln(n);
  writeln(f(n):0:0); readln
end.
```

Fig. 5.5

```
program joseph1;
var k,n,s,x:real;

begin
  write('n,k,s=');
  readln(n,k,s);
  x:=k*s;
  while x>n do
    x:=int(((k*(x-n)-1)/(k-1)));
  writeln('the ',s:0:0,
    '-th elim. person has #',
    x:0:0);
  readln
end.
```

Fig. 5.6

For arbitrary k, the number x of the s-th eliminated person can be found by means of the program in Fig. 5.6; the proof appears at the end of this section. For $k = 2$, $k/(k-1) = 2$, and the 5-th line of the program becomes x:=2*(x-n)-1. Here we explain another new feature of Turbo Pascal:

$$\text{write}('n, k, s =');\ \text{readln}(n, k, s);$$

The first statement causes $n, k, s =$ to appear on the screen. Now we must type in n, then press the space bar, type in k, press the space bar again, and finally type s with carriage return.

For $k = 2$, there is a simple expression for the number $f(n)$ of the last survivor. First we note that $f(n) = 1$ for $n = 2^m$. For arbitrary n, let m be the largest integer such that $2^m \leq n$. We write $n = 2^m + (n - 2^m)$. Now remove the persons numbered $2,4,6,\ldots,2(n-2^m)$, leaving 2^m persons in the circle. By the above result, the first one of these 2^m persons will survive. The place number of this one is $2(n - 2^m) + 1$. Hence, for $k = 2$, the number of the last survivor is

$$f(n) = 2(n - 2^m) + 1,$$

where 2^m is the largest power of $2 \leq n$. For example, if $n = 1988 = 1024 + 964$, then $f(n) = 2 * 964 + 1 = 1929$.

In the binary representation, $f(n)$ can be computed by transferring the first digit of n to the end:

$$n = 1b_1b_2 \ldots b_m = 2^m + (n - 2^m) \rightarrow 1 = 2(n - 2^m) + 1 = f(n)$$

Exercises for Sections 1-5:

1. The sequence $f(n)$ is defined by $f(1) = f(2) = f(3) = f(4) = 1$ and $f(n) = 4f(n-1) + 3f(n-2) + 2f(n-3) + f(n-4)$ for $n > 4$.

    a) Write a program which prints $f(20)/f(19)$.

    b) By means of the program **bis** (Fig. 3.4) find the maximal root of $x^4 - 4x^3 - 3x^2 - 2x - 1 = 0$, and compare with $f(20)/f(19)$. [Hint for a): Use the transformation $(a, b, c, d) \leftarrow (b, c, d, a + 2b + 3c + 4d)$.]

2. Rewrite the program **bis** in Fig. 3.4, so that it works for functions increasing or decreasing from a to b. Try also to remove recursion. [Hint: The absolute value of x is **abs(x)**. It can be applied to integers and reals.]

3. Find by means of the program **bis** the smallest positive solution of
    a) $x - \cos x = 0$;   b) $x \ln x = 1$;   c) $xe^x = 1$;   d) $\tan x = x$.

4. The equation $x^3 - 3x + 1 = 0$ has three real roots. Find them using program **bis**.

5. For this and the following problems we set $\mathrm{fib}(n) = f(n)$ and we define the Lucas sequence by means of $g(0) = 2$, $g(1) = 1$, $g(n) = g(n-1) + g(n-2)$, $n \geq 2$.

    a) Write a program which for input n finds $g(n) - 5f^2(n)$.

    b) Conjecture and prove, for instance by induction.

6. Write a program which for input n tests if the following formulas are true:
    a) $g(n) = f(n-1) + f(n+1)$;
    b) $f^2(n+1) - f(n)f(n+2) = (-1)^n$.

7. Let $f(x)$ be the probability of eventual success in the bold gamble, starting with fortune x. Find exactly
    a) $f(1/3)$ b) $f(2/5)$ c) $f(1/1984)$ for $p = q = 1/2$.

8. Here is another solution of the Josephus problem for any k.

    The number t of the s-th eliminated person can be found by means of the following algorithm: Set $x = 1 + k(n - s)$ and set $q = k/(k-1)$. Then generate the "integer" geometric sequence $x(1) = \lceil x \rceil$, $x(2) = \lceil qx(1) \rceil$, $x(3) = \lceil qx(2) \rceil, \ldots$, where $\lceil y \rceil$ denotes the ceiling of y, i.e. the smallest integer $\geq y$. If a is the largest term $\leq nk$ then $t = 1 + kn - a$. Translate this algorithm into a program.

9. The number $f(n)$ in the Josephus problem for $k = 2$ can also be found as follows: Write n in the binary system, replace each "0" by "-1" and you get $f(n)$. For instance, $n = 50 = 110010$ implies $f(n) = 11 - 1 - 11 - 1 = -1 + 2 - 4 - 8 + 16 + 32 = 37$. Show this.

10. Show that for $k = 2$ in the Josephus problem the last remaining person has the number $f(n) = 1 + 2n - 2^k$ with $k = \lfloor \log_2 n \rfloor + 1$

11. *Duplication Formula for the Fibonacci Sequence.* Let $u(1) = 0$, $u(2) = 1$, $u(n) = u(n-1) + u(n-2)$ for for $n \geq 3$. That is, $u(n) = \text{fib}(n-1)$. Prove by induction

   (i) $u(2n) = u^2(n) + u^2(n+1)$, $n > 1$,

   (ii) $u(2n+1) = 2u(n)u(n+1) + u^2(n+1)$, $n \geq 1$.

   Construct a fast recursive program `fastfib` based on (i) and (ii). Use type `real` for the function values. It requires only $O(\log n)$ steps instead of $O(n)$ in `fib1` or `fibit` and $O(c^n)$ (with $c > 1$) in `fib`. Compare the speeds of `fibit` and `fastfib`.

   Remark: By means of matrices one can show that similar duplication formulas exist for all linear difference equations.

12. Write a program which prints the sequence $J(n, k)$ of eliminations in the Josephus problem. For instance, $J(8, 3) = (3, 6, 1, 5, 2, 8, 4, 7)$. This sequence is called the *Josephus permutation.*

13. Let $H_n = 1 + 1/2 + 1/3 + \ldots + 1/n$ be the harmonic numbers.
    a) Find H_{10000} by adding terms from left to right.
    b) Find the same sum by adding terms from right to left.
    c) Which sum is more accurate and why?

14. Find by experimentation the sum
$$\text{round}(n/2) + \text{round}(n/4) + \text{round}(n/8) + \ldots$$
   for positive integers n. (For a definition of round(x) see the summary of Pascal.) Try to prove your observations.

15.   a) Find all pairs (x, y) of positive integers from $1 \ldots 100$, which satisfy
$$|x^2 - xy - y^2| = 1.$$
    b) Try to guess how one might generate all solutions.

16. How many of the 6-digit blocks 000000 to 999999 have the property that the sum of the first three digits equals the sum of the last three digits. (The idea here is to reduce computation as much as possible.)

17. *Dragon Curves.* The sequence $a(n)$ is defined as follows:
$a(4n+1) = 1$, $a(4n+3) = 0$ for $n \geq 0$, and $a(2n) = a(n)$ for $n \geq 1$.
    a) Write a program which prints n bits of this sequence.
    b) Use these digits to draw a curve as follows: Start at the origin of the plane lattice (graph paper) and go one step to the right. If the next bit is

"1" then turn left by 90° and go one step forward. If the next bit is "0" then turn right by 90 and go one step forward. You get a strange curve with many regularities, which is called a "dragon curve".

c) Show that the sequence $a(n)$ is not periodic. (Easy Olympiad level problem.)

18. Change the programs `faciter1` and `faciter2` so that they are also valid for $n = 0$.

19. The *stutter* function. Translate the functional equation

$$f(0) = 0, \quad f(1) = 1, \quad f(x) = \begin{cases} f(2x)/4 & \text{for } 0 < x < \frac{1}{2}, \\ \frac{3}{4} + f(2x-1)/4 & \text{for } \frac{1}{2} \le x < 1. \end{cases}$$

into a recursive program. (Theoretically, the program should stop only if x is a finite binary fraction, although the equation determines $f(x)$ uniquely for all rational x. It does not determine $f(x)$ for irrational x unless we specify the additional condition that f be monotonic. After studying Section 6 you will be able to translate x and $f(x)$ into binary notation and you will see the reason for the function's name.

Appendix: Proof of the Josephus Algorithm (Fig. 5.6)

Consider any person P whose first count is x_1, and whose i-th count is x_i. We claim that if P is not eliminated at the i-th count, then

$$(1) \qquad x_{i+1} = x_i + n - \left\lfloor \frac{x_i}{k} \right\rfloor.$$

Indeed, the number of persons eliminated at the count x_i is $\lfloor x_i/k \rfloor$. There remain $n - \lfloor x_i/k \rfloor$ persons. After counting off these persons we get back to P.

Now we form the sequence $x_1, x_2, \ldots$, until we reach an x_j which is divisible by k. Then we have reached the (x_j/k)-th eliminated person. Thus the last count of the s-th eliminated person is ks. We solve (1) for x_i, so that we can go backwards to find his original number. Now

$$(2) \qquad x_i/k = \lfloor x_i/k \rfloor + e, \quad \text{with } 1/k \le e \le (k-1)/k;$$

here $e \ne 0$, else x_i would be a multiple of k, and the person would have been eliminated at the i-th round. From (1) and (2) we get

$$x_i = \frac{(x_{i+1} - n)k}{k-1} - \frac{k}{k-1}e = \frac{(x_{i+1} - n)k - 1}{k-1} - \frac{k}{k-1}\left(e - \frac{1}{k}\right)$$

By (2) the term subtracted on the right is between 0 and $\frac{k-2}{k-1}$. Since x_i is an integer, it is the greatest integer part of the first term:

$$(3) \qquad x_i = \left\lfloor \frac{(x_{i+1} - n)k - 1}{k-1} \right\rfloor = x_{i+1} - n + \left\lfloor \frac{x_{i+1} - n - 1}{k-1} \right\rfloor.$$

To find the original number of the s-th eliminated person, set $x_{i+1} = ks$ and apply (3) repeatedly until we get a number $\leq n$; this is the number we are looking for.

The reader should do one example by hand by forming the sequence $x_1, x_2, \ldots$ and the inverse sequence by means of (3). Suggestion: use $n = 18$, $k = 3$, $x_1 = 7$.

CHAPTER 2

Algorithms in Number Theory

6. Greatest Common Divisor

6.1 Euclid's Algorithm. Let a and b be integers. We denote their greatest common divisor by $\gcd(a,b)$. If we define $\gcd(0,0)=0$, then for all integers a, b

$$\gcd(a,1) = 1, \quad \gcd(a,a) = |a|, \quad \gcd(a,0) = |a|, \quad \gcd(a,b) = \gcd(b,a).$$

If $\gcd(a,b)=1$, we say that a and b are *relatively prime*, or that one is prime to the other.

Let us assume $a \geq b > 0$. Then a can be uniquely represented by b in the form

$$(1) \qquad\qquad a = bq + r, \qquad 0 \leq r < b.$$

The integers q and r are the quotient and remainder on division of a by b. Thus $q = \lfloor a/b \rfloor, \quad r = a - b\lfloor a/b \rfloor$, see Fig. 6.1.

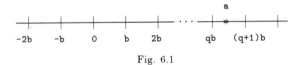

Fig. 6.1

In Pascal these operations are denoted by

$$\text{q} := \text{a div b}, \qquad \text{r} := \text{a mod b}.$$

Now any divisor of a and b is also a divisor of $ax + by$ for any integers x, y. Thus in (1)

$$d \mid a, \; d \mid b \to d \mid r \quad \text{and} \quad d \mid b, \; d \mid r \to d \mid a.$$

Hence

$$(2) \qquad\qquad \gcd(a,b) = \gcd(b,r) = \gcd(b, a \bmod b).$$

By means of (2) we can replace the pair (a,b) by the smaller pair $(b, a \bmod b)$ with the same gcd. Repeating this step, we obtain ever smaller pairs, until finally a pair $(g,0)$ is reached. Then

$$\gcd(a,b) = \gcd(g,0) = g.$$

We get immediately the recursive function in Fig. 6.2 and its iterative version in Fig. 6.3.

```
function gcd(a,b:integer):
begin                   integer;
  if b=0 then gcd:=a
  else gcd:=gcd(b,a mod b)
end;
```

Fig. 6.2

```
function gcd(a,b:integer):
                    integer;
var r:integer;
begin
  while b>0 do
  begin r:=a mod b;
    a:=b; b:=r
  end;
  gcd:=a
end;
```

Fig. 6.3

Since these functions often occur as parts of bigger programs, we will save 6.2 and 6.3 under the names **gcdrec** and **gcditer**, respectively. We shall insert them into programs as needed. This can usually be done by means of the **include** command, or by copying with a text editor.

6.2 Euler's ϕ-function. We define $\phi(n) = $ number of elements in 1 .. n which are prime to n.

```
program Euler;
var i,n,phi:integer;
function gcd(a,b:integer):integer;
begin if b=0 then gcd:=a
      else gcd:=gcd(b,a mod b)
end;
begin
  write('n=');readln(n);phi:=0;
  for i:=1 to n do if gcd(n,i)=1
  then phi:=phi+1;
  writeln('phi(',n,')=',phi)
end.
```

Fig. 6.4

The program in Fig. 6.4 computes $\phi(n)$ from its definition. With my AT I got phi(30000)=8000 in five seconds. Then I replaced gcdrec by gcditer and ran it with the same data. This time the AT required 3 seconds.

One could compute $\phi(n)$ faster from the formula

$$\phi(n) = n\frac{p-1}{p} \cdot \frac{q-1}{q} \cdots, \text{ where}$$

$p, q, ..$ are the *distinct* prime factors of n. This formula follows from the Chinese Remainder Theorem.

6.3 Extended Euclidean Algorithm.   The gcd of a and b can be written as a linear combination of a and b with integer coefficients x, y. That is, there exist integers x, y such that

$$\gcd(a,b) = ax + by.$$

This fact and the values of x and y can be obtained by keeping track of what happens as we carry out Euclid's algorithm. We set up two columns to record the contributions of a and b to the remainders we compute. We start with the two lines

(2) a 1   0   (i.e. $a = 1 * a + 0 * b$),

(3) b 0   1   (i.e. $b = 0 * a + 1 * b$).

Let $q = a$ div b. By subtracting from row (2) q-times row (3) we get

(4) r 1 $-q$ (i.e. $r = a * 1 + (-q) * b$).

Now we cross out row (2) and repeat the step with the remaining rows. The last line before r becomes zero will be

(5) $\gcd(a, b) = ax + by$.

The program in Fig. 6.5 prints $\gcd(a, b)$, x, y in this order.

```
program lincom;
var a,b:integer;

procedure dio(a,u,v,b,x,y:integer);
var q:integer;
begin if b=0 then writeln(a,' ',u,' ',v)
   else begin q:=a div b; dio(b,x,y,a mod b,u−q*x,v−q*y) end
end;

begin
   write('a,b='); readln(a,b); dio(a,1,0,b,0,1)
end.
```

Fig. 6.5

6.4 Visible Points in the Plane Lattice. Let L_2 denote the plane lattice, i.e. the set of points (x, y) with integral coordinates. A point (x, y) in L_2 will be called *visible* (from the origin) if $\gcd(x, y)=1$. Otherwise, the point will be called *invisible*. Let $s(n)$ be the number of visible lattice points in the square $1 \leq x, y \leq n$. Then the proportion of visible lattice points in the square is $p(n) = s(n)/n^2$. What happens to $p(n)$ for $n \rightarrow \infty$? Each visible point (x, y) obscures infinitely many points (kx, ky), $k = 2, 3, 4, \ldots$.. Hence we may guess that $p(n) \rightarrow 0$. We write a program which computes $p(n)$ for input n.

The amazing stability of $p(n)$ for various values of n in Fig. 6.7 suggests that on the contrary $\lim_{n \to \infty} p(n) = p$ exists, and $p \approx 0.61$. But $n = 200$ is not particularly large. Let us take $n = 30000$. Now we must test $n^2 = 9 \cdot 10^8$ points, a huge task for a PC. So we settle for a random sample of $m = 10000$ points and find the proportion of visible points in the sample. In Turbo Pascal **random(n)** chooses at random, with equal probabilities, one of the integers $0 .. n - 1$. So $1 +$ **random(n)** chooses each of the integers $1 .. n$ with the same probability $1/n$. The program **cheb** in Fig. 6.8 gives 0.6098 as an estimate of $p(n)$. This program uses constants $m = 10000$, $n = 30000$. They must be declared before variable declarations. In addition it uses the new instruction **randomize**, which is explained in A Short Summary of Turbo Pascal.

```
program visible;
var i,j,n,s:integer; p:real;
function gcd(a,b:integer):
                      integer;
begin if b=0 then gcd:=a
  else gcd:=gcd(b,a mod b)
end;
begin write('n=');
    readln(n); s:=0;
  for i:=1 to n do
  for j:=1 to n do
  if gcd(i,j)=1 then s:=s+1;
  p:=s/n/n;
  writeln('s=',s,'    p=',p:0:5)
end.
```

n	$s(n)$	$p(n)$
10	63	0.63000
20	255	0.63750
40	979	0.61187
50	1547	0.61880
100	6087	0.60870
120	8771	0.60910
150	13715	0.60956
200	24463	0.61157

Fig. 6.7

Fig. 6.6

```
program cheb;
const m=10000; n=30000;
var a,b,i,count:integer;
{$I gcdrec}   {Here Turbo Pascal inserts the file gcdrec}
begin randomize;
  count:=0;
  for i:=1 to m do
  begin
    a:=1+random(n); b:=1+random(n);
    if gcd(a,b)=1 then count:=count+1
  end;
  writeln(count)
readln; end.
```

Fig. 6.8

We now sketch two different ways of obtaining the value of the probability $p(n)$ for large values of n. For any finite n the equations we use are only approximately true. To make the proof rigorous we would have to deal with that. We would also have to prove that p approaches the same limit for shapes other than squares, e.g. circles, otherwise the result would not have much meaning. For the basic concepts of probability, see the Crash Course.

a) Let A_d be the event $\gcd(x, y) = d$. A_d is equivalent to the occurrence of three independent events: x is a multiple of d, y is a multiple of d, $\gcd(x/d, y/d) = 1$. Thus the probability of the event A_d is

$$P(A_d) = \frac{1}{d} \cdot \frac{1}{d} \cdot p = \frac{p}{d^2} \ .$$

Since one of the disjoint events $A_1, A_2, A_3, \ldots$ must occur, we have

$$p/1 + p/4 + p/9 + \ldots = 1 ,$$

and using Euler's result $\sum_{i=1}^{\infty} 1/i^2 = \pi^2/6$, we find that $p = 6/\pi^2 \approx 0.6079271019$.

b) Let q and r be different primes. In the sequence of natural numbers we consider the subsequences $a_n = qn$, $b_n = rn$, $c_n = qrn$ of all multiples of q, r, qr. A randomly chosen natural number falls into the subsequences a_n, b_n, c_n with probabilities $1/q$, $1/r$, $1/qr$, respectively. Since $1/(qr) = (1/q)(1/r)$, we can say that divisibility by q and r are independent events. If we choose the natural numbers x, y at random, then both will fall into a_n with probability $1/q^2$, and q is not a common factor of x and y with probability $1 - 1/q^2$. The events that neither x nor y is divisible by q are independent for different primes q. So the probability for $\gcd(x,y)=1$ is

$$p = \prod \left(1 - \frac{1}{q^2}\right) \quad \text{(over all primes } q) ,$$

$$\frac{1}{p} = \frac{1}{1 - 1/2^2} * \frac{1}{1 - 1/3^2} * \frac{1}{1 - 1/5^2} * \cdots ,$$

$$= \sum_{n \geq 0} \frac{1}{2^{2n}} * \sum_{n \geq 0} \frac{1}{3^{2n}} * \sum_{n \geq 0} \left(\frac{1}{5^{2n}} * \cdots = \sum_{n \geq 1} \frac{1}{n^2} = \frac{\pi^2}{6} ;\right.$$

(By multiplying out the brackets we got the reciprocal of each square exactly once. This follows from the unique factorization theorem.) Hence

$$p = \frac{6}{\pi^2} .$$

How can we find the sum

$$s = \sum_{n \geq 1} \frac{1}{n^2}$$

to the maximum accuracy of **real** arithmetic in Turbo Pascal, 11 digits, without using Euler's result? Let

$$s = 1 + 1/4 + \cdots + 1/n^2 + E(n) .$$

We want to estimate the error $E(n) = 1/(n+1)^2 + 1/(n+2)^2 + \cdots$. For large values of k, k^2 is relatively close to $(k - \frac{1}{2})(k + \frac{1}{2})$ and hence

$$\frac{1}{k^2} \approx \frac{1}{(k - \frac{1}{2})(k + \frac{1}{2})} = \frac{1}{k - \frac{1}{2}} - \frac{1}{k + \frac{1}{2}} .$$

Consequently

$$E(n) = \frac{1}{(n+1)^2} + \frac{1}{(n+2)^2} + \cdots$$

$$\approx \left(\frac{1}{n + \frac{1}{2}} - \frac{1}{n + \frac{3}{2}}\right) + \left(\frac{1}{n + \frac{3}{2}} - \frac{1}{n + \frac{5}{2}}\right) + \cdots = \frac{1}{n + \frac{1}{2}} .$$

The right side is $> E(n)$, but by a factor $\leq (n+1)^2/((n+1)^2 - 1/4)$. Having found a good estimate of $E(n)$, we compute $s(n) = 1 + \ldots + 1/n^2$ starting at the end (to minimize rounding errors) and add the correcting term $1/(n + 0.5)$. Now we test $n = 500, 1000, 1500, \ldots$ until there is no change any more. The results are shown in Fig. 6.8 a,b.

```
program Zeta2;
var i,n:integer; sum:real;
begin
  write('n='); readln(n); sum:=0;
  for i:=n downto 1 do
        sum:=sum+1/i/i;
  sum:=sum+1/(n+0.5);
  writeln(sum)
end.
```

n	Zeta2(n)
500	1.6449340675
1000	1.6449340669
1500	1.6449340669

Fig. 6.8a   Fig. 6.8b

6.5 The Binary GCD-Algorithm. The binary gcd-algorithm is based on the obvious relations

$$u \text{ even}, v \text{ even} \rightarrow \gcd(u, v) = 2 * \gcd(u \operatorname{div} 2, v \operatorname{div} 2)$$
$$u \text{ even}, v \text{ odd} \rightarrow \gcd(u, v) = \gcd(u \operatorname{div} 2, v)$$
$$u > v \rightarrow \gcd(u, v) = \gcd(u - v, v)$$
$$u \text{ odd}, v \text{ odd} \rightarrow u - v \text{ even}, |u - v| < \max(u, v) .$$

In machine language this algorithm can be somewhat faster than the standard algorithm of Euclid. This is because division by 2 on a binary computer is just a shift by one place. Without access to machine language this algorithm does not pay off. In Pascal there is on purpose no easy access to machine language. We program this algorithm not because of its speed, but because it is instructive. Again, the recursive version is especially simple. In Pascal odd(x) is true if x is odd, but there is no corresponding boolean function **even**. This is no handicap since even(x)=odd$(x - 1)$. Fig. 6.9 shows the recursive version **rebingcd** and Fig. 6.10 shows the iterative version **bingcdit**.

Let us rewrite **bingcdit** into "machinelike" language. The functions **pred**(a) and **succ**(a) may be faster than $a - 1$ and $a + 1$, respectively. Here **pred** and **succ** stand for *predecessor* and *successor*, respectively. In addition we replace a div 2 by a *shr* 1 and $2 * a$ by a *shl* 1. Here a *shr* b and a *shl* b shift the binary representation of a by b places to the right and left, respectively. So a *shr* $b = a$ div 2^b, a *shl* $b = a * 2^b$. See the figure and Exercise 26.

```
program rebingcd;
var u,v:integer;
function gcd(u,v:integer):
                    integer;
begin
  if u=v then gcd:=u;
  if odd(u-1) and odd(v-1)
  then gcd:=2*gcd(u div 2,
                  v div 2);
  if odd(u+v) then if odd(v)
  then gcd:=gcd(u div 2,v)
  else gcd:=gcd(u,v div 2);
  if odd(u) and odd(v)
          then if u>v
  then gcd:=gcd(u-v,v)
  else if v > u then
      gcd:=gcd(u,v-u)
end;
begin write('u,v=');
        readln(u,v);
  writeln(gcd(u,v))
end.
```

Fig. 6.9

```
program bingcdit;
var gcd, x,y,u,v,k:integer;
begin write('u,v=');
        readln(u,v);
  x:=u; y:=v; k:=1;
  while odd(x-1) and
          odd(y-1) do
  begin
    x:=x div 2;y:=y div 2;
                  k:=2*k
  end;
  repeat
    while odd(x-1)
      do x:=x div 2;
    while odd(y-1)
      do y:=y div 2;
    if x > y then x:=x-y;
    if y > x then y:=y-x
  until x=y;
  gcd:=x*k;
        writeln('gcd=',gcd)
end.
```

Fig. 6.10

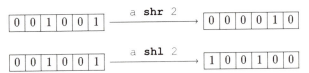

6.6 GCD and LCM.

We give an algorithm which computes $\gcd(a,b)$ and $\text{lcm}(a,b)$ at the same time. It is due to Stanley Gill. ($\text{lcm}(a,b)$ = least common multiple of $a,b=ab/\gcd(a,b)$).

We start with $x=a$, $y=b$, $u=a$, $v=b$ and move as follows:

If $x < y$ we set $y \leftarrow y - x$ and $v \leftarrow v + u$.

If $y < x$ we set $x \leftarrow x - y$ and $u \leftarrow u + v$.

The algorithm ends with $x = y = \gcd(a,b)$ and $(u+v)/2 = \text{lcm}(a,b)$. The invariants of these transformations are

$$P: \qquad \gcd(x,y) = \gcd(x - y, y) = \gcd(x, y - x),$$

$$Q: \qquad xv + yu = 2ab,$$

$$R: \qquad x > 0, \ y > 0.$$

P and R are obviously invariant. We show the invariance of Q. At the beginning we have $ab + ab = 2ab$, which is obviously correct. At the next step the left side of Q becomes either

$$x(v + u) + (y - x)u = xv + yu \quad \text{or} \quad (x - y)v + y(u + v) = xv + yu .$$

Thus the left side of Q does not change. At the end we have

$$x = y = \gcd(a, b) \quad \text{and} \quad \frac{u + v}{2} = \frac{ab}{\gcd(a, b)} = \operatorname{lcm}(a, b) .$$

The program can be found in Fig. 6.11.

```
program gcdlcm;
var x,y,u,v,gcd,lcm:integer;
begin write('u,v='); readln(u,v);
  x:=u;y:=v;
  while x<>y do
  begin
    if x<y then begin y:=y-x; v:=u+v end
    else begin x:=x-y; u:=u+v end
  end;
  gcd:=x; lcm:=(u+v) div 2;
  writeln(gcd,' ',lcm)
end.
```

Fig. 6.11

6.7 GCD for Numbers Beyond Maxint. In Turbo Pascal an integer can be at most $2^{15} - 1$ or 32767. In Turbo 4.0 and later there is a standard data type LongInt with range $-2^{31} + 1$ to $2^{31} - 1$ for which div and mod are defined.

In IBM Turbo 5 and later even larger integers can be represented using the variable types extended or comp. They become available if we enter the {$N+$}; (numeric processing mode) compiler directive in the second line of the program. (In Turbo Pascal for the Macintosh no directive is needed.) These types have 63-bit mantissas. The operations mod and div are not defined for these types.

```
function gcd(a,b:real):real;
begin
  if b=0.0 then gcd:=a
  else gcd:=gcd(b,a-b*int(a/b))
end;
```

Fig. 6.12

As an example of how to get by without mod and div, let us consider the use of reals which can represent integers up to $2^{31} - 1$. Then we can replace $a \operatorname{div} b$ by $\operatorname{int}(a/b)$ and $a \bmod b$ by $a - b*\operatorname{int}(a/b)$. The function gcd in Fig. 6.12 will do the job. It will be stored under the name gcdreal.

Exercises for Section 6:

1. Let $a=8991$, $b=3293$.
    a) Find $\gcd(a, b)$.
    b) Find a solution (x, y) of $\gcd(a, b) = ax + by$ using `lincom`.
    c) Find $\gcd(a, b)$ using the program `rebingcd`.

2. Find $\gcd(a, b)$ and $\mathrm{lcm}(a, b)$ for $a=2431$ and $b=1309$ using the program `gcdlcm`.

3. Let a, b be integers with at most 11 digits. Using the function `gcdreal` find $\gcd(a, b)$ for
    a) $a=987654321$, $b=123456789$
    b) $a=9753197531$, $b=2468008642$
    c) $a=12345678901$, $b=10234567891$.

4. Rewrite the program `lincom` into `relincom` which accepts up to 11 digits and use the inputs from Exercise 3 to get $\gcd(a, b)$ as a linear combination of a and b.

5. Find the last three (four) digits of a) 7^{9999} b) 7^{99999} c) 7^{999999}

6. Make the program `visible` in Fig. 6.6 twice as efficient using that, if $t(n)$ is the number of visible points in the triangle $0 < y < x \le n$, then $s(n) = 2t(n) + 1$.

7. Write an iterative and a recursive program to find $1 + 2 + \ldots + n$.

8. Write a recursive and an iterative program to find the binary representation of n.

9. Write a program which computes the digital sum $d(n)$ of n.

10. Let $d(n)$ be the digital sum of n. Write a program which finds $d(d(n)), d(d(d(n))), \ldots$ until a single digit number $f(n)$ is reached. Show that $f(mn) = f(f(m)f(n))$. This was formerly used as a computational check.

11. Write a program which reverses the digits of an integer. For instance, $1988 \mapsto 8891$. Change the program so that you can reverse integers up to 11 digits.

12. Write a slow recursive program for $C(n, s)$, the number of s-subsets of an n-set, based on the recursion
    $$C(n, s) = C(n - 1, s - 1) + C(n - 1, s), \quad C(n, 0) = C(n, n) = 1.$$

13. Write a fast recursive program for $C(n, s)$, which is based on the recursion
    $$C(n, s) = C(n - 1, s - 1) * n/s, \quad C(n, 0) = 1.$$

14. Introduce in Fig. 2.1 a variable count, which counts the number of function calls. Find a recursion for count(n) and try to express count(n) by means of fib(n).

Mathematical Explorations:

15. Write a program which prints fib$(n+1)$/fib(n) for $n = 1$ to 30. What do you see? Can you prove your observations?

16. Print those terms of the Fibonacci sequence, which are divisible by some fixed positive integer d. Experiment with different values of d, make observations and conjectures. Can you prove your conjectures ?

17. The Fibonacci sequence mod m will eventually become periodic.
    a) Why ?
    b) A periodic sequence can be either purely periodic or periodic with a tail, as in Fig. 6.13 or 6.14, respectively. Show that the Fibonacci sequence mod m is always purely periodic.
    c) Write a program which finds the period $L(m)$ for the modulus m.

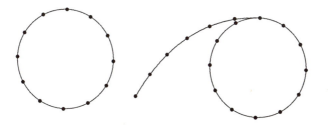

Fig. 6.13   Fig. 6.14

18. Experiment with the program in 17c) and make conjectures about
    a) $L(p^n)$ for any prime p.
    b) $L(p_1^{a_1} p_2^{a_2} \ldots p_r^{a_r})$ for primes p_1 to p_r.
    By means of b) conjecture a formula for $L(10^n)$.

19. Try to discover theorems about the length L of the period of the Fibonacci sequence mod p, if the prime p has the form
    a) $p = 5k \pm 1$ b) $p = 5k \pm 2$ c) $p = 5$

20. The Tribonacci sequence is defined by $t(1) = t(2) = t(3) = 1$, $t(n) = t(n-1) + t(n-2) + t(n-3)$, $n > 3$.
    a) Why is this sequence purely periodic mod m?
    b) Find the period $L(m)$ for the modulus m and discover some theorems about $L(m)$.

21. Find Zeta3 $= 1 + 1/2^3 + 1/3^3 + 1/4^3 + \ldots$, in a way similar to Fig. 6.8. Show that this time, for $n=200$ we get an accuracy of 11 digits.

22. Define a lattice point (x, y, z) to be visible if $\gcd(x, y, z)=1$. Rewrite the programs `visible` and `cheb` for space and call them `visible3` and `cheb3`. Guess a formula for the probability that $\gcd(x, y, z)=1$. Can you give a heuristic argument for this formula?

23. Guess a formula for the probability that k random integers have $\gcd=d$. Check this formula empirically with the computer. Give a heuristic argument for this formula.

24. *Morse-Thue Sequence.* This infinite sequence of binary digits can be constructed recursively: start with 0; to each initial segment append its complement: 0, 01, 01 10, 0110 1001, 01101001 10010110, ... Thus, having constructed the first $1 + 1 + 2 + 4 + 8 + 16 + \cdots + 2^n = 1 + 2^{n+1}$ digits, the next block of $1 + 2^{n+1}$ is determined.

    a) Show that the sequence is aperiodic.

    b) The sequence has a much more interesting property: it is self-similar. Striking every second digit reproduces the sequence. Prove this.

    c) Let the digits of the sequence be $x(0), x(1), x(2), \ldots$. Show that $x(2n) = x(n)$, $x(2n + 1) = 1 - x(2n)$. Write a recursive algorithm to find $x(n)$ based on these recurrences.

    d) Show that $x(n) = 1 - x(n - 2^k)$, where 2^k is the largest power of 2 contained in n, i.e. the largest power of 2 which is $\leq n$. Write an algorithm to find $x(n)$ based on this property.

    e) Write the nonnegative integers in binary notation: 0, 1, 10, 11, 100, 101, 110, 111, ... Now replace each number by the sum of its digits mod 2. You get the sequence 01101001..., which is again the Morse-Thue sequence. Prove this and write an algorithm for finding the n-th digit $x(n)$ based on this idea.

25. The infinite binary *Keane Sequence*
    $x(0)\,x(1)\,x(2)\ldots\ =\ 001\,001\,110\,001\,001\,110\,110\,110\,001\ldots$ is formed as follows. (Commas separate successively longer initial segments.)

    0, 001, 001001110, $\ldots, A,\ AAC(A),\ \ldots$ where $C(A)$ is the digitwise complement of the binary block A. Construct an algorithm which finds the n-th digit $x(n)$. Hints: a) Translate $0,1,2,3,\ldots$ into the ternary system. Do you see a pattern? b) Show that $x(3n) = x(n)$, $x(3n + 1) = x(n)$, $x(3n + 2) = 1 - x(n)$. Write a program based on these recurrences.

26. Rewrite `bingcd` in Fig. 6.10 into "machinelike" language, as explained in the text. Incorporate this new program into a program `cheb1` and test its speed versus `cheb` in Fig. 6.8.

27. a) Print one row of Pascal's triangle, i.e., the sequence $C(n, 0) \ldots C(n, n)$. Use a vector $p[0..n]$ generated by addition, as in

1	5	10	10	5	1		
	1	5	10	10	5	1	
1	6	15	20	20	15	6	1

b) Modify a), so that rows 0 to n of Pascal's triangle are printed.

c) Write a program which prints Pascal's triangle mod 2 and study the recursive pattern.

28. a) A woman walks to the market with n eggs in her basket. She collides with a car, causing all eggs to be broken. The driver offers to pay for every one of her n eggs. But she did not know n exactly. She only knew that by taking them 2,3,4,5,6 at a time, one egg was left, but by taking them 7 at a time, it came out even. Find all numbers ≤ 3000 satisfying this property.

b) Which is the most likely number if one egg weighs about 56 grams?

c) Generalize.

29. Consider again the linear recurrence $v_0 = 3$, $v_1 = 0$, $v_2 = 2$, $v_n = v_{n-2} + v_{n-3}$, $n > 2$. Write a program which prints a table of those numbers satisfying $3 \leq n \leq 4001$ and $n \mid v_n$. Compare this table with the table printed by the program in Fig. 10.2. For the moment you may type in Fig.10.2 without understanding it. Do you think there is now enough evidence for the theorem $n \mid v_n \Leftrightarrow n$ is prime? We will return to this problem again.

30. *Analysis of an Algorithm.* Start with two piles of a and b chips, respectively. You may double the number of chips in the smaller pile at the expense of the larger one. From (c, c) you move to $(0, 2c)$. Stop as soon as the first component is reduced to 0. In the language of algorithms this can be summarized as follows:

```
while  a > 0
do if  a < b, then  (a, b) ← (2a, b − a),
            else  (a, b) ← (a − b, 2b) od.
```

a) For what initial positions does the algorithm stop, and after how many steps?

b) When do you get a pure cycle, and how long is it ?

c) When do you get a cycle with a tail, and how long is the tail ?

d) Suppose a and b are positive rationals. Answer a) to c) in this case.

e) Let a and b be positive reals. Answer a) to c) in this case. Hint: Simplify the algorithm substantially by the observation that $a + b$ is an invariant of the transformation.

31. Find a recursive equation for the number $f(n)$ of ones in the binary representation of n, i.e. express $f(n)$ for $n > 0$ in terms of values of f for smaller arguments and predefined functions of Pascal.

32. By systematic search find a number n with the property
$\phi(n) = \phi(n+1) = \phi(n+2)$. See Section 6.2.

33. The function f from the positive integers to the positive integers is defined
by $f(1) = 1$, $f(2n) = f(n) + g(n-1)$, $f(2n+1) = f(n+1) + g(n/2)$ for
all positive integers n. Here $g(x)$ is the smallest power of $2 > x$, x a real.
Write a recursive program which tests that $f(f(n)) = n$ for any input n.
A function with this property is called an *involution*.

7. All Representations of n in the Form $x^2 + y^2$

We want to generate all nonnegative integer solutions of

$$x^2 + y^2 = n$$

with $\sqrt{n} \geq x \geq y \geq 0$. It pays to subdivide the task into three parts (Fig.7.1):

1. Initialization.
2. Go from A to B.
3. Go from B to C.

Initialization: x:=0; y:=0;

Go from A to B:

```
repeat x:=x+1; y:=y+1
until x*x+y*y>=n;
```

Go from B to C:

```
repeat
   go down in unit steps
   until you are below or on the circle x² + y² = n;
   if you are on, print the current position (x, y).
   go one step to the right
until x * x > n.
```

Fig. 7.1

Thus we get the program quadsum1.

```
program quadsum1;
var n,x,y:integer;
begin write('n=');readln(n);
  x:=0; y:=0;
  repeat
    x:=x+1; y:=y+1
  until x*x+y*y>=n;
  repeat
    while x*x+y*y>n do y:=y-1;
    if x*x+y*y=n then
        writeln(x,' ',y);
    x:=x+1
  until x*x>n
end.
```

Fig. 7.2

```
program quadsum2;
var n,x,y:integer;
begin
    write('n=');readln(n);
    x:=trunc(sqrt(n/2)); y:=x;
    repeat
       while x*x+y*y>n do y:=y-1;
       if x*x+y*y=n then
           writeln(x,' ',y);
       x:=x+1;y:=y-1
    until x*x>n
end.
```

Fig. 7.3

In program quadsum2 we go from A to B in one step. We also utilize that the part of the circle we are concerned with goes down more steeply than 45°. Consequently, if $(x, y + 1)$ is outside the circle then so is $(x + 1, y)$. That allows us to bypass the latter point and go directly to $(x + 1, y - 1)$. These refinements do not speed up the computation by much, but `quadsum2` gives us an opportunity to introduce two new Pascal functions: `sqrt(n)` is the square root of n and it is of the type `real`; `trunc(u)` is the same as `int(u)`, but is of type `integer`, whereas `int(u)` is of type `real`. The assignment `x:=int(sqrt(n/2))` would be a syntax error since x was declared to be an `integer`.

8. Pythagorean Triples

We will extend the program **quadsum2** so that it prints a table of primitive Pythagorean Triples, i.e. all positive integer solutions (x, y, z) of

$$x^2 + y^2 = z^2 \,, \ \gcd(x, y) = 1 \,, \ z \leq max \,.$$

```
program pyttrip1;
var z,n,x,y,max: integer;
{$I gcdrec}

begin
  write('max=');readln(max);
  z:=1;
while z<=max do
begin z:=z+2; n:=z*z;
  x:=trunc(sqrt(n/2)); y:=x;
  repeat
    while x*x+y*y>n do y:=y-1;
    if x*x+y*y<n then x:=x+1
    else begin
      if gcd(x,y)=1 then
      writeln(y:4,x:4,z:5);
      x:=x+1;y:=y-1
    end
  until x*x>n
 end
end.
```

Fig. 8.1 Prints all triples with $z \leq$ max.

```
program pyttrip2;
var a,b,max:integer;
{$I gcdrec}        begin
  write('max=');readln(max);
  a:=2; b:=1;
repeat
  while b>0 do
  begin
    if gcd(a,b)=1 then
    writeln(a*a-b*b:4,
            2*a*b:4,a*a+b*b:5);
    b:=b-2
  end;
  b:=a; a:=a+1
until a>max
end.
```

Fig. 8.2

y	3	5	8	7	20	12	9	28	11	33	16	48	36	13	39	65	20
x	4	12	15	24	21	35	40	45	60	56	63	55	77	84	80	72	99
z	5	13	17	25	29	37	41	53	61	65	65	73	85	85	89	97	101

First we must read in the function **gcdrec** at the appropriate place. In Turbo Pascal we can use the compiler directive {$I gcdrec}. Now z must be odd; for, if z were even, x and y would have the same parity. If both are even, the triple is not primitive; if both are odd, the equation $x^2 + y^2 = z^2$ does not hold mod 4. Hence we will run through the odd values of z. And we print the triple only if $\gcd(x, y) = 1$. We get the program `pyttrip1`, see Fig. 8.1, which prints all triples with $z \leq$ max. Primitive Pythagorean Triples (x, y, z) can also be represented in the form

$$x = a^2 - b^2, \qquad y = 2ab, \qquad z = a^2 + b^2,$$
$$a > b, \qquad \gcd(a, b) = 1, \qquad a + b \text{ odd};$$

see, e.g., Invitation to Number Theory by Oystein Ore, NML vol. 20, MAA 1967.

Fig. 8.2 shows the simpler program `pyttrip2`, which prints all triples with $a \leq$ max. It is based on these formulas.

9. Counting the Lattice Points in a Ball

Let $\text{lat}(s, n)$ be the number of lattice points in the closed n-ball

$$x_1^2 + x_2^2 + \cdots + x_n^2 \leq s.$$

In Fig. 9.1 we use $n = 3$ for illustration. The equatorial section of the n-ball is an $(n-1)$-ball with radius $\sqrt{s}$. By definition there are $\text{lat}(s, n-1)$ lattice points in this ball. To these we must add the contributions of the latitudinal sections at heights $h = 1, 2, \ldots, \lfloor \sqrt{s} \rfloor$ from the equatorial section. The latitudinal section with height h is an $(n-1)$-ball with radius $\sqrt{s - h^2}$, and it contributes $\text{lat}(s - h^2, n-1)$ lattice points to the total count. For symmetry reasons we ignore the sections in the "southern hemisphere" and count instead the contributions of the sections above the equator twice. Thus we get the program fragment

```
count   ← lat(s,n-1)
for h   ← 1 to ⌊√s⌋ do
count   ← count + 2 * lat(s-h²,n-1).
```

The cleanest breakoff condition would be

(1)   if $n = 0$ then lat $\leftarrow 1$.

That is, the 0-ball contains just the origin. We could "help" the program by using $\text{lat}(s, 1) \leftarrow 1 + 2\lfloor \sqrt{s} \rfloor$ (Fig. 9.2). Then the breakoff condition would be

(2)   if $n = 1$ then lat $\leftarrow 1 + 2\lfloor \sqrt{s} \rfloor$.

The program `lattice` in Fig. 9.3 is based on (2). It works also for integers beyond `maxint`, since we have declared the relevant variables to be `reals`. The table in Fig. 9.4 was computed by means of this program.

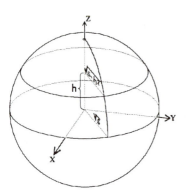

Fig. 9.2. The 1-ball $x^2 \leq s$

Fig. 9.1

```
program lattice;
var n:integer; s:real;

function lat(s:real;n:integer):real;
var h,count: real;
begin
  if n=1 then lat:=1+2*int(sqrt(s))
  else
  begin
   count:=lat(s,n-1); h:=1;
   while h*h<=s do
   begin
    count:=count+2*lat(s-h*h,n-1);
    h:=h+1
   end;
   lat:=count
  end
end;

begin
  write('s,n='); readln(s,n);
  writeln( lat(s,n):0:0 )
end.
```

Fig. 9.3

s	$\mathrm{lat}(s,2)$
10	37
100	317
1000	3149
10000	31417
100000	314197
1000000	3141549
10000000	31416025
100000000	314159053
1000000000	3141592409
10000000000	31415925457

Fig. 9.4

Fig. 9.5

The table shows the conceptually simplest way to compute π, just by counting. Unfortunately, there is no simple way to find error bounds. Note that in going from $s = 10000$ to $s = 100000$ the estimate we get for π becomes worse.

Let

$$f(s) = \text{lat}(s, 2) = \pi s + e(s),$$

where $e(s)$ is the error. Gauss found around 1800 the following crude estimate: To each lattice point in or on the circle he assigned the unit square "north-east" of the point. The sum of the areas of these squares is $f(s)$. This is, however, not quite equal to the area of the circle. Some squares protrude beyond the circle and there is, on the other hand some unfilled area in the circle (Fig. 9.5). However, the circle about O with radius $\sqrt{s} + \sqrt{2}$ covers these squares completely and the circle about O with radius $\sqrt{s} - \sqrt{2}$ is completely covered by these squares. Thus we have

$$\pi(\sqrt{s} - \sqrt{2})^2 < f(s) < \pi(\sqrt{s} + \sqrt{2})^2$$

which implies

(3) $$|f(s) - \pi s| < 2\pi(\sqrt{2s} + 1).$$

We can write this less precisely, for large s, as

(4) $$|f(s) - \pi s| < C\sqrt{s}, \quad \text{where } C \text{ is some constant.}$$

It took over 100 years to find a better estimate. In 1906 W. Sierpinski proved the surprising result

(5) $$|f(s) - \pi s| < C\, s^{1/3}, \quad \text{where } C \text{ is some constant.}$$

On the other hand G. H. Hardy and E. Landau proved that there is no constant C such that

(6) $$|f(s) - \pi s| < Cs^{1/4}.$$

It is believed that, for each $t > 1/4$, there is a number C_t such that

(7) $$|f(s) - \pi s| < C_t s^t,$$

but so far this has been proved only for $t > 7/22$, and even that was very difficult.

Exercises for Sections 7-9:

1. Find the number of representations as a sum of two squares of the following integers: 5, 13, 17, 29, 5*13=65, 5*13*17= 1105, $5 * 13 * 17 * 29 = 32045$, $5^2 * 13 = 325$, $5^3 * 13 = 1625$, $5^2 * 13^2 = 4225$, $5^3 * 13^3$. Try to guess theorems about the number of representations.

2. Write a program which finds all lattice points on the circle $x^2 + y^2 = n$ by starting at C and going to B (Fig. 7.1).

3. If $n \bmod 4 = 3$ then $x^2 + y^2 = n$ has no solutions. Show this. Use this fact to write a brute force algorithm `pyttrip3` for finding all Pythagorean triples. The variable a runs from 2 to $c - 1$. Print a, b, c if $b = \sqrt{c^2 - a^2}$ is an integer. How do you test if b is an integer? Only one of a, b, c and b, a, c should be printed.

4. Make `quadsum2` more efficient by using the fact that
$$(x + 1)^2 = x^2 + (2x + 1), \quad (y - 1)^2 = y^2 - (2y - 1).$$

5. There is another formula for the number $f(s)$ of lattice points in the circle $x^2 + y^2 \le s$:

(8) $f(s) = 1 + 4(\lfloor s \rfloor - \lfloor s/3 \rfloor + \lfloor s/5 \rfloor - \lfloor s/7 \rfloor + \lfloor s/9 \rfloor - \ldots).$

Find by means of this formula   a) $f(10000)$ b) $f(1000000)$.
Hint for b): With $s = 500000$ we have $\lfloor 2s/(s + 1) \rfloor - \lfloor 2s/(s + 3) \rfloor + \ldots + \lfloor 2s/(2s - 3) \rfloor - \lfloor 2s/(2s - 1) \rfloor = +1 - 1 + 1 - 1 + \cdots + 1 - 1 = 0$. Using similar relations reduce the number of terms from 500000 to about 50000. See the references for Chapter II.

6. Evaluate lat(10000,3) and give an estimate of π based on this number. Compare with the estimate based on lat(10000,2).

7. The volume of the 4-ball is $v_4(r) = c_4 r^4$, where c_4 is the volume of the unit ball in 4-space. Estimate the constant c_4 by counting lattice points lat(100,4) and lat(1000,4), which give estimates for c_4.
   a) Can you guess the exact value of c_4?
   b) Divide the estimate by π. Can you now guess c_4?
   c) Multiply the estimate by 2 and divide by π. Can you now guess c_4?

8. The circle occupies $\pi/4$ (roughly 3/4) of the circumscribed square. The 3-ball occupies $\pi/6$ (roughly 1/2) of the circumscribed cube. Roughly what proportion of the circumscribed cube does the 4-, 5-, 6-ball occupy?

9. The following formula was known to Gauss:
$$\mathrm{lat}\,(s, 2) = 1 + 4\lfloor \sqrt{s} \rfloor + 4\lfloor \sqrt{s/2} \rfloor^2 + 8 \sum_{i=a}^{b} \lfloor \sqrt{s - i^2} \rfloor$$
$$\text{with} \quad a = \lfloor \sqrt{s/2} \rfloor + 1 \text{ and } b = \lfloor \sqrt{s} \rfloor.$$

Prove this formula. Write a program `lattice1` based on it, and compare its speed with that of `lattice`.

10. From (8) in Exercise 5 we can derive the formula of Leibniz
$$\frac{\pi}{4} = 1 - \frac{1}{3} + \frac{1}{5} - \frac{1}{7} + \cdots.$$

But this formula was known at least a century earlier in India with three successively better error estimates:

$$\frac{\pi}{4} = 1 - \frac{1}{3} + \frac{1}{5} - \cdots + \frac{1}{(n-3)} - \frac{1}{(n-1)} + f(n);$$

$$f(n) \approx \frac{1}{2n}, \quad f(n) \approx \frac{n}{2n^2+2}, \quad f(n) \approx \frac{n^2+4}{2n^3+10n}.$$

Find π using the above sum and the third remainder formula with $n = 20$ and with $n = 50$.

The Indian sources can be found in the references for Chapter II.

10. Sieves

From a set we strike all elements which do not have a certain property. This is called *sieving*. Sieves play a fundamental role in computer science and mathematics. We treat some examples of sieves from number theory.

10.1. Square-free Integers An integer is called *square-free*, if it is not divisible by the square of an integer > 1. Let $A(n)$ be the number of square-free integers in the set $1 \ldots n$. Then $q(n) = A(n)/n$ is the proportion of square-free integers in $1 \ldots n$.

```
program square_free;
var i,k,m,n,s:integer; x:array[1..15000] of 0..1;
begin
  write('n='); readln(n);
  m:=trunc(sqrt(n)); s:=0;
  for i:=1 to n do x[i]:=1;
  for i:=2 to m do
  begin k:=0;
   repeat
     k:=k+i*i;
     if k<=n then x[k]:=0
   until k>=n
  end;
  for i:=1 to n do
  begin
    s:=s+x[i];
    if i mod 500 = 0 then
      writeln(i:5,s/i:10:5)
  end
end.
```

500	0.61200	8000	0.60812
1000	0.60800	8500	0.60812
1500	0.61000	9000	0.60811
2000	0.60750	9500	0.60842
2500	0.60920	10000	0.60830
3000	0.60800	10500	0.60810
3500	0.60914	11000	0.60791
4000	0.60825	11500	0.60783
4500	0.60800	12000	0.60775
5000	0.60840	12500	0.60768
5500	0.60764	13000	0.60777
6000	0.60767	13500	0.60770
6500	0.60800	14000	0.60779
7000	0.60786	14500	0.60772
7500	0.60813	15000	0.60800

Fig. 10.1

If

$$q = \lim_{n \to \infty} q(n)$$

exists, we call it the *density* of square-free integers. We computed $q(500)$ to $q(15000)$ in steps of 500 to get some idea of q.

We set up an array $x[1], x[2], \ldots, x[n]$ and initialize each $x[i]$ to be 1. Then we sieve with square factors p up to $p = \sqrt{n}$ by setting the corresponding $x[i] = 0$. We get the program in Fig. 10.1. The output suggests that $q \approx 0.608$. Conjecture: $q = 6/\pi^2$. See exercises.

10.2. Sieve of Eratosthenes The following algorithm is called *Sieve of Eratosthenes*:

1. Set $p := 2$.
2. Strike out all multiples of $p \geq p^2$.
3. Set $p :=$ the first integer beyond p not yet struck out, and go to 2.

```
program eratsiev;
var d,i,k,n:integer;
  x:array[1..16000] of 0..1;
begin
write('n='); readln(n);
for i:=1 to n do x[i]:=1;
d:=trunc(sqrt(n+n+1));
k:=3;
while k<=d do
begin
  if x[(k-1) div 2]=1 then
  begin
    i:=k*k;
    while i<=n+n+1 do
    begin
      x[(i-1) div 2]:=0;
      i:=i+k+k
    end
  end;
  k:=k+2
end;
for k:=1 to n do if x[k]=1
  then write(k+k+1:6)
end.
```

Fig. 10.2

It uses the fact that a proper divisor of a number must precede it but cannot be 1. After step 2 the first number not struck out is a prime, otherwise it would have been struck out as a multiple of its smallest divisor. Every prime sieves out first its own square. E.g., 7 first sieves out 7^2, since $2*7, 3*7, \ldots, 6*7$ have been struck already as multiples of 2, 3, or 5. When sieving using p, one first strikes p^2 and then every pth number. But since $p^2 + p, p^2 + 3p, p^2 + 5p, \ldots$ are even, they are already struck out. Hence we may proceed in steps of $2p$. The remaining numbers are primes. In the program **eratsiev** after completing the sieving we have $x[k] = 1$ or 0 if $2k + 1$ is prime or not, respectively. This memory saving trick makes **eratsiev** hard to comprehend. If you do not understand it then rewrite the program by giving up this trick. The sieve is very fast. On my AT the primes from 3 to 31991 are sieved out in less than a second. The printing takes longer.

We suggest that you run this program for $n = 2000$ and print the corresponding table of primes. You should save the page of primes from 3 to 4001. It will come in handy on many occasions.

10.3. A Closed Formula for an Irregular Sequence The sequence
a_n: 1, 3, 4, 6, 8, 9, 11, 12, 14, 16, 17, 19, 21, 22, 24, 25, 27, 29, 30, ...
was produced by the following sieve:

Take $a_1 = 1$ and delete $a_1 + 1 = 2$.
Keep the next integer $a_2 = 3$, but delete $a_2 + 2 = 5$.
Keep the next integer $a_3 = 4$, but delete $a_3 + 3 = 7$, etc.

We first make an outline of the program which generates the sequence.

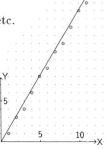

```
1. for i:=1 to n do x[i]:=1;
2. i:=1; j:=1; a[j]:=1;
3. delete i+j by setting x[i+j]:=0;
4. repeat i:=i+1 until x[i]>0;
5. set j:=j+1; a[j]:=1 and goto 3;
```

Fig. 10.3

Our aim is to find a closed formula for a_n. A look at the graph of this highly irregular sequence shows that a_n grows almost linearly. Thus $a_n \approx tn$, where t may be some irrational number, which accounts for the small but erratic fluctuation about a straight line. To find a good approximation for t we must evaluate a_n/n for large n.

The program sieve in Fig.10.4 gives $t \approx 1.6176$ for $n = 2000$. This is very close to the golden ratio $\frac{1}{2}(1 + \sqrt{5}) = 1.61803\ldots$. We can convince ourselves that if a_n/n approaches a limit t then t is indeed the golden ratio as follows. The number of integers in the complementary sequence up to $a_n + n$ is n. (Two sequences of natural numbers are called *complementary* if together they contain each natural number exactly once.) Hence the number of numbers in the sequence itself up to that point is a_n. So $a_{a_n} \approx a_n + n$. Assuming that $a_n \approx tn$, we get $t^2 n \approx tn + n$ and hence $t^2 = t + 1$. Since a_n/n is always smaller than t, we conjecture that $a_n = \lfloor tn \rfloor$.

```
program sieve;
var i,j,n:integer;
    a,x:array[1..4000] of
                    integer;
begin
  write('n='); readln(n);
  for i:=1 to n do x[i]:=1;
  for j:=1 to n do a[j]:=0;
  i:=1; j:=1; a[j]:=1;
  while i<n do
  begin
    x[i+j]:=0;
    repeat i:=i+1 until
                    x[i]>0;
    j:=j+1; a[j]:=i
  end;
  writeln(a[j]/j:10:8)
end.
```

Fig. 10.4

A small change in the program checks if this formula is true for $1 \ldots n$. We introduce a boolean variable b and replace `writeln(a[j]/j)` by the following lines

```
b:=true; t:=(1+sqrt(5))/2;
for i:=1 to j do if a[i]<>trunc(t*i) then b:=false;
writeln(b);
```

The beautiful theorem of Sam Beatty enables us to verify our conjecture quickly. It says that if $\alpha > 1$ is irrational then the complementary sequence of $\{\lfloor \alpha n \rfloor\}$ is $\{\lfloor \beta n \rfloor\}$, where β is the number such that $1/\alpha + 1/\beta = 1$. (For proofs see the solution of ex. 2, section 48 and Honsberger's Ingenuity in Mathematics, NML24.)

The golden ratio number t satisfies $1/t + 1/t^2 = 1$, hence the complementary sequence of $\{\lfloor tn \rfloor\}$ is $\{\lfloor t^2 n \rfloor\}$. But $t^2 = t + 1$, hence $\lfloor t^2 n \rfloor = \lfloor tn + n \rfloor = \lfloor tn \rfloor + n$. So the sequence $\{\lfloor tn \rfloor\}$ obeys the relation which defines our sequence.

10.4. An Olympiad Problem. Do there exist 1983 distinct positive integers ≤ 100000 with no three in arithmetic progression? (Problem 5 of IMO XXIV, Paris 1983.)

To get some numerical evidence we construct a tight sequence a_j with no three in arithmetic progression by means of the following "greedy" algorithm: $a_0 = 0$, $a_1 = 1$, $a_j =$ the smallest positive integer, which does not form an arithmetic progression with any two previous terms.

We start with an x-array and an a-array, and we set $a[0] = 0$, $a[1] = 1$, $x[i] = 1$ for $i = 2$ to 10000. The terms $a < b < c$ are in arithmetic progression if $2b = a + c$, or $c = 2b - a$. Of the numbers $i = 2$ to 10000 we successively sieve out those which are in arithmetic progression with any two of the preceding elements $a[0], \ldots, a[j-1]$. The first i that remains is $a[j]$.

Sieve out all elements which form an arithmetic progression with any two of $a[0]$ to $a[j-1]$:

```
for k:=0 to j-1 do x[2*i-a[k]]:=0;
```

Find the first element of the x-array not sieved out:

```
repeat i:=i+1 until x[i]=1;
```

The first i with $x[i] = 1$ is $a[j]$:

```
j:=j+1; a[j]:=i;
```

This is repeated until we have $j = 256$, say. Then the elements $a[0]$ to $a[256]$ are printed out. We get the program **greedy** in Fig. 10.5 and the sequence tabulated in Fig. 10.6. By looking closely at this sequence, we observe ever longer gaps — between 1 and 3, 4 and 9, 13 and 27, 40 and 81, etc. — and the upper endpoints of these gaps are successive powers of 3: 3, 9, 27, 81, 243, 729, 2187. We also observe that the upper endpoint, U, of a

```
program greedy;
var i,j,k:integer; a,x:array[0..8000] of integer;
begin
  for i:=0 to 8000 do x[i]:=1;
  a[0]:=0; a[1]:=1; i:=1; j:=1;
  repeat
    for k:=0 to j−1 do x[2*i−a[k]]:=0;
    repeat i:=i+1 until x[i]=1;
    j:=j+1; a[j]:=i;
  until j=256;
  for i:=0 to 256 do write(a[i]:6)
end.
```

Fig. 10.5

0	1	3	4	9	10	12	13	27	28	30	31
36	37	39	40	81	82	84	85	90	91	93	94
108	109	111	112	117	118	120	121	243	244	246	247
252	253	255	256	270	271	273	274	279	280	282	283
324	325	327	328	333	334	336	337	351	352	354	355
360	361	363	364	729	730	732	733	738	739	741	742
756	757	759	760	765	766	768	769	810	811	813	814
819	820	822	823	837	838	840	841	846	847	849	850
972	973	975	976	981	982	984	985	999	1000	1002	1003
1008	1009	1011	1012	1053	1054	1056	1057	1062	1063	1065	1066
1080	1081	1083	1084	1089	1090	1092	1093	2187	2188	2190	2191

Fig. 10.6

gap is one greater than twice its lower endpoint, L, i.e. $U = 2L + 1$. This leads to an alternative algorithm for generating arbitrarily long sequences with the desired property. Beginning with any finite sequence

$$(1) \qquad\qquad a_1, \ a_2, \ \ldots, \ a_k$$

having the property, it adds not just the next term, but a whole block of k new terms. These are obtained by translating each term in (1) by $2a_k + 1$ units. We now have the sequence of $2k$ terms

$$(2) \qquad a_1, \ a_2, \ \ldots, a_k, \ a_{k+1} = 2a_k + 1 + a_1, \ \ldots, \ a_{2k} = 2a_k + 1 + a_k \ .$$

Exercise 1. a) Prove that this sequence of $2k$ terms inherits the desired property from the sequence of k terms.

b) Beginning with the sequence 0, 1, write a program for this algorithm, and observe that it generates the same sequence as the greedy algorithm.

c) Prove that $a_{2^j+1} = 3^j$.

Exercise 2. a) Rewrite the program **greedy** so that it prints the output in base 3, and observe that all numbers appearing in the sequence have only ternary digits 0 or 1, and that all such numbers appear in increasing order, i. e., 0, 1, 10, 11, 100, 101, 110, ... up to $3^j = 10...0$ (j zeros).

b) Prove that the finite sequence generated by the greedy algorithm consists exactly of the numbers whose ternary digits are 0's and 1's. (Hint: Use induction.)

c) Prove that the sequence generated by the alternative algorithm starting with 0, 1 consists exactly of the numbers whose ternary digits are 0's and 1's. (Note that b) and c) together prove that your observation in Exercise 1 b) persists for arbitrarily long sequences.)

To answer the question posed in the Olympiad problem, we need only verify that the first $2^{11} = 2048 > 1983$ terms of our sequence are all $\leq 10^5$. The largest of them is $a_{2^{11}}$, the lower endpoint of the gap with upper endpoint 3^{11}. Its value, because of the relation $U = 2L + 1$, or $3^{11} = 2a_{2^{11}} + 1$, is $a_{2^{11}} = (3^{11} - 1)/2 = 88573 < 10^5$. We conclude that our sequence of 2048 integers from 0 to 88573 contains no three in arithmetic progression.

10.5. Ulam's Sequence At the 1963 number theory conference in Boulder, Colorado, S. Ulam proposed the following sequence U : $u_1 = 1, u_2 = 2$, $u_n =$ the smallest integer that can be uniquely represented as a sum of two different preceding terms.

There is an obvious, but slow, program for this sequence (Exercise 2 below). There is also a sophisticated and vastly superior program which uses a "double sieve" to compute the elements of $U \leq n$. It uses two boolean arrays $x[1..n]$ and $y[1..n]$. For $i > 2$

x[i]=true iff i is representable as a sum in at least one way.

y[i]=true iff i is representable as a sum in at most one way.

The counter k is incremented until it reaches the first integer beyond the previous member of U that is representable in exactly one way. This integer is in U, i.e. x[k] and y[k] =true, and all integers $k + i$ for $1 < i < k + 1$ must have their x[k+i], y[k+i] updated. Fig. 10.7 describes an algorithm which identifies all elements of U which are $\leq n$. Change it into a Pascal program. Several exercises are based on it.

```
x[1..n]:=[true,true,false,...,false]
y[1..n]:=[true,   ...,  true]
k:=1
while k<n do
begin
  k:=k+1
  while not (x[k] and y[k]) and (k<n) do k:=k+1;
  if k-1<n-k then min:=k-1 else min:=n-k;
              {computes min(k-1,n-k)}
  for i:=1 to min do
  begin
    y[k+i]:= y[k+i] and not(x[k+i] and x[i] and y[i])
    x[k+i]:= x[k+i] or (x[i] and y[i])
  end
end;
```

Fig. 10.7

Exercises for Section 10:

1. In the central prison of Sikinia there are n cells numbered 1 to n, each
occupied by a single prisoner. The state of each cell can be changed from
closed to open and vice versa by a half-turn of the key. To celebrate the
Centennial Anniversary of the Republic it was decided to grant a partial
amnesty. The president sent an officer to the prison with the instruction

```
for i:=1 to n do
turn the keys of cells i, 2i, 3i,... .
```

A prisoner was freed if at the end his door was open. Which prisoners are
set free? Remark: Do not think, just sieve!

2. a) Write an "obvious" program for the Ulam sequence and compare its
speed with that based on Fig. 10.7.
The following questions should be answered by means of Fig. 10.7.

   b) In the interval 1..32000 find all solutions of $u_i + u_{i+1} = u_k$.

   c) In 1..32000 find all pairs of consecutive numbers which are U-num-
bers.

   d) Among the distances $g_i = u_{i+1} - u_i$ between successive U numbers
in the interval 1..32000, find the largest.

   e) What proportion of U-numbers are twins with distance 2?

   f) Find the frequency distribution of the distances g_i.

   g) Does the asymptotic distribution of U-numbers parallel that of the
distribution of primes?

   h) What sequence do you get if the two preceding terms need not be
different?

3. Construct a sequence $0 < a_1 < a_2 < a_3 < \ldots$ of integers with the property that each nonnegative integer n can be uniquely represented in the form $n = a_i + 2a_j$, where i and j need not be distinct. Collect numerical data, conjecture, prove.

4. Consider the sequence of the first digit in
   a) fib(n) for $n = 1$ to 10000,
   b) 2^n for $n = 1$ to 10000,
   c) 3^n for $n = 1$ to 10000.

   Find the frequencies $x[1], \ldots, x[9]$ of the digits 1 to 9 in each of the sequences found in parts a), b), c) above. How do you avoid overflow? (In this range instances when the roundoff error can affect the first digit are sure to be so rare that you need not consider this posibility.)

   The result is surprising, but it can be shown that the frequency of the digit n is $\log_{10}(1 + 1/n)$ for $n = 1, \ldots, 9$.

5. Prove that the density of square-free integers is $p = 6/\pi^2$. Use the procedures of Section 6.4.

6. **The Frobenius Problem.** Let $a_1, \ldots, a_k$ be natural numbers with $\gcd(a_1, \ldots, a_k) = 1$. The natural number N can be represented by $a_1, \ldots, a_k$ if there are nonnegative integers $x_1, \ldots, x_k$, such that $N = a_1 x_1 + \cdots + a_k x_k$. It is not hard to show that all sufficiently large numbers can be so represented. The Frobenius Problem is to find the largest number $g(a_1, \ldots, a_k)$ without such a representation. The problem has many interpretations. E.g., given coins with denominations $a_1, \ldots, a_k$, find the largest amount which cannot be paid with these coins.
   a) For $k = 2$ there is a simple solution. Find it by experimenting with the computer.
   b) For $k > 2$ only a few partial results are known. Write a program for $k = 3$, which collects experimental data.
   c) Write an efficient program which for any input $a_1, \ldots, a_k$ finds $g(a_1, \ldots, a_k)$. (Very difficult problem.)

7. **The $3n + 1$ Problem.** The following algorithm for generating a sequence is quite old, yet it is not even known if it always stops:
   i) Start with a natural number n.
   ii) If $n = 1$ then stop.
   iii) If n is odd then set $n := 3n + 1$.
   iv) If n is even then set $n := n$ div 2 and go to ii)
   a) Write a program, which for input n prints the sequence $n, f(n)$, $f(f(n)), \ldots, 1$, where

$$f(n) = \begin{cases} 3n + 1 & \text{if} \quad n \text{ is odd} \\ n/2 & \text{if} \quad n \text{ is even.} \end{cases}$$

b) Write a program which for input n counts the number $t(n)$ of terms of the sequence.

c) Print a table $(i, t(i))$ for $i = a, a+1, \ldots, b$.

d) For input n print the triple $(n, t(n), \max(n))$, where $\max(n)$ is the largest term of the sequence.

e) The algorithm can stop only if it encounters a power of 2. Write a program, which for input n prints the exponent of this power of 2.

f) Write a program which for i between a and b, prints the largest $t(i)$ and the values of i for which it occurs.

8. Suppose you want to test, for all numbers from 2 to b, whether or not the $3n + 1$ algorithm stops.

a) Show that you only need to test odd numbers.

b) Show that if you start with n you may stop as soon as you encounter a number smaller than n.

c) Show that you need to test only numbers of the form $4n - 1$.

d) Show that you need not test numbers of the form $16k + 3$ or $128k + 7$.

9. Consider once again the $3n + 1$ algorithm. Let us study the record-breaking numbers, i.e. those numbers which require more steps to reach 1 than any preceding number. Write a program which prints these numbers, the number of steps to reach 1 and the maximum height reached. Partial answer for checking:

serial number	record-breaking number	steps	largest term	serial number	record-breaking number	steps	largest term
1	2	1	2	16	327	143	9232
2	3	7	16	17	649	144	9232
3	6	8	16	18	703	170	250504
4	7	16	52	19	871	178	190996
5	9	19	52	20	1161	181	190996
6	18	20	52	21	2223	182	105504
7	25	23	88	.	.	.	.
8	27	111	9232	.	.	.	.
9	54	112	9232	.	.	.	.
10	73	115	9232	51	5649499	612	1017886660
11	97	118	9232	52	6649279	664	15208728208
12	129	121	9232	53	8400511	685	159424614880
13	171	124	9232	54	11200681	688	159424614880
14	231	127	9232	55	14934241	691	159424614880
15	313	130	9232	56	15733191	704	159424614880

10. **Fibonacci squares: An empirical study.** Are there squares among the Fibonacci numbers? Obviously fib(1)=fib(2)=1 and fib(12)= 144 are squares, and also fib(0)=0, but not everybody starts with 0. Are there more squares in the sequence? Our aim is to sieve out the nonsquares among the Fibonacci numbers from 1 to 10000. We first set x[i]:=1

for i:=1 to 10000. These are our potential Fibonacci squares. The idea is to consider the sequence modulo a prime p. Now any square is also a square mod p. Equivalently, a nonsquare mod p is not a square. So we sieve out the nonsquares mod $3,5,7,11,13,\ldots$. Take the Fibonacci sequence mod 3: $1,1,2,0,2,2,1,0,1,1,\ldots$. It has period 8. The only nonsquare mod 3 is 2. So we can sieve out all terms with indices $8n+3$, $8n+5$, $8n+6$ by setting the corresponding x[i] := 0. Now take the sequence mod 5: $1,1,2,3,0,3,3,1,4,0,4,4,3,2,0,2,2,4,1,0,1,1,\ldots$. This sequence has period 20. The nonsquares mod 5 are 2 and 3. So we set x[20n+r]:=0 for $r = 3,4,6,7,13,14,16,17$. Go on in this way until all the elements except x[1], x[2], x[12] are zero. See references at the end of this book.

11. Explore the sequence

$$f(n) = \begin{cases} 3n - 1 & \text{if } n \text{ is odd} \\ n/2 & \text{if } n \text{ is even.} \end{cases}$$

The sequence always seems to run into one of four cycles. Find them and prove this for $n < 10000$. Make also some checks with large numbers.

12. **Conway's permutation sequences.** A sequence is defined by

$$f(n) = \begin{cases} \lfloor (3n + 1)/4 \rfloor & \text{if } n \text{ is odd} \\ 3n/2 & \text{if } n \text{ is even} \end{cases}$$

or, more lucidly $2m \mapsto 3m$, $4m - 1 \mapsto 3m - 1$, $4m + 1 \mapsto 3m + 1$, from which it is clear that the operation is invertible. So, if we connect n with $f(n)$, $n = 1, 2, \ldots$, the resulting graph consists only of disjoint cycles and doubly infinite chains. Explore some sequences. It is conjectured that there are only four cycles. But there is no conjecture about the number of infinite chains. What is the status of the sequence containing the number 8?

13. Proceeding as in Ex. 10, find by successive sieving the triangular Fibonacci numbers, i.e., Fibonacci numbers of the form $m(m + 1)/2$, below 1000.

11. Rotation of an Array

How do you rotate an array v of n elements left by m positions? For instance, with $n = 10$ and $m = 3$ the array 0123456789 should be transformed into 3456789012. Can you do this with little extra storage? The problem makes sense if m and n are both large. Of course, you could copy the first m array elements in a second array $y[1..m]$, moving the remaining elements left m places and appending the first m elements from the temporary array y back to the end of the array v. But what if m is so large that not enough space for copying is left?

We first consider an easy to understand and quite efficient solution. Rotation means switching two blocks $AB \mapsto BA$. Suppose we have a procedure **reverse**, which reverses a block: $A \mapsto A^R$. Then we form $(A^R B^R)^R$ and this is BA. For our example with $n = 10$, $m = 3$ we get

$$0123456789 \mapsto 2109876543 \mapsto 3456789012.$$

Thus we must call the procedure **reverse(a,b)** (reverse the part of the array from the a-th to the b-th elements) three times:

```
reverse (1,m):    210 3456789
reverse (m + 1, n):  210 9876543
reverse (1, n):   3456789 012.
```

The program **rotate** in Fig. 11.1 is based on this idea.

There is a completely different algorithm which may be somewhat more efficient if $\gcd(m, n) = 1$ or some small number. This so called **dolphin-algorithm** consists of moving array elements into their new places one by one, with the next move being to fill the space that has been vacated. First we look at the case $m = 3$, $n = 10$, with $\gcd(m, n) = 1$. We save a copy of $v[0]$ by $x \leftarrow v[0]$. Now we move $v[3]$ to its final position $v[0]$ leaving $v[3]$ free to be occupied by $v[6]$, etc.:

$$x \leftarrow v[0] \leftarrow v[3] \leftarrow v[6] \leftarrow v[9] \leftarrow v[2] \leftarrow v[5]$$
$$\leftarrow v[8] \leftarrow v[1] \leftarrow v[4] \leftarrow v[7] \leftarrow x.$$

In assigning $v[i] \leftarrow v[i+3]$ we reduce mod 10 as soon as $i + 3$ becomes 10 or more.

Let now $m = 6$, $n = 15$ with $\gcd(m, n) = 3$. In $v[i] \leftarrow v[i+6]$ we reduce $i + 6$ mod 15 and get

$$x \leftarrow v[0] \leftarrow v[6] \leftarrow v[12] \leftarrow v[3] \leftarrow v[9] \leftarrow x$$
$$x \leftarrow v[1] \leftarrow v[7] \leftarrow v[13] \leftarrow v[4] \leftarrow v[10] \leftarrow x$$
$$x \leftarrow v[2] \leftarrow v[8] \leftarrow v[14] \leftarrow v[5] \leftarrow v[11] \leftarrow x.$$

The number of cycles is 3, or, generally $\gcd(m, n) = \gcd(m, n - m)$. The corresponding Pascal program is shown in Fig. 11.2.

```
program rotate;
var k,m,n:integer;
    v:array[0..200] of
                    integer;
procedure reverse(l,r:
                     integer);
var i,j,t:integer;
begin
  i:=l; j:=r;
  while j-i>0 do
  begin
    t:=v[i]; v[i]:=v[j];
                  v[j]:=t;
    i:=i+1; j:=j-1
  end
end;

begin
  write('m,n=');readln(m,n);
  for k:=1 to n do v[k]:=k;
  reverse(1,m);
  reverse(m+1,n);
  reverse(1,n);
  for k:=1 to n do
   write(v[k],' ')
end.
```

Fig. 11.1

```
program dolphin;
var d,i,j,k,m,n,x:integer;
    v:array[0..200] of
                      integer;
{$I gcditer}
begin write('m,n=');
  readln(m,n);
  for i:=0 to n-1 do
    v[i]:=i;
  d:=gcd(m,n-m);
  for i:=0 to d-1 do
  begin
    j:=i; x:=v[i];
    repeat
      k:=j; j:=j+m;
      if j>=n then j:=j-n;
      v[k]:=v[j]
    until j=i;
    v[k]:=x
  end;
  for i:=0 to n-1 do
    write(v[i],' ')
end.
```

Fig. 11.2

12. Partitions

Let $p(n)$ be the number of different partitions of n into natural parts. For instance,

$$5 = 4+1 = 3+2 = 3+1+1 = 2+2+1 = 2+1+1+1 = 1+1+1+1+1.$$

Thus $p(5) = 7$. In addition let $f(m,n)$ be the number of partitions of m with each part at most n. Then $p(n) = f(n,n)$. But $f(m,n)$ can be found recursively:

(1) $f(1,n) = f(m,1) = 1$ for all m,n;

(2) $f(m,n) = f(m,m)$ for $m \leq n$;

(3) $f(m,m) = 1 + f(m,m-1)$;

(4) $f(m,n) = f(m,n-1) + f(m-n,n)$ for $n < m$.

We can combine (2) and (3) into

$$f(m, n) = 1 + f(m, m - 1) \text{ for } m \leq n .$$

Thus we get the recursive program parts in Fig. 12.1. It gives $p(39) = 31185$. But $p(40) >$ maxint. We could declare the function values as reals, but we will not get very far in this way. We have a tree recursive computational process and the computation time grows exponentially with n. After some deliberation we succeed in writing the very fast iterative algorithm **partiter** in Fig. 12.2, which is easy to comprehend without comment. In this program we have put $f(0, m) = f(0, 0) = p(0) = 1$. Then $f(m, n) = f(m, n - 1) + f(m - n, n)$ is also valid for $m = n$. By storing $f(j, i)$ in $f(j, i - 1)$ we get a one-dimensional problem. The recursion $f(j, i) = f(j, i - 1) + f(j - i, i)$ turns into $p[j] \leftarrow p[j] + p[j - i]$. We must start at i and proceed up to n.

```
program parts;
var n:integer;
function f(m,n:integer):
                    integer;
begin
  if (m=1) or (n=1) then f:=1
  else if m<=n then
    f:=1+f(m,m-1)
  else f:=f(m,n-1)+f(m-n,n)
end;

begin write('n=');readln(n);
  writeln('p(',n,')=',f(n,n))
end.
```

Fig. 12.1

```
program partiter;
var i,j,n:integer;
    p:array[0..200] of real;
begin
  write('n='); readln(n);
  for i:=0 to n do p[i]:=1;
  for i:=2 to n do
    for j:=i to n do
      p[j]:=p[j]+p[j-i];
  writeln('p(',n,')=',p[n]:0:0);
readln;end.
```

Fig.12.2

With the program **partiter** we found

$$p(127) = 3\,913\,864\,295, \qquad p(159) = 97\,662\,728\,555,$$
$$p(138) = 12\,292\,341\,831, \qquad p(160) = 107\,438\,159\,470.$$
$$p(149) = 37\,027\,355\,200,$$

We are not completely sure that the last digit is correct, but Ramanujan has proved the congruences

$$p(5m + 4) \equiv 0 \pmod 5, \qquad p(7m + 5) \equiv 0 \pmod 7,$$
$$p(11m + 6) \equiv 0 \pmod{11}.$$

Of the above values of n, 159 is of the form $7m + 5$ and the others have the form $11m + 6$. So $p(159)$ should be divisible by 7 and the other values we obtained should be divisible by 11. This is indeed the case except for $p(160)$, which is not divisible by 11. But this number has 12 digits and Turbo Pascal gives only 11-digit accuracy.

13. The Money Changing Problem

In how many different ways can you change one dollar? That is, in how many different ways can you pay 100 cents using five different kinds of coins: cents, nickels, dimes, quarters and half-dollars (worth 1,5,10,25, and 50 cents, respectively)? We generalize slightly and ask: In how many ways can you pay a sum of n cents with US-coins? Other countries have other coins. So we generalize again and ask:

A country has coins with denominations $D = \{1, d_2, d_3, \ldots, d_k\}$ units. In how many ways can you pay a sum of n units ?

We want to have $d_1 = 1$, so we can pay any amount. Let $a(n, k)$ be the number of ways to pay a sum of n units using coins from D. Then we have

$$a(n, 1) = 1,$$
$$a(n, k) = 0 \text{ for } n < 0,$$
$$a(n, k) = a(n, k - 1) + a(n - d_k, k).$$

Indeed, a payment of the sum n either contains the coin d_k (at least once) or it does not. There are $a(n, k-1)$ ways to pay without using coin d_k and $a(n - d_k, k)$ ways to pay using coin d_k. The recursion for $a(n, k)$ is implemented by the program in Fig. 13.1; before running it, one has to add a line assigning the values of d[1],..,d[k].

```
program change;
var i,k,n:integer;
     d:array[1..10] of integer;
function a(n,k:integer):
                    integer;
begin
  if n<0 then a:=0
  else if k=1 then a:=1
  else a:=a(n,k-1)+a(n-d[k],k)
end;

begin
  write('n,k=');readln(n,k);
  for i:=1 to k do
               readln(d[i]);
  writeln;
  writeln('# of ways to
               pay=',a(n,k))
end.
```

```
program change1;
const d:array[1..5] of
      integer=(1,5,10,25,50);
var i,k,n:integer;
function a(n,k:integer):
                         real;
begin
  if n<0 then a:=0
  else if k=1 then a:=1
  else
     a:=a(n,k-1)+a(n-d[k],k)
end;

begin
  write('n,k=');
  readln(n,k);
  writeln('# of ways to
               pay=',a(n,k):0:0)
end.
```

Fig. 13.1   Fig. 13.2

For U. S. coins we get

n	100	200	300	400	500
$a(n,k)$	292	2435	9590	26517	−5960

The negative number for $n = 500$ indicates that a(n,k) > maxint. It should be $32767+(32768-5960+1) = 59576$, but we are not sure since we do not know if there was just one "wraparound". A slight change of the program will do. We declare the function values $a(n, k)$ to be reals, and we make an insignificant change in the output: writeln(a(n,k):0:0) instead of writeln(a(n,k)). That is enough. But in Turbo Pascal not only constant numbers, but also constant arrays can be declared. So we declare the array of U. S. coins to be a constant, as shown in Fig. 13.2. If we use another set of coins this array must be retyped. Indeed, for $n = 500$, we get the result $a(n, k) = 59576$. The computation for $n = 500$ takes about 7 seconds on my AT, twice as long as it did with integers.

We are dealing here with a tree recursive process. Let us construct an iterative program to deal with larger n or with other countries. The direct translation of Fig. 13.1 or 13.2 using a matrix $a[n, k]$ is too space consuming. Instead we will compute the table $a[n, k]$ row-wise using the recurrence. We get the program change2.

```
program change2;
var i,j,k,n:integer;
    d:array[1..8] of integer;
    a:array[0..600] of real;
begin
  write('n,k=');readln(n,k);
  for j:=1 to k do
    readln(d[j]);
  for i:=0 to n do a[i]:=1;
  for j:=2 to k do
  for i:=d[j] to n do
    a[i]:=a[i]+a[i-d[j]];
  writeln('a(n,k)=',
          a[n]:0:0)
end.
```
                Fig. 13.3

```
program change3;
var d,i,j,k,n:integer;
    a:array[0..3000] of real;
begin
  write('n,k=');readln(n,k);
  for i:=0 to n do a[i]:=1;
  for j:=2 to k do
  begin
    readln(d);
    for i:=d to n do
      a[i]:= a[i]+a[i-d];
  end;
  writeln('a(n,k)=',a[n]:0:0)
end.
```
                Fig. 13.4

The program change2 computes $a(n, k) = 59576$ for n=500 in a tiny fraction of a second. Even more elegant and efficient is change3. In this program you should not start with the input $d = 1$. Instead you should start with the second coin in the 9-th line in the program.

With the program change3 we get for US-coins

n	600	700	800	900	1000	2000	3000
a(n,k)	116727	207530	343145	53633	801451	11712101	57491951

The computation time is negligible.

14. An Abstractly Defined Set

A set S is defined by means of three axioms:

A1. $0 \in S$

A2. $x \in S \rightarrow 2x + 1 \in S$ and $3x + 1 \in S$

A3. S is the minimal set satisfying axioms A1 and A2.

This is an instructive example of recursion and backtracking. We first write a boolean function which recognizes if a number n belongs to S. Which of the numbers 511, 994, 995, 996, 997, 998, 999 do belong to S? The trees in Fig. 14.1 show that only 511, 994, 999 belong to S. From these numbers there is a descending path leading to 0.

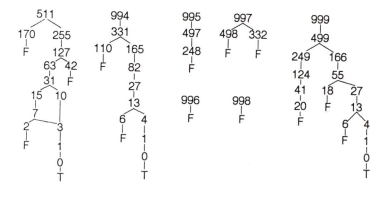

Fig. 14.1

Study the programs inset and setiter until you understand them completely. They are not difficult. If you run inset for $n = 1000$ it will print the table in Fig. 14.4. The program setiter computes and counts the elements of the same set iteratively. It is a sieve in reverse.

```
program inset;
var i,n:integer;
function inS(n:integer):
                 boolean;
begin
  n:=n-1;
  if n<=0 then inS:=true
  else if (n mod 2<>0) and
          (n mod 3<>0)
  then inS:=false
  else if n mod 3<>0 then
           inS:=inS(n div 2)
  else if n mod 2<>0 then
           inS:=inS(n div 3)
  else inS:=inS(n div 2) or
            inS(n div 3)
end;
begin
  write('n='); readln(n);
  for i:=0 to n do if inS(i)
  then write(i:5)
end.
```

Fig. 14.2

```
program setiter;
var a,b,i,n,count:integer;
   x:array[0..10000] of 0..1;
begin
  write('n=');readln(n);
  count:=0;
  for i:=0 to n do x[i]:=0;
  i:=0; x[i]:=1;
  repeat
    if x[i]=1 then
    begin a:=2*i+1;b:=3*i+1;
      if a<=n then x[a]:=1;
      if b<=n then x[b]:=1
    end;
    i:=i+1
  until i>n div 2;
  for i:=0 to n do
    if x[i]=1 then
    begin count:=count+1;
    write(i:5) end;
  writeln('count=',count)
end.
```

Fig. 14.3

0	1	3	4	7	9	10	13	15	19	21	22
27	28	31	39	40	43	45	46	55	57	58	63
64	67	79	81	82	85	87	91	93	94	111	115
117	118	121	127	129	130	135	136	139	159	163	165
166	171	172	175	183	187	189	190	193	202	223	231
235	237	238	243	244	247	255	256	259	261	262	271
273	274	279	280	283	319	327	331	333	334	343	345
346	351	352	355	364	367	375	379	381	382	387	388
391	405	406	409	418	447	463	471	475	477	478	487
489	490	495	496	499	511	513	514	517	519	523	525
526	543	547	549	550	559	561	562	567	568	571	580
607	639	655	663	667	669	670	687	691	693	694	703
705	706	711	712	715	729	730	733	735	742	751	759
763	765	766	769	775	777	778	783	784	787	811	813
814	819	820	823	837	838	841	850	895	927	943	951
955	957	958	975	979	981	982	991	993	994	999	1000

Fig. 14.4

15. A Recursive Program with a Surprising Result

Given a pile of n chips, $n > 1$. This pile is randomly split into two piles with k and $n - k$ chips, and the product $k(n - k)$ is computed. Then each of the two piles, provided it has more than one chip, is randomly split. The product of the number of chips in its parts is added to the preceding product. This procedure is repeated until all piles are reduced to 1.

a) Write a recursive program to find the sum of the $n-1$ products.

b) Surprisingly, the sum does not seem to depend on the way of splitting.

c) Can you prove this result?

The program sumprod yields the following table:

```
program sumprod;
var n:integer;

function f(n:integer):
                      integer;
var k:integer;
begin
  if n<3 then f:=n-1 else
  begin
    k:=1+random(n-1);
    f:=k*(n-k)+f(k)+f(n-k)
  end
end;

begin
  write('n='); readln(n);
  writeln(f(n))
end.
```

Fig. 15.1

n	2	3	4	5	6	7	8	9	10
$f(n)$	1	3	4	10	15	21	28	36	45

Thus we guess $f(n) = \binom{n}{2}$. For a proof let us tie each chip to every other chip by a string. Initially there will be $n(n - 1)/2$ strings. Splitting a pile into piles X, Y with x and y chips, respectively, we cut all strings connecting X to Y. These are xy strings. The total number of strings cut will be $n(n - 1)/2$. This is also the total sum of products, no matter how we split the piles.

16. Sequence 56 in Sloane's Handbook of Integer Sequences

We define a function f on the nonnegative integers by means of the recursion
$$f(0) = 0, \quad f(1) = 1, \quad f(2n) = f(n), \text{ and}$$
$$f(2n + 1) = f(n) + f(n + 1) \text{ for all } n > 0.$$

Let $n = 2^s + e_{s-1}2^{s-1} + \cdots + e_0$ be the binary representation of n. If we reverse the digits of the binary representation (e.g. $10100 \to 00101$), we get $u = e_0 2^s + e_1 2^{s-1} + \cdots + e_{s-1} 2 + 1$. One can prove that f has the same value at n as it has at the integer obtained by reversing the binary digits of n, i.e., always $f(n) = f(u)$. The proof is not simple. Let us write a program, which prints $f(n)$ and the truth value of the relation $f(n) = f(u)$.

```
program reverse1;
var n,r,u: integer;

function f(n:integer):
                  integer;
begin if n<2 then f:=n
   else if odd(n-1) then
     f:=f(n div 2)
   else f:=f((n-1) div 2)
          + f((n+1) div 2)
end;
begin write('n='); readln(n);
  r:=n; u:=0;
  repeat
    u:=u*2+r mod 2;
    r:=r div 2
  until r=0;
  writeln(f(n),'   ',
          f(n)=f(u))
end.
```
Fig. 16.1

```
program reverse2;
var n,r,u:real;

function f(n:real):real;
begin if n<2 then f:=n
   else if n/2=int(n/2) then
                  f:=f(n/2)
   else
     f:=f((n-1)/2)+f((n+1)/2)
end;
begin write('n='); read(n);
  r:=n; u:=0;
  repeat
    u :=u *2+r-2*int(r/2);
    r:=int(r/2)
  until r=0;
  r:=f(n);
  write(r:1:0,'   ',r=f(u))
end.
```
Fig. 16.2

With the program **reverse1** we check at random integers up to $n = 32767$ and we always get the answer TRUE. To check integers beyond maxint we must rewrite the program as in Fig. 16.2. But if we try the input $n = 1234567$ we must wait over 45 seconds to get the result **6279 TRUE** (with my AT).

The process is tree recursive, but the depth of the tree is merely about $\log_2 n$. So we can get very far, but the number 123456789 is almost out of our reach. It requires 8.75 minutes to get **83116 TRUE**. Let us transform the recursion to speed up the computation. For better understanding we compute one function value in detail:

$$f(85) = f(42) + f(43) = f(21) + f(21) + f(22)$$
$$= 2 * f(21) + f(22) = 2 * f(10) + 3 * f(11)$$
$$= 5 * f(5) + 3 * f(6) = 5 * f(2) + 8 * f(3)$$
$$= 13 * f(1) + 8 * f(3) = 13 * f(0) + 21 * f(1) = 21.$$

Let us introduce the function

$$g(n, i, j) = i * f(n) + j * f(n + 1).$$

Then we have

$$g(2n, i, j) = i * f(2n) + j * f(2n + 1)$$
$$= (i + j) * f(n) + j * f(n + 1) = g(n, i + j, j),$$
$$g(2n + 1, i, j) = i * f(2n + 1) + j * f(2n + 2)$$
$$= i * f(n) + (i + j) * f(n + 1) = g(n, i, i + j).$$

Thus we get for g the recursions

$$g(2n, i, j) = g(n, i + j, j)$$
$$g(2n + 1, i, j) = g(n, i, i + j).$$

Now $f(n) = g(n, 1, 0)$. By repeated application of the recursions for g we can get to $g(0, i, j) = j$. Then $f(n) = j$.

The program **reverse3** is based on this idea. With this program we get $f(n)$ for any n up to 11 digits in a tiny fraction of a second.

```
program reverse3;
var n:real;
function g(n,i,j:real):real;
begin
  if n=0 then g:=j
  else if n/2<>int(n/2)
  then g:=g((n-1)/2,i,i+j)
  else g:=g(n/2,i+j,j)
end;
begin write('n=');readln(n);
  write('f(n)=',g(n,1,0):1:0)
end.
```

Fig. 16.3

Exercises for Sections 11-16:

1. Here is yet another method for rotating an array. Initially we have $V = AB$. This initial condition must be transformed into $V = BA$. Suppose that the block A is not longer than B. Then we can write $B = B_0 B_1$ with block B_1 having the same length as A. Blocks of equal length are easy to swap and we get $V = B_1 B_0 A$. Now A is in its final position and we must transform $B_1 B_0$ into $B_0 B_1$. But this is the same task with shorter blocks B_1, B_0. The case that the block B is not longer than A is treated analogously. Write a program based on this idea.

2. Let $b(0) = 1$ and for $n \geq 1$ let $b(n)$ denote the number of ways of writing n as a sum of powers of 2 (order does not count). Thus

$$7 = 4 + 2 + 1 = 4 + 1 + 1 + 1 = 2 + 2 + 2 + 1 = 2 + 2 + 1 + 1 + 1$$
$$= 2 + 1 + 1 + 1 + 1 + 1 = 1 + 1 + 1 + 1 + 1 + 1 + 1$$

and so $b(7) = 6$. Show that

(1) $b(2n + 1) = b(2n),$ (2) $b(2n) = b(2n - 2) + b(n).$

a) Construct a recursive program, which computes $b(n)$.

b) Construct an iterative program, which computes $b(n)$. Partial result:

$$b(132) = 31196.$$

3. Let $r(n, m)$ be the number of partitions of n into parts, the smallest being at least m. We define $r(0, m) = 1$. Obviously $p(n) = r(n, 1)$, with $p(n)$ as defined in § 12.

a) Show that $r(n, m) = r(n, m + 1) + r(n - m, m)$

b) Write a recursive program parts2, which computes $p(n)$.

c) Write an iterative program partit, which is similar to partiter.

4. Let the number of partitions of n into parts less than or equal to m be denoted by $p(n, m)$. Prove that $p(n, m) = p(n, m - 1) + p(n - m, m)$ (Euler). Find boundary conditions and write a recursive program parts3, which computes p. Compare its speed with that of parts.

5. Let $p(u, n, m)$ be the number of partitions of u into n parts each $\leq m$. Show that $p(u + 1, n + 1, m) = p(u, n + 1, m) + p(u, n, m) - p(u - m, n, m)$ (Euler). Find boundary conditions and write a program to compute p.

6. The list of all German coins consists of $D = \{1, 2, 5, 10, 50, 100, 200, 500\}$ Pfennigs. In how many ways can you change

a) 1 DM into smaller coins

b) 5 DM using all coins? (1 DM=1 German Mark=100 Pfennigs.)

7. The list of Russian coins consisted of $D = \{1, 2, 3, 5, 10, 15, 20\}$ kopeks. In how many ways could you change 1 Ruble (=100 kopeks) into coins?

8. The list of Swiss coins less than 1 SF (Swiss Franc) consists of $D = \{1, 2, 5, 10, 20, 50\}$ Rappen. In how many ways can you change 1 SF into smaller coins ? (1 SF= 100 Rappen.)

9. In how many ways can you change 1 English Pound into coins? A few years ago the list of English coins was $D = \{0.5, 1, 2, 5, 10, 20, 50, 100\}$ Pence. How do you handle 0.5 Pence? (The half-Penny coin is no longer in circulation, and the Pound is 100 Pence.)

10. You have a set $a_1 < a_2 < \ldots < a_s$ of integer weights. In how many different ways can the weight n be composed of these weights ? That is, write a program which finds the number C_n of solutions of

$$a_1 e_1 + a_2 e_2 + \cdots + a_s e_s = n , \qquad e_i \in \{0, 1\}.$$

11. Solve the preceding problem if the weights may be placed on both pans of the scales. That is, write a program which finds the number D_n of solutions of

$$a_1 e_1 + a_2 e_2 + \cdots + a_s e_s = n , \qquad e_i \in \{-1, 0, 1\}.$$

In Exercises 10 and 11, experiment with different sets of weights, especially the set $1, 2, 2^2, \ldots, 2^{s-1}$ in 10, and the set $1, 3, 3^2, \ldots, 3^{s-1}$ in 11.

12. A set S is defined by means of three axioms:

   A1: $1 \in S$.   A2: $x \in S \to 2x \in S, \; 3x \in S, \; 5x \in S$.

   A3: S is the minimal set satisfying axioms A1 and A2.

   a) Write a boolean function, which recognizes if $n \in S$.
   b) Show that S contains all elements of the form $2^a 3^b 5^c$ with natural numbers a, b, c.
   c) Write a program which prints the elements of S in order.

13. For the result in Section 15 there is a beautiful geometrical solution. Try to discover it.

14. The set S is the minimal set satisfying the two axioms:

   A1: $0 \in S$ A2: $x \in S \to 2x + 1 \in S, \; 3x + 1 \in S, \; 5x + 1 \in S$.

   Write recursive and iterative programs which print and count all elements of $S \leq n$.

15. Does the set $\{1, 2, \ldots, 3000\}$ contain a subset A of 2000 elements such that $x \in A$ implies $2x \notin A$? (10th Austrian-Polish Mathematics Competition.) This problem was generalized by E. T. H. Wang (Ars Combinatoria 1989, pp. 97-100) as follows: A set S of integers is called double-free (D.F.) if $x \in S \to 2x \notin S$. Let $N_n = \{1, 2, ..., n\}$ and $f(n) = max\{|A| : A \in N_n$ is D.F.$\}$. He proves $f(n) = \lceil n/2 \rceil + f(\lfloor n/4 \rfloor)$. Write a recursive program for f and solve the original problem.

The following problems 16 to 19 refer to Section 16.

16. Write a program which finds the three smallest n such that $f(n) = x$ for $x = 8, 10, 20, 30$.

17. Write an iterative program for $f(n)$. Hint: The relation $f(n) = af(m) + bf(m + 1)$ is correct for $a = 1$, $b = 0$, $m = n$. Its correctness can be maintained. Indeed: for $m = 2k$ we have $f(n) = (a + b)f(k) + bf(k + 1)$ and for $m = 2k + 1$ we have $f(n) = af(k) + (a + b)f(k + 1)$. So for even m we set $a := a + b$; $m := m$ div 2, and for odd m we set $b := a + b$; $m := m$ div 2. As soon as $m = 1$ we have $f(n) = b$.

18. A function f defined on the positive integers is given by

$$f(1) = 1, \qquad f(3) = 3, \qquad f(2n) = f(n),$$
$$f(4n + 1) = 2f(2n + 1) - f(n), \qquad f(4n + 3) = 3f(2n + 1) - 2f(n)$$

for all positive integers n. Determine the number of positive integers ≤ 1988 for which $f(n) = n$. This was a difficult problem of the IMO XXIX(1988). But with a computer it becomes a routine problem.

a) Write a recursive function which computes $f(n)$.

b) Write a program which solves the problem by brute force.

c) Compute a table of $f(n)$ and try to guess what the function does.

d) Write a program which for input n finds the binary representations of n and $f(n)$. What does the function do?

e) Find the number of positive integers $\leq 2^k$ for which $f(n) = n$.

f) Prove that f reverses the binary representation of an odd integer.

g) Compute iteratively the array $f[k]$ for every $k \in 1..n$.

19. Show that in Exercise 17 $f(n) = f(u)$. Hint: Analyze Ex. 17 or 18 f).

20. a) With the program **inset** in Fig.14.3 find all set elements ≤ 10000.

b) What proportion of the differences of consecutive numbers are 1's?

c) What proportion of the differences of consecutive numbers are 2's?.

d) Find the number of elements in $1...n$ for $n = 1000, 2000, ..., 10000$. Observe the strange fluctuations in the ten "millenia". Does this behavior persist ?

21. In topic 11 we learned how to swap the blocks AB into BA. How would you transform ABC into CBA? That is, two nonadjacent blocks A, C are to be swapped.

22. **The Frobenius Problem revisited.** Given coins with denominations $a[1], ..., a[s]$. Let $b[k]$ be the number of solutions of the equation $a[1]x_1 + \cdots + a[s]x_s = k$, x_i nonnegative integers. Write a program which stores $b[1], ..., b[n]$ into the array $b[0...n]$.

23. Do Exercise 16 in Exercises to Sections 1-5 by analogy with the preceding problem. Call the program **equisum6**.

17. Primes

Statisticians tell us that we should use only real life data. But gathering sufficiently extensive data is costly and can lead to boredom if you have no special interest in the data. The natural numbers are an excellent statistical data base with easy access. They are free, they are as real as any other data base, and their properties are fascinating. In this section we will deal with primes and their distribution. This topic belongs to number theory proper.

17.1. How to Recognize a Prime. A natural number is a prime if it has exactly two divisors. If $n > 1$ is not a prime it can be decomposed into non-trivial factors

$$n = d_1 d_2 , \quad 1 < d_1 < n , \quad 1 < d_2 < n .$$

The divisors d_1, d_2 cannot both be $> \sqrt{n}$, otherwise $d_1 d_2 > n$. That is, a composite number n has non-trivial divisors $d \leq \sqrt{n}$. In other words:

If n has no divisors d in $1 < d \leq \sqrt{n}$, then n is a prime.

We want to write a program which tests if a number n is prime. We do not know anything about n since it may have been picked at random by the computer. So we must think of every possibility. Such a recursive function can be found in Fig. 17.1. This function returns the values **true** or **false** for primes and non-primes, respectively. In the main program we must enter n and call **prime(n,3)**. We can test integers up to **maxint**. The function in Fig. 17.2 recognizes primes up to 11 digits.

```
function prime(n,d:integer):boolean;
begin
  if n<2 then prime:=false
  else if n=2 then prime:=true
  else if n mod 2=0 then  prime:=false
  else begin
    if d*d>n then prime:=true
    else if n mod d=0 then prime:=false
    else prime:=prime(n,d+2)
  end
end;
```

Fig. 17.1

```
function prime(n,d:real): boolean;
begin if n<2.0 then prime:=false
  else if n=2.0 then prime:=true
  else if n/2.0=int(n/2.0) then prime:=false
  else begin
    if d*d>n then prime:=true
    else if n/d=int(n/d) then prime:=false
    else prime:=prime(n,d+2.0)
  end
end;
```

Fig. 17.2

Very often we know that a number is odd and greater than one . In this case we can use the simpler functions in Fig. 17.3 and 17.4 for integers up to maxint and up to 11 digits, respectively. They will be stored under the names natprime and realprime. Then we can read them from the file at appropriate places.

```
function prime(n,d:integer): boolean;
begin
  if d*d>n then prime:=true
  else if n mod d=0 then prime:=false
  else prime:=prime(n,d+2)
end;
```

Fig. 17.3

```
function prime(n,d:real): boolean;
begin
  if d*d>n then prime:=true
  else if n=d*int(n/d) then prime:=false
  else prime:=prime(n,d+2.0)
end;
```

Fig. 17.4

Let us now write a program, which for odd integers a, b with $1 < a < b$ counts all primes in the interval $[a, b]$ (Fig. 17.5 and 17.6). The functions "natprime" and "realprime" were read from the file. With these programs we count the numbers of primes in ten consecutive "centuries" above 10000, 30000, 100000, 1000000 and get the counts tabulated for each century:

```
program primes;                      program prime_count;
var a,b,x,count:integer;             var a,b,x,count:real;
{$I natprime}                        {$I realprime}
begin                                begin
write('a,b='); readln(a,b);            write('a,b='); readln(a,b);
  x:=a; count:=0;                      x:=a; count:=0;
  repeat                               repeat
    if prime(x,3) then                   if prime(x,3) then
      count:=count+1;                         count:=count+1;
    x:=x+2                               x:=x+2
  until x>b;                           until x>b;
  writeln('primes in [a,b]=',         writeln('primes in [a,b]=',
                     count)                            count:0:0)
end.                                 end.
```

Fig. 17.5   Fig. 17.6

above	10000:	11	12	10	12	10	8	12	11	10	10	Mean 10.6
above	30000:	9	12	9	9	9	9	9	8	13	8	Mean 9.5
above	100000:	6	9	8	9	8	10	8	7	6	10	Mean 8.1
above	1000000:	6	10	8	8	7	7	10	5	6	10	Mean 7.7

That is, in the interval $[a, b] = [10001, 10100]$ there are 11 primes, etc.

The density of primes in the neighborhood of a large number x is roughly $1/\ln x$. This has been observed around 1800 by looking at tables. More precise statements of the "prime Number Theorem" and their difficult proofs can be found in some of the more advanced books on number theory.

According to the prime Number Theorem the number of primes among a hundred integers starting with a is, on the average, $100/\ln a$. For our four intervals the expected numbers given by this formula are 10.8, 9.7, 8.7 and 7.2.

The distribution of primes looks fairly random although it has obvious non-random features. For example, two consecutive numbers are never both primes except for 2 and 3. We can obtain one measure of how random the distribution of primes is on a larger scale as follows. Take a long interval starting with some large integer x and select integers at random from this interval until we have selected as many as there are primes in it. Then compare the numbers of primes in selected centuries of the interval with the number of randomly picked numbers in it.

The probability of having k selected integers in a century is given by the "Poisson distribution", discussed in books on probability and statistics. A result of such a comparison is that in each of our sets of ten centuries there are fewer centuries with exceptionally many or exceptionally few primes than one would have in a random distribution with the same mean. Primes are left over after certain arithmetic progressions are removed by the sieve of Eratosthenes. Arithmetic progressions are far from random and it is not too surprising that the result of the process seems to be somewhat more evenly distributed than a randomly generated sequence would be.

17.2. Empirical Study of the Goldbach Conjecture.

The Goldbach Conjecture is a deep number theoretic problem, still unsolved after more than 200 years. It states that every even number, starting with 6, can be represented as the sum of two odd primes. We will investigate the number of ways it is possible to represent an even number e as the sum of two odd primes. We call this number the *Goldbach count*. First we want to see the representations (Fig. 17.7), then we concentrate on the count alone (Fig.17.8).

```
program goldbach;
var g,x,count:integer;
 {$I natprime}
begin
  write('g='); readln(g);x:=3;
                    count:=0;
  repeat
    if prime(x,3) then
    if prime(g-x,3) then
    begin write(x,'+',
      g-x,' '); count:=count+1
    end;
    x:=x+2
  until x>g-x;
  writeln;
  writeln('goldbach count=',
                    count)
end.
```

Fig. 17.7

```
program goldbach_count;
var g,x,count:integer;
 {$I natprime}
begin
  write('g='); readln(g);
  x:=3; count:=0;
  repeat
    if prime(x,3) then
    if prime(g-x,3) then
    count:=count+1;x:=x+2
  until x>g-x;
  writeln('goldbach count=',
                    count)
end.
```

Fig. 17.8

Which numbers have particularly low counts?
Which numbers seem to have particularly high counts?
Do you see any pattern?
From around 36 on, numbers divisible by 3 seem to have a high count. Powers of 2 and numbers $2p$ with p a large prime seem to have low counts. Divisibility by small odd primes seems to raise the Goldbach count on the

average. Take $n = 3*5*7 = 105$. So $2n = 210$ should have a high count.
Indeed:

e	200	202	204	206	208	210	212	214	216	218	220	222	224	226	228
count	8	9	14	7	7	19	6	8	13	7	9	11	7	7	12

The reason why multiples of 3, say, have a higher than average Goldbach
count is as follows. Tables indicate that about half of all primes up to
any given limit are $\equiv 1 \pmod 3$ and the other half are $\equiv -1 \pmod 3$.
This has been proved and generalized as the Prime Number Theorem for
Arithmetic Progressions. Consequently about half the sums $p + q$ of two
primes are multiples of 3 while the other two residue classes mod 3 each
have only about a fourth of these numbers. Hardy and Littlewood, in
Acta Mathematica (1923), gave probabilistic arguments, but not a proof,
for the validity of the formula

$$P(2n) \sim C \left(\prod_{\substack{\text{odd prime} \\ \text{factors of } n}} \frac{p-1}{p-2} \right) \frac{2n}{(\ln 2n)^2}$$

where

$$C = 2 \prod_{\text{odd primes}} \left(1 - \frac{1}{(p-1)^2} \right) \approx 1.320324 .$$

The actual Goldbach counts tend to differ from these expected values
by several percent even for five-digit numbers.

17.3 Prime Twins, Triples, Quadruples

Prime couples of the form $(p, p+2)$ are called *twin primes*. For example:
(3,5), (5,7), (11,13), (17,19), (29,31), ... Starting with $p = 5$ they are all
of the form $(6n - 1, 6n + 1)$, since all primes starting with 5 are contained
in these two arithmetic progressions.

A prime triple of the form $(p, p+2, p+6)$ is a 2-4 *triple prime*. A 4-2
triple prime has the form $(p, p+4, p+6)$.

A 2-4-2 quadruple prime has the form $(p, p+2, p+6, p+8)$. A 4-2-4
quadruple prime has the form $(p, p+4, p+6, p+10)$. By *quadruple prime*
we shall mean those of the first form.

We want to write a program which prints all twin primes between $a > 8$
and b. Let x be the first integer $\geq a$ of the form $6n - 1$. Then $y =
6*\text{int}(a/6) + 5$. Fig. 17.9 is the sketch of such a program.

```
program twins;
var a,b,x:integer;
{$I natprime}
begin
  write('a,b=');readln(a,b);
  x:=6*trunc(a/6)+5;
  repeat
    if prime(x,5) then
    if prime(x+2,5) then
      writeln(x,'   ',x+2);
    x:=x+6
  until x>b-2
end.
```

Fig. 17.9

```
program twin_primes;
label 0;
var a,b,x,y:real;
begin
  write('a,b=');readln(a,b);
  writeln; x:=6*int(a/6)+5;
  repeat y:=5;
    repeat
      if (x=y*int(x/y)) or
         (x+2=y*int((x+2)/y))
      then goto 0;
      y:=y+2
    until y>sqrt(x)+1;
    writeln(x:0:0,'   ',
                  x+2:0:0);
0:  x:=x+6
  until x>=b
end.
```

Fig. 17.10

1000037	1000039	1000859	1000861	1002257	1002259
1000211	1000213	1000919	1000921	1002341	1002343
1000289	1000291	1001087	1001089	1002347	1002349
1000427	1000429	1001321	1001323	1002359	1002361
1000577	1000579	1001387	1001389	1002719	1002721
1000619	1000621	1001549	1001551	1002767	1002769
1000667	1000669	1001807	1001809	1002851	1002853
1000721	1000723	1001981	1001983	1002929	1002931
1000847	1000849	1002149	1002151		

Fig. 17.11

We want to go far beyond maxint. Then it may be a good idea to test both x and $x + 2$ simultaneously, because small prime factors are more frequent than large ones. The program in Fig. 17.10 uses a goto to break out of a loop prematurely. It does not use the function natprime. The label must be declared.

The program twin_primes was first run for $a = 1000000$ and $b = 1001000$. There are 11 twins in this millenium. Let us split this interval evenly into 10 centuries. Just two centuries are empty, the second and the fourth. Suppose we toss 11 balls at random into 10 cells. What is the expected number of empty cells? Any one cell remains empty with probability 0.9^{11}. The expected number of empty cells is $10 * 0.9^{11} = 3.138$. We have too few data to show that twin primes are too evenly spread out to be called random, although we suspect it. (Fig. 17.11). Next the program was run for two more millenia. The combined data look now more random. Distances between twin primes are much larger than distances between primes. There is little dependence between them.

Let us now construct a program for quadruple primes (those of the 2-4-2 kind). The program prime_quadruple is easy to understand. Let us look at the output for $a = 5$, $b = 20000$.

program prime_quadruple;	5	7	11	13
	11	13	17	19
var a,b,p:integer;	101	103	107	109
{$I natprime}	191	193	197	199
	821	823	827	829
begin	1481	1483	1487	1489
writeln('a,b');	1871	1873	1877	1879
readln(a,b);	2081	2083	2087	2089
p:=6*trunc(a/6)+5;	3251	3253	3257	3259
repeat	3461	3463	3467	3469
if prime(p,3) then	5651	5653	5657	5659
if prime(p+2,3) then	9431	9433	9437	9439
if prime(p+6,3) then	13001	13003	13007	13009
if prime(p+8,3) then	15641	15643	15647	15649
writeln(p,' ',p+2,' ',	15731	15733	15737	15739
p+6,' ',p+8);	16061	16063	16067	16069
p:=p+6	18041	18043	18047	18049
until p>b	18911	18913	18917	18919
end.	19421	19423	19427	19429

Fig. 17.12

Apart from the first atypical quadruple all the others end up in 1, 3, 7, 9. This is easy to prove. (Exercise 12.) So the steps must be multiples of 10 as well as of 6, i.e. 30-steps. That is, as soon as we find the first quadruple in an interval we can proceed in steps of 30. Suppose I want all quadruples between a and b. Where do I start? With the first term of the sequence $30q + 11$ which is $\geq a$. This term is $30\,\mathrm{trunc}((a + 18)/30) + 11$. Check this!

So we can replace the formula for p in Fig. 17.12 by

```
p:= 30*trunc((a+18)/30)+11;
```

and the two lines before the final end by

```
p:=p+30;
until p>b−8;
```

This speeds up the program by a factor of about 4.

17.4 Factoring

We want to factor integers up to 11 digits. First we assume that n is odd and greater than 1. We start with the trial divisor $d = 3$. The pseudo-program in Fig 17.13 is self-explanatory. It can be translated into the Pascal program in Fig.17.14.

```
d:=3
while d*d<=n do
  while n=d*int(n/d) do
    write(d); n:=n/d
  d:=d+2
write(n)
```

Fig. 17.13

```
program factor;
var n,d: integer;
begin
write( 'n=');readln(n);
                      d:=3;
while d*d<=n do
begin
  while n=d*int(n/d) do
  begin
    write(d:0,'*');n:=n div d
  end;
  d:=d+2
end;
if n>1 then writeln(n:0:0)
end.
```

Fig. 17.14

```
program factor1;
var n,d:real; s:integer;
begin
  write('n='); readln(n);
  while n=2*int(n/2) do
  begin
    write(2,'*'); n:=n/2
  end;
  while n=3*int(n/3) do
  begin
    write(3,'*');n:=n/3
  end;
  d:=5; s:=2;
  while d*d<=n do
  begin
    while n=d*int(n/d) do
    begin
      write(d:0:0,'*'); n:=n/d
    end;
    d:=d+s; s:=6-s
  end;
  if n>1 then writeln(n:0:0)
end.
```

Fig. 17.15

Let us factor some large numbers with this program:

$$123456789 = 3 * 3 * 3607 * 3803$$
$$1111111111 = 11 * 41 * 271 * 9901$$
$$11111111111 = 21649 * 513239$$
$$111111111111 = 3 * 3 * 11 * 13 * 37 * 101 * 9901$$
$$999999999999 = 3 * 3 * 3 * 7 * 11 * 13 * 37 * 101 * 9901$$
$$11111111111111 = 3 * 5 * 7 * 11 * 13 * 37 * 101 * 19802$$

All the numbers except the last one are correctly factored. The last would require more binary digits than Turbo Pascal uses for reals; it has been rounded to 1 111 111 111 110. The program has factored this number correctly, except that the last factor, 19 802, is interpreted as a prime since the program never divides by 2.

All primes beyond 3 lie in one of the two arithmetic progressions $6n - 1$ and $6n + 1$. That is, from 5 on the trial divisor d may be incremented alternatively in steps of 2 and 4. The program factor1 is valid for any natural number. It first divides out the factors 2 and 3. Then it starts with $d = 5$ and step size $s = 2$. The assignment $s \leftarrow 6 - s$ assures that s alternates between 2 and 4.

18. Representation of n as a Sum of Four Squares

Inspired by Diophantus' classical treatise, Bachet conjectured (1621) that every natural number is the sum of at most four squares. He verified this up to $n = 351$. The first proof was given by Lagrange (1770). We can tell how difficult this theorem was from the fact that Euler, over the years, settled several special cases, but failed to prove it.

As in the case of the Goldbach conjecture, we consider the number of representations. We observe that this number increases enormously with n. One can even try to find patterns in this sequence of numbers. See, e.g., G. Pólya's Mathematics and Plausible Reasoning, vol. 1, Ch. IV.

The following program finds all representations of n as $x^2 + y^2 + z^2 + u^2$ with $x \geq y \geq z \geq u$:

```
program four_squares;
var x,y,z,u,n,r:integer;

begin
  write( 'n=' );readln(n);
  r:=trunc(sqrt(n)); writeln;
  for x:=r downto r div 2 do
    for y:=x downto 0 do
      for z:=y downto 0 do
        for u:=z downto 0 do
          if x*x+y*y+z*z+u*u=n then
            write(x,' ',y,' ',z,' ',u,'   ');
  writeln
end.
```

Fig. 18.1

For $n = 169$, 207 and 399 the program gives us the following representations:

169:	13 0 0 0	12 5 0 0	12 4 3 0	11 4 4 4	10 8 2 1
	10 7 4 2	9 6 6 4	8 8 5 4		
207:	14 3 1 1	13 6 1 1	13 5 3 2	11 9 2 1	11 7 6 1
	11 6 5 5	10 9 5 1	10 7 7 3	9 9 6 3	

399:															
19	6	1	1	19	5	3	2	18	7	5	1	18	5	5	5
17	10	3	1	17	9	5	2	17	7	6	5	15	13	2	1
15	11	7	2	15	10	7	5	14	13	5	3	14	11	9	1
13	13	6	5	13	11	10	3	13	10	9	7	11	11	11	6

Exercises for Sections 17 and 18

1. Let us look at powers of 2. Denote by $G(2^n)$ the Goldbach count of 2^n. Check the following table:

n	3	4	5	6	7	8	9	10	11	12
$G(2^n)$	1	2	2	5	3	8	11	22	25	53

n	13	14	15	16	17	18	19	20	21	22
$G(2^n)$	76	151	244	435	749	1314	2367	4239	7471	13705

2. Plot $\ln(G(2^n))$ versus n. What do you get for large n?

3. Write a program which checks if a number n satisfies the Goldbach Conjecture. That is, it must print TRUE as soon as it finds a representation of e as a sum of two odd primes $a + b$ with $a \le b$. Find also the quotient $\ln(e)/\ln(a)$.

4. Write a program which prints a and e only if a is larger than all the a's of the preceding e's. Find again $\ln(e)/\ln(a)$.

5. Rewrite the program in Fig. 17.12, so that it works for numbers beyond maxint.

6. Find all positive integer solutions (x_n, y_n) of $x^2 - dy^2 = 1$ for $d = 2, 3, 5, 7, 10$ in a sufficiently large interval. From the data guess recurrences $x_{n+1} = f(x_n, y_n)$, $y_{n+1} = g(x_n, y_n)$ and prove them by induction.

7. Write programs which compute prime triples of type 2-4 as well as 4-2.

8. Find the first prime p beyond max so that $2p + 1$ is also a prime. Use max $= 1000, 10000, 100000, 1000000, 10000000, 100000000$.

9. Are there quadratic polynomials which assume unusually many prime values? Count the primes of the form $x^2 + 1$, $x^2 + x + 41$, $x^2 + x + 1$ up to $x = 4000$.

10. Rewrite the function prime more efficiently as follows: Starting with 5 you can proceed alternately in steps of 2 and 4. If we set initially $s \leftarrow 2$ then the counter $s \leftarrow 6 - s$ oscillates between 2 and 4.

11. Rewrite the function prime in Exercise 10 iteratively and find out if there is a speedup.

12. Starting with (11, 13, 17, 19) all prime quadruples lie in the sequence $30q + 11$. Why?

13. Write programs for computing 4-2-4 prime quadruples.

14. Explain the line before the last in the programs `factor` and `factor1` (Figs. 17.14, 15). What is the reason for "`if` n>1 `then` "? Why not just "`writeln(n:0:0)`"?

15. Factor the Fermat-number $F_5 = 2^{32} + 1 = 4294967297$ with programs `factor` and `factor1` and compare the run times. Eliminate the searches for factors 2 and 3 in the program `factor1` and compare the run times again.

16. Compute a list of natural numbers which are not the sums of three squares (0 allowed) and make a conjecture about the form of these numbers.

17. Prove that the numbers of the form identified in ex. 16 are indeed not the sums of three squares. **Remark:** It is a very deep theorem that every positive integer not of the form given in the solution of ex. 16 is the sum of three squares.

18. a) Prove that a number of at least four digits cannot be equal to the sum of the cubes of its digits.
    b) Find all numbers with at most three digits which are equal to the sum of the cubes of their digits.

19. *Wheels.* To factor a large number n we first divide it by $2, 3, 5, 7$. From then on we can use step sizes $4, 2, 4, 2, 4, 6, 2, 6$ of period 8 to get the trial divisors $11, 13, 17, 19, 23, \ldots$ which contain all the remaining primes. Use this to write a factoring program.

20. Find the smallest integer which can be represented as a sum of three 4th powers in two different ways.

21. Sieve out the positive integers which cannot be represented in the form
    a) $x + y + xy$ b) $x + y + 2xy$ with positive integers x, y. Comment.

22. Find all numbers n up to 1000 so that all numbers of the form $n - 2^k$ with $2 \le 2^k < n$ are prime. The six numbers you will find are the only known numbers with this property.

Additional Exercises for Sections 1 to 18

1. If n runs from 1 to a^2, the function $f(n) = \lfloor n + \sqrt{n} + 0.5 \rfloor$ runs up to $a^2 + a$, leaving out exactly a numbers.
   a) Write a program which prints the skipped values.
   b) Make a conjecture and try to prove it.

2. If n runs from 1 to some upper bound max, then $f(n) = \lfloor n + \sqrt{2n} + 0.5 \rfloor$ will skip some values.
   a) Write a program which prints the skipped values.
   b) Make a conjecture and try to prove it.

3. a) For $k = 2, 3, 4, \ldots$ investigate empirically the values left out by the functions $f_k(n) = \lfloor n + \sqrt{n/k} + 0.5 \rfloor$.

b) Try to guess a closed formula for these values.

c) Prove your conjecture.

4. a) Find the values skipped by $f(n) = \lfloor n + \sqrt{kn} + 0.5 \rfloor$. Try to guess a closed formula for the skipped values and to prove your conjecture.

b) Find the values skipped by $f(n, p, q) = \lfloor n + \sqrt{np/q} + 0.5 \rfloor$ for positive integers p, q.

5. *Stirling Numbers of the Second Kind.* Let $S(n, k)$ be the number of partitions of the set $\{1, 2, \ldots, n\}$ into exactly k nonempty subsets. Show that

$$S(n, 1) = S(n, n) = 1 , \quad S(n, k) = 0 \text{ for } k > n,$$
$$S(n, k) = S(n - 1, k - 1) + kS(n - 1, k).$$

Write recursive and iterative programs for $S(n, k)$.

6. *Stirling Numbers of the First Kind.* Let $f(n, k)$ be the number of permutations of $\{1, 2, \ldots, n\}$ with exactly k cycles. Show that

$$f(n, k) = f(n - 1, k - 1) + (n - 1)f(n - 1, k).$$
$$f(n, 1) = (n - 1)! \qquad f(n, n) = 1 \qquad f(n, k) = 0 \text{ for } k > n.$$

Write recursive and iterative programs for $f(n, k)$.

7. Find all representations of n in the form $x^3 + y^3 + z^3$ with nonnegative integers x, y, z. Explain why numbers of the form $9k \pm 4$ cannot be sums of three cubes.

8. Find all representations of n as a sum of 4 nonnegative integer cubes.

9. Write a program for listing the integers from 1 to max which can be represented in the form $x^2 + 2y^2$.

10. Show empirically that any number from 1 to 10000 can be represented as a sum of at most three triangular numbers, i.e. numbers of the form $\binom{n}{2}$. Do it by sieving.

11. *The mode (plateau problem).* Let a[1..n] be an array of integers sorted in increasing order. The mode m is the most frequent value. Construct an algorithm which finds m and its frequency f. The program should also find the length of the longest plateau (stretch of equal values) for any array, even if not sorted.

12. *Common elements of two sequences.* We are given two arrays f[0..m-1] and g[0..n-1] of natural numbers in increasing order. Find the terms which occur in both sequences and the number k of common terms.

13. A sequence $g(n)$ is defined by $g(0) = 0$ and $g(n) = n - g(g(n - 1))$ for $n > 0$.

a) Construct a recursive program for $g(n)$ and compute its values up to $g(40)$.

b) Write an iterative program which uses an array g[0..1000].

c) Plot the sequence and try to guess a closed formula for $g(n)$.

d) Prove your conjecture.

14. A sequence $h(n)$ is defined by
$$h(0) = 0 \text{ and } h(n) = n - h(h(h(n-1))) \text{ for } n > 0.$$
   a) Write a recursive program for $h(n)$ and find its values up to $h(40)$.
   b) Write an iterative program which uses an array h[0..1000].
   c) Plot the sequence and make conjectures about a closed formula for $h(n)$.

15. A sequence $q(n)$ is defined by
$$q(1) = q(2) = 1 \text{ and } q(n) = q(n - q(n-1)) + q(n - q(n-2)) \text{ for } n > 2.$$
   a) Write a recursive program and compute the values of $q(n)$ up to $q(30)$.
   b) Write an iterative program using an array $q[1\ldots2000]$ and compute the values of q up to $q[2000]$.
   c) Write all integers 7, 13, 15, 18, ... that get left out. Are there infinitely many of these? (This is a hard unsolved problem.) See also Ex. 23, below.

16. Find all 4-2-4-2-4 sextuples of primes below 50000.

17. Find numerical evidence for the following theorem due to Euler:
   Every number 2^n for $n \geq 3$ can be represented in the form $2^n = 7x^2 + y^2$ with odd x and y.
   From the numerical evidence deduce a proof. (Moscow Mathematics Olympiad 1985.)

18. Consider again Perrin's sequence v_n defined by $v_0 = 3$, $v_1 = 0$, $v_2 = 2$, $v_n = v_{n-2} + v_{n-3}$, $n \geq 3$. See Section 3. Write a program which for each n from 3 to maxint, tests if $n|v_n$. If the test is positive and n is not prime, it should print n. Do you now have enough evidence for the conjecture $n|v_n \Leftrightarrow n$ is a prime?

19. On visiting Ramanujan, Hardy mentioned that he took taxi #n, but this number did not look remarkable to him. "On the contrary", replied Ramanujan, "it is the smallest natural number which can be represented as a sum of two cubes in two different ways". Find n. (Next problem is the continuation.)

20. Hardy asked if Ramanujan knows the answer to the problem for fourth powers. After some deliberation Ramanujan replied: "I cannot find an example, I think the first such number must be large". Indeed, Euler has found $635318657 = x^4 + y^4 = u^4 + v^4$, $x < y$, $u < v$, $x \neq u$. Find x, y, u, v. Is this the smallest such number?

21. *The smallest common element of three ordered arrays.* Given are three increasingly ordered arrays F, G, H. Find the smallest common element. Since the arrays are ordered, we need the smallest i, such that $F[i] = G[j] = H[k]$. Apply the algorithm to the Fibonacci sequence F, to the squares G, and to the multiples H of 9. Print i, j, k, and $F[i]$.

22. A sequence is defined by means of

$$f(1) = f(2) = 1, \qquad f(n) = f(f(n-1)) + f(n - f(n-1)) \text{ for } n > 2.$$

The table below shows some values of this sequence. We see that $f(13) = \ldots = f(16) = 8$.

n	1	2	3	4	5	6	7	8	9	10	11	12	13	14	15	16	17	18	19	20	21
$f(n)$	1	1	2	2	3	4	4	4	5	6	7	7	8	8	8	8	9	10	11	12	12

This is an *interval of constancy* of length 4 starting at 13. On the other hand, $f(16)$ to $f(20)$ is a *run* of length 5 starting at 16. In 1..max (depending on your computer) find
a) the leftmost (rightmost) longest interval of constancy,
b) the leftmost (rightmost) longest run.

23. Let us return to the sequence in Ex. 15., defined by $q(1) = q(2) = 1$ and $q(n) = q(n - q(n-1)) + q(n - q(n-2))$ for $n > 2$. Hofstadter [1979] calls its growth pattern chaotic. But this applies only to local growth. Find a simple global regularity in its growth by going as far out as your PC permits.

24. *Floyd's Function.* A function G is defined on the integers as follows:

$$G(n) = \begin{cases} n - 10 & \text{for} \quad n > 100 \\ & \text{else} \quad G(G(n + 11)). \end{cases}$$

Write a recursive program which finds $G(n)$.

25. A function Q is defined on the integers by

$$Q(a, b) = \begin{cases} 0 & \text{if} \quad a < b \\ Q(a - b, b) + 1 & \text{if} \quad a \geq b. \end{cases}$$

Find $Q(a, b)$.

26. There is a conjecture that every large integer n is either a square or the sum of a prime and a square. Write a program which finds all numbers from 0 to 32000 which are not representable in this way. Is the conjecture plausible? Write also a program, which for input n prints the number of representations $n = m^2 + p$. For $n = 11$ it should print 3, since $11 = 0^2 + 11 = 2^2 + 7 = 3^2 + 2$.

27. Find the function L defined on the integers by

$$L(n) = \begin{cases} 0 & \text{if} \quad n = 1 \\ L(n \text{ div } 2) + 1 & \text{if} \quad n > 1. \end{cases}$$

28. What does the program **product** (Fig. 18.2) do for natural inputs x, y?

```
program product;
var x,y:integer;

function prod(x,y:integer):integer;
begin
  if y=1 then prod:=x
  else if odd(y) then
  prod:=x+prod(x,y-1)
  else prod:=prod(x+x,y div 2)
end;

begin
  write('x,y=');readln(x,y);
  writeln(prod(x,y))
end.
```

Fig. 18.2

29. Find a permutation of the set $\{1, 2, \ldots, 9\}$ so that the number consisting of the first n digits is divisible by n for every n from 1 to 9. The idea is to reduce the work a little by thinking and then proceed with the computer. (By hard thinking you can also find the unique number without a computer.)

30. Let B_n be the number of partitions of an n-set. The numbers B_n can be computed by means of the "Bell triangle" below. The first column is $B_0, B_1, B_2, \ldots$ and the last entry in each row is the next Bell number. Here $B_0 = 1$ by definition. How is this triangle formed? Write a program which for input n prints the Bell triangle up to line n.

```
1
1    2
2    3    5
5    7    10   15
15   20   27   37   52
52   67   87   114  151  203
203  255  322  409  523  674  877
877
```

31. Write a program based on $S(n, k)$, the Stirling Numbers of the Second Kind which for input n prints B_n.

32. a) Write a program which prints

```
1
2 2
3 3 3
4 4 4 4
. . . . . . . . . .
```

b) Show that $a_n = \lfloor \sqrt{2n} + 0.5 \rfloor$ is a closed expression for the n-th term of this sequence 1, 2, 2, 3, 3, 3, 4, 4, 4, 4, ...

c) It is claimed that $a_n = \lfloor \frac{1}{2}(1 + \sqrt{8n - 7}) \rfloor$ $a_n = \lceil \frac{1}{2}(\sqrt{8n + 1} - 1) \rceil$ are also closed expressions for the n-th term. Check if these claims are true for $n = 1 \ldots 1000$.

33. a) Write a program which prints the sequence

$$1$$
$$2 \quad 4$$
$$5 \quad 7 \quad 9$$
$$10 \quad 12 \quad 14 \quad 16$$
$$\cdots \cdots \cdots \cdots \cdots \quad .$$

b) Show that $a_n = 2n - \lfloor \sqrt{2n} + 0.5 \rfloor$ is a closed expression for the n-th term of the sequence.

34. Let X and Y be natural numbers. We say that X is contained in Y if the binary representation of Y goes over into that of X when some (possibly none) digits are omitted. For example, $X = 1010$ is contained in $Y = 1001100$. Construct an algorithm which, for given natural numbers A, B, finds the largest C contained in A as well as in B. (First International Programming Contest, 1987.)

35. A sequence is defined by $a_1 = 1$, $a_n = \lfloor \sqrt{a_1 + \cdots + a_{n-1}} \rfloor$, $n > 1$. Explore this sequence.

36. *Ulam's lucky number sieve.* From the list 1, 2, 3, 4,... of all positive integers remove every second number, leaving 1, 3, 5, 7, 9,.... Since 3 is the first number (above 2) that has not been used as a sieving number, we remove every third number from the remaining numbers, yielding 1, 3, 7, 9, 13, 15, 19, 21,.... Now every seventh number is removed, leaving 1, 3, 7, 9, 13, 15, 21,.... Numbers that are never removed are considered to be "lucky". Write a program which prints the lucky numbers up to max.

19. The Best Rational Approximation

Given a real number $r > 0$, we want to find a sequence of ever better rational approximations p/q with $q \le qmax$. The following algorithm is probably the simplest:

```
Start with p = 0, q = 1.
If p/q  <  r then set p := p + 1.
If p/q  =  r then stop.
If p/q  >  r then set q := q + 1 and stop if q  ≥  qmax.
```

We measure the distance between r and p/q by means of $d = |r - p/q|$. Initially we have $d = r$. Each time we get a smaller distance than the

preceding ones, the fraction p/q is printed. If we run the program in
Fig. 19.1 with

$$r = \tau = (1 + \sqrt{5})/2 \approx 1.6180339887498948432,$$
$$r = \sqrt{2} \approx 1.4142135623730955049,$$
$$r = \pi \approx 3.141592653589793238$$

we get most of the entries presented in the table; to get the largest ones
we need to use the integer type `longint` and the real-type `extended` for
the variables.

```
program ratap;
var p,q,qmax:integer;
    d, r, min: real;
begin
  write('r,qmax=');
  readln(r,qmax);
  p:=0; q:=1; min:=r;
  repeat
    if p/q<r then p:=p+1
      else q:=q+1;
    d:=abs(r-p/q);
    if d<min then
    begin
      min:=d;
      write(p:7,'/',q)
    end
  until (q>=qmax) or (d=0)
end.
```

τ	$\sqrt{2}$	π
1/1	1/1	1/1
2/1	3/2	2/1
3/2	4/3	3/1
5/3	7/5	13/4
8/5	17/12	16/5
13/8	24/17	19/6
21/13	41/29	22/7
34/21	99/70	179/57
55/34	140/99	201/64
89/55	239/169	223/71
144/89	577/408	245/78
233/144	816/577	267/85
377/233	1393/985	289/92
610/377	3363/2375	311/99
987/610	4756/3363	333/106
1597/987	8119/5741	355/113
2584/1597	19601/13860	52163/16604
4181/2584	27720/19601	52518/16717

Fig. 19.1

Some such sets of numbers can be characterized in interesting ways: In
the first case we get pairs of successive Fibonacci numbers. In the second
case it turns out that the p's and q's are the solutions of $p^2 - 2q^2 = \pm 1$
and $p^2 - 2q^2 = 2$. In the third case we get all famous approximations
of π, like 22/7, 355/113. We will return to this problem when we treat
continued fractions.

Exercise:

1. *A Project.* Run the program **ratap** for

   a) $r = \sqrt{3} \approx 1.7320508076$,   b) $r = \sqrt{5} \approx 2.2360679775$
   c) $r = \sqrt{7} \approx 2.6457513110$,   d) $r = \sqrt{10} \approx 3.1622776602$
   e) $r = \sqrt{13} \approx 3.6055512754$,   f) $r = e \approx 2.7182818285$
   g) Numbers with randomly selected digits.

(It is easier to let Pascal compute the constants by modifying the program slightly. If you use variables of type **extended** you will get the constants to 19-digit accuracy.)

Try to make sense of the data. How is a line computed from preceding lines? If needed, you should use $qmax > 1000$. Try to strike some lines to get a simpler rule. Try to find rules for the stricken lines. Two different rules may be intertwined.

20. The Maximum of a Unimodal Function

A function which is strictly increasing up to a point in an interval and strictly decreasing past that point, as in Figs. 20.2 and 20.3, is called *unimodal* in the interval. We want to find the maximum of a unimodal function f on $[a, b]$.

We give a method suitable when we have no information about the derivative of f and finding values of the function is laborious. We confine the maximum to smaller and smaller intervals by repeating the following procedure: Pick in the interval two points x, y where we will evaluate f. Let $x - a = s(b - a)$, $y - a = t(b - a)$. (If $a = 0$ and $b = 1$ then $x = s$ and $y = t$.) We are going to choose s, t so that one of the points x, y can be used again at the next iteration.

Consider first the case when $f(x) > f(y)$. Then our new interval of uncertainty about the location of the maximum is $[a, y]$, because f must be decreasing somewhere to the left of y, and once it starts to decrease it continues to do so. In order to use the old x as the new y we must have

$$x - a = t(y - a), \text{ or } s(b - a) = t(y - a) = t^2(b - a) \Rightarrow s = t^2.$$

Similarly, if $f(x) \leq f(y)$, the new interval of uncertainty about the location of the maximum is $[x, b]$. In order to use the old y as the new x we must have $y - x = s(b - x)$, or $(t - s)(b - a) = s(1 - s)(b - a)$. Hence $t - s = s - s^2$. Using $s = t^2$ we get $t^4 - 2t^2 + t = 0$ or $t(t-1)(t^2 + t - 1) = 0$. Hence $t = \frac{1}{2}(\sqrt{5} - 1) \approx 0.618$ (golden section).

We write an iterative program to locate the maximum of a function f which is unimodal in $[a, b]$. If we set $f(x) = \sin(x/2)$ and take $a = 0$, $b = 6$ we get the program **unimaxit**.

It prints **max=3.1415898**. Only 5 decimals of **max** are correct. We cannot expect more. See exercise below.

In standard Pascal the function f in this procedure could be a function-parameter, i.e. a dummy which is replaced by a particular function when the routine is called, but this cannot be done in versions of Turbo Pascal available so far.

For a preassigned number n of function evaluations one can narrow down the interval where the maximum must be by somewhat more than the factor $1/\tau^{n-1}$, where $\tau = 1.61\ldots$ denotes the golden ratio. J. Kiefer showed that the following procedure is optimal if one makes no assumption about the derivatives of f.

Let fib(n) be the nth Fibonacci number. Divide the interval $[a, b]$ into fib(n) equal parts, and in the first two steps evaluate $f(x)$ at the subdivision points fib($n-1$) and fib($n-2$); see Fig. 20.4. Whatever the function values are, the maximum is now confined to an interval consisting of fib($n-1$) segments, and we can repeat the procedure. The last step is different. Then we have two equal intervals. We evaluate $f(x)$ at a second point very close to the middle to determine whether $f(x)$ is increasing or decreasing there, and that tells us on which side the maximum lies. (See *Proc. Amer. Math. Soc.* **4** (1953), 502.)

The ratio fib($n-1$)/fib(n) rapidly approaches the golden ratio which was used to divide the intervals in unimaxit. That is why unimaxit is close to optimal.

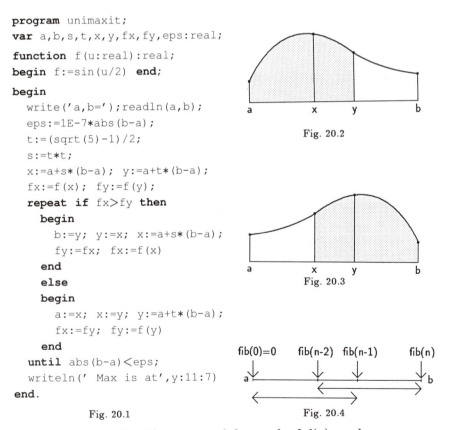

```
program unimaxit;
var a,b,s,t,x,y,fx,fy,eps:real;

function f(u:real):real;
begin f:=sin(u/2) end;

begin
  write('a,b=');readln(a,b);
  eps:=1E-7*abs(b-a);
  t:=(sqrt(5)-1)/2;
  s:=t*t;
  x:=a+s*(b-a); y:=a+t*(b-a);
  fx:=f(x); fy:=f(y);
  repeat if fx>fy then
    begin
      b:=y; y:=x; x:=a+s*(b-a);
      fy:=fx; fx:=f(x)
    end
  else
    begin
      a:=x; x:=y; y:=a+t*(b-a);
      fx:=fy; fy:=f(y)
    end
  until abs(b-a)<eps;
  writeln(' Max is at',y:11:7)
end.
```

Fig. 20.2

Fig. 20.3

fib(0)=0 fib(n-2) fib(n-1) fib(n)

Fig. 20.1

Fig. 20.4

We should note that if segments of the graph of $f(x)$ can be approximated by parabolas as well as is usually the case, then one gets to the maximum faster by substituting into f the x-value of the vertex of a parabola through points of the curve we have already computed, than by using an interval-dividing strategy.

Exercise: Suppose f has a maximum at $x = c$ and $f''(x) \approx -3$ near $x = c$. Suppose the error in evaluating $f(x)$ can be as large as 10^{-9}. Then there is an interval I around $x = c$ such that the computed value of $f(x)$ for points in this interval may be greater than the computed value of $f(c)$. Find the length of I.

We can expect that unimax will become unreliable when we have narrowed our search to I.

CHAPTER 3

Probability

Before tackling this chapter, it may be advisable to read the Appendix: *A Crash Course in Probability.*

21. The Random Number Generator (RNG)

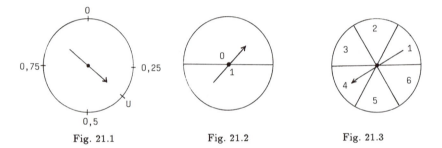

Fig. 21.1 Fig. 21.2 Fig. 21.3

Turbo Pascal provides two RNG's, which constitute the raw material from which we can fabricate other RNG's. These are:

random   generates a random variable U which is uniformly distributed in the interval (0,1) (Fig. 21.1). It has type **real**. *Uniformly distributed* means U falls into a subinterval $[a,b)$ of $[0,1)$ with a probability equal to the length of that interval, i.e. $b - a$. More generally, a point U generated by a random process is uniformly distributed in some subset Ω of the plane or space if it falls into a subset A of Ω with probability

$$P(A) = \text{volume of } A / \text{volume of } \Omega$$

(We consider only subsets which have volume.)

random(n)   generates each of the integers $0, 1, \ldots, n-1$ with probability $1/n$. It has type **integer** and is bounded above by **maxint**.

random(2)   generates tosses of a fair coin with faces labeled 0, 1 (Fig. 21.2).

`1+random(6)`   generates rolls of a fair die (Fig. 21.3).

`2*random(2)-1`   generates steps 1 or −1 of a symmetric random walk on the line (Fig.21.4).

`trunc(random+p)` generates spins of the spinner in Fig. 21.5. It generates a real number uniformly distributed in an interval which extends a distance p above 1, as shown in Fig. 21.6. Truncating it gives 1 with probability p. It has type **integer**.

`a+random(b-a+1)` generates each integer from the set `[a..b]` with probability $1/(b - a + 1)$, provided that $b \geq a$. If $n > maxint$, we must use `int(n*random)` of type real instead of `random(n)`. If $b >$ maxint, we use `a+int((b-a+1)*random)` to get $a, \ldots, b$ of type real.

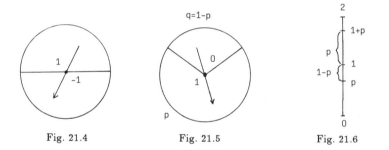

Fig. 21.4   Fig. 21.5   Fig. 21.6

The pseudo codes in Fig.21.7 to 21.9 perform random walks on the line, in the plane, and in space, starting at the origin.

```
1. x←0                 1. x[1]←0; x[2]←0      1. x[1]←x[2]←x[3]←0
2. x←2*random(2)-1     2. a←1+random(2)       2. a←1+random(3)
3. goto 2              3. x[a]←x[a]+          3. x[a]←x[a]+
                           2*random(2)-1          2*random(2)-1
                       4. goto 2              4. goto 2

    Fig. 21.7              Fig. 21.8              Fig. 21.9
```

22. Finding Geometric Probabilities by Simulation

In this section we find some geometric probabilities and expectations. The exact solutions are quite complicated and require multiple integrals. Instead we will repeat a random experiment 10000 times and get an estimate of the probability or expectation. This is called *simulation*.

Problem 1. Two points are chosen at random in a unit square. Find by simulation the expected distance between the two points. The coordinates x, y, z, u in Fig. 22.1 are chosen by means of Turbo Pascal's `random` function. Then the points (x, y) and (z, u) are uniformly distributed in the unit square. We do not assign the values of x, y, z, u to variables since we need them only to compute dx and dy. The variable `sum` accumulates all the n distances and the average `sum/n` is a good estimate E_{est} of the expectation E. The corresponding program in Fig. 22.2 was run for $n = 10000$.

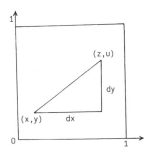

Fig. 22.1

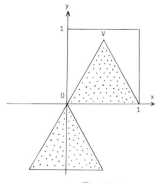

Fig. 22.4

```
program expdist1;
var i,n: integer;
    dx,dy,sum: real;
begin
write('n=');  readln(n);
sum:=0;
for i:=1 to n do
begin
   dx:=random-random;
   dy:=random-random;
   sum:=sum+sqrt(dx*dx+dy*dy)
end;
writeln('expected dist=',
                sum/n:9:6)
end.
```

Fig. 22.2

```
program expdist2;
var i,n:integer;
    x,y,z,u,sum,dist,d:real;
begin write('n=');readln(n);
sum:=0; d:=sqrt(3);
for i:=1 to n do
begin
  repeat x:=random; y:=random
  until y<d*(0.5-abs(x-0.5));
  repeat z:=random; u:=random
  until u<d*(0.5-abs(z-0.5));
  dist:=sqrt((sqr(x-z))
                 +sqr(y-u));
  sum:=sum+dist
end;
writeln('exp. dist.=', sum/n)
end.
```

Fig. 22.3

As an estimate for the expected distance we get $E_{est} = 0.52267$.

The exact value, given in L.A. Santalo [1976], is

$$E = \frac{\sqrt{2} + 2 + 5\ln(1 + \sqrt{2})}{15} \approx 0.52141.$$

E_{est} is a very good estimate of E.

Problem 2. Two points are chosen at random inside an equilateral triangle of side 1. Find by simulation the expected distance between them.

The idea is to choose the two points inside the unit square and count them only if they also lie inside the triangle in Fig. 22.4. If we translate the angle with vertex V to the origin O we get $y < -\sqrt{3}|x|$ for the shaded area. Now we translate back to $V = (\frac{1}{2}, \frac{1}{2}\sqrt{3})$ and get

$$y - \sqrt{3}/2 < -\sqrt{3}|x - \tfrac{1}{2}|.$$

For $n = 10000$ the program **expdist2** gives the estimate $E_{est} = 0.36108$. In his formula (4.17) Santalo gives the exact value

$$E = \tfrac{1}{5} + 3\ln\tfrac{3}{20} \approx 0.36479.$$

We get E_{est} with an error of 1%.

Problem 3. Two points are chosen at random in a unit circle. Find by simulation the expected distance.

We must choose (x, y) and (z, u) at random in $(-1,1)$ (Fig. 22.5). This is accomplished by x ← 2*random−1, etc. But only points inside the unit circle about O are suitable. The program expdist3 is based on this idea. For $n = 10000$ it gives $E_{est} = 0.90091$. Santalo gives the exact value

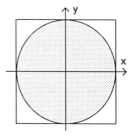

Fig. 22.5

$$E = \frac{128}{45\pi} \approx 0.90541.$$

This time the simulated value is within about 0.5% of the exact value.

```
program expdist3;                          repeat z:=2*random-1;
var i,n:integer;                                  u:=2*random-1
        x,y,z,u,sum,dist:real;             until z*z+u*u<=1;
begin randomize;                           dist:=sqrt(sqr(x-z)
write('n=');readln(n);                                   +sqr(y-u));
sum:=0;                                    sum:=sum+dist
for i:=1 to n do                           end;
begin                                      writeln('exp. dist.=',
   repeat x:=2*random-1;                                sum/n:9:6)
      y:=2*random-1                        end.
   until x*x+y*y<=1;
```

Fig. 22.6

Problem 4. Buffon's Needle Problem (1777), the first problem in geometric probability.

In the plane, consider an infinite set of parallel lines spaced a units apart. Onto the plane we throw "at random" a curve C of arbitrary shape (Fig. 22.8). (Buffon used a needle of length a.) Let S be the number of intersections with the parallels. We want to find $E(S)$, the expected number of intersections. For a line segment AB of length L we set $E(S) = f(L)$ with an as yet unknown function f. Throw the broken line PQR onto the parallels. If PQ has S_1 intersections and QR has S_2 intersections, then $S = S_1 + S_2$. The expectation of the sum of two random variables is the sum of their expectations, even if the variables are not independent. Thus we have $E(S) = E(S_1 + S_2) = E(S_1) + E(S_2)$, or

$$（1）\qquad f(L_1 + L_2) = f(L_1) + f(L_2),$$

i. e. f is *additive*. A function f which is additive and also increasing must be the linear function

$$（2）\qquad f(L) = L * f(1) .$$

The additive property of expectations tells us that the expected number of intersection with any polygon is also given by (2), where L is the length of the polygon. Any "decent" curve can be approximated by a broken line to any accuracy. Thus (2) is valid for a curve of any shape and length L. Now take a circle of diameter a and length $L = \pi a$. It has always two intersections (Fig. 22.9), i.e.

$$f(\pi a) = \pi a f(1) = 2 \rightarrow f(1) = \frac{2}{\pi a}$$

$$（3）\qquad\qquad f(L) = \frac{2L}{\pi a}$$

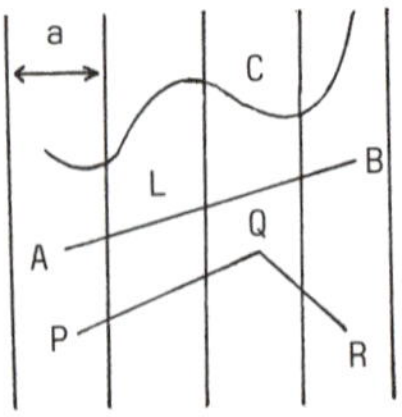

Fig. 22.8 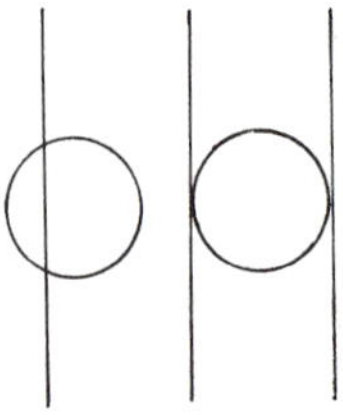   Fig. 22.9 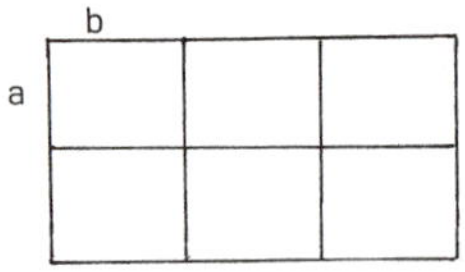   Fig. 22.10

Now suppose the plane is divided into congruent rectangles with sides a, b. Onto this plane we throw a curve of length L. If there are S_1 intersections with the verticals and S_2 intersections with the horizontals, with $S = S_1 + S_2$, then

$$E(S) = E(S_1) + E(S_2) = \frac{2L}{\pi a} + \frac{2L}{\pi b} = \frac{2L}{\pi}\left(\frac{1}{a} + \frac{1}{b}\right).$$

For $a = b = L = 1$, we get

$$E(S) = 4/\pi, \text{ or } \pi = 4/E(S)$$

If we estimate $E(S)$ by means of

$$E_{est} = \frac{\text{intersections}}{\text{throws}} = \frac{S}{T}$$

then we get for π the estimate

$$\pi_{\text{est}} = 4T/S.$$

Fig. 22.11

Let us write a program which finds an estimate of π. One problem is that in Turbo Pascal the function int is the integer part of x only for $x \geq 0$. For negative x it rounds up and not down. But we will write our own function intp which also rounds correctly for negative x.

The interpretation of throwing a needle "at random" onto the chessboard is contained in the assignments

```
x:= random, y:= random, a:=π*random
```

They tell us that the coordinates of the center of the needle are random numbers with a uniform distribution between 0 and 1, and its angle with the northern direction, counted to the right, is random between 0 and π. The periodic pattern of the chess board ensures that the result will be the same as if we had picked the center with a uniform distribution from a larger number of squares. One may think that a proper idealization of a random throw is that the center could land with equal probability in any of the infinitely many squares of the plane. However, attempts to give a mathematical formulation to that notion lead to contradictions.

Study the program BUFFON until you understand it completely. In fact it does not count some intersections, but the probability of their occurrence is zero. Input $T = 10000$ resulted in $\pi_{\text{est}} = 3.1375$ with an error of 0.13%.

Throwing a needle on a chessboard will give better approximations to π than the usual experiment based on formula (3). An even better needle experiment is described in exercise 7.

Of course BUFFON is good only for simulating Buffon's experiment. It is worthless for computing π. It uses π to get a very crude estimate of π. Yet Buffon's needle problem has some interesting applications in stereology. See H. Solomon [26].

Buffon's needle problem has some interesting applications in stereology.
See H. Solomon [26].

```
program BUFFON;
var a,b,c,x,y:real; i,t,s:integer;
function intp(x:real): integer;
begin
  intp(x):=int(x+2)-2
end;
begin
  write('t='); readln(t);s:=0;
  for i:=1 to t do begin
    x:=random; y:=random;
    a:=pi*random; b:=cos(a)/2; c:=sin(a)/2;
    if intp(x-b)<>intp(x+b)then s:=s+1;
    if intp(y-c)<>intp(y+c)then s:=s+1
  end;
  writeln('PiRoof=',4*t/s)
end.
```

Fig. 22.12

Exercises for Section 22

1. Fig. 22.7 shows another way of selecting pairs points for calculating the
   average distance of two randomly selected points in the unit circle. Using
   this method instead of the one in Fig. 22.6 makes the program about 25%
   faster. Try to understand it.

```
repeat                          z:=random; u:=random
  x:=random; y:=random;         until z*z+u*u<=1;
until x*x+y*y<=1;               if random(2)=0 then z:=-z;
repeat                          if random(2)=0 then u:=-u;
```

Fig. 22.7

2. In each of two squares in Fig. 22.13 a point is chosen at random. Find
   by simulation the mean distance between the two points. (One can show
   that the value is $E \approx 1.08814$.)

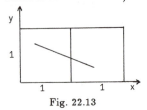

Fig. 22.13

Fig. 22.14

4. Inside a unit cube four points P_1, P_2, P_3, P_4 are chosen at random. Find by simulation the expected volume of the tetrahedron $P_1 P_2 P_3 P_4$. In this case I do not know the exact value of the expected volume.

5. Abel suggests to Cain: Let us choose at random two points P and Q inside a unit circle. If $|PQ| \leq 1$ then I win, otherwise you win. What is Abel's chance of winning? The random choice is to be performed as in Fig. 22.6 or 22.7. Theory predicts that Abel wins with probability $1 - (3\sqrt{3}/4)/\pi \approx 0.5865$.

6. Pick two random angles `alpha:=2*π*random`, `beta:=2*π*random`. From the center, move in the `alpha` and `beta`-directions by `random` distances to P_1 and P_2 respectively, see Fig. 22.15. Find the expected distance betwen P_1 and P_2. It does not agree with that found in *Problem 3*. Why not?

   If, instead of averaging the distance

$$|P_1 P_2| = \sqrt{r_1^2 + r_2^2 - 2 r_1 r_2 \cos(\beta - \alpha)},$$

we find the average of the quantity $\sqrt{r_1 + r_2 - 2\sqrt{r_1 r_2} \cos(\beta - \alpha)}$, we get the result of *Problem 3*. Why?

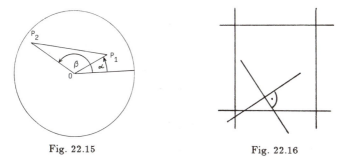

Fig. 22.15   Fig. 22.16

7. Two needles of unit length are tied at their centers, so as to form a cross. This cross is thrown on a chessboard with unit squares. In this way we get an even better estimate of π. Write a simulation program based on this idea, similar to the program BUFFON. (See Fig. 22.16.)

8. Find the mean distance between two points chosen at random in the unit cube.

   Theory predicts (see American Mathematical Monthly 1978, p. 278)

$$E = \frac{4 + 17\sqrt{2} - 6\sqrt{3} + 21\ln(1 + \sqrt{2}) + 42\ln(2 + \sqrt{3}) - 7\pi}{105} = 0.661707.$$

23. Random Choice of an s-Subset from an n-Set

The random choice of an s-sample from an n-population is an interesting and important statistical problem. A deep probe into this problem alone would serve as an excellent introduction to Computer Science. We first consider some elementary algorithms. By definition an s-sample is chosen at random from an n-population if each s-sample has the same probability $1/\binom{n}{s}$ to be chosen.

a) The algorithm sample1 is one of the simplest and cleanest solutions. For each i from 1 to n it calls random(n). If random(n) < s, then it prints i and decreases s by 1. For each call it also decreases n by 1. If n is the number of elements still to be considered and s the number of elements still to be chosen, then the next number will be printed with probability s/n.

```
program sample1;
var i,n,s:integer;
begin
   write('s,n=');
   readln(s,n); i:=1;
   repeat
     if random(n)<s then
     begin
       s:=s-1;write(i,'   ')
     end;
     i:=i+1; n:=n-1;
   until n=0;
end.
```

Fig. 23.1

```
program sample2;
var i,n,r,s:integer;
        x:array[1..10000] of byte;
begin
   write('s,n=');readln(s,n);
   for i:=1 to n do x[i]:=0;
   for i:=1 to s do
   begin
     repeat r:=1+random(n)
     until x[r]=0;
     write(r,'   '); x[r]:=1;
   end;
end.
```

Fig. 23.3

Fig. 23.2

The algorithm has time complexity $O(n)$ since random(n) is called n times. But it gives the sample in sorted order which is often an advantage.

The algorithm is not easy to understand. The number 1 is obviously printed with the correct probability $p_1 = s/n$. It requires some computation to see that the number 2 is printed with the same probability. From Fig.23.2 we get

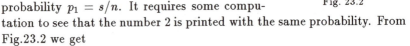

b) The algorithm **sample2** is more efficient if s is much smaller than n, and it is easy to understand. We choose at random a number r from 1 to n. Then we test if it was already chosen. If so, in which case $x[r] = 1$, we repeat the random choice. Otherwise r is printed and we set $x[r] = 1$ as a reminder that r was already chosen. But the algorithm gives the numbers in random order. **Sample3** is an easy change, which gives sorted output.

```
program sample3;
var i,n,r,s:integer;
  x:array[1..10000] of byte;
begin
  randomize;
  write('s,n=');
  readln(s,n);
  for i:=1 to n do x[i]:=0;
  for i:=1 to s do
  begin
    repeat r:=1+random(n)
    until x[r]=0;
    x[r]:=1
  end;
  for i:=1 to n do if x[i]=1
  then write(i,'  ')
end.
```

Fig. 23.4

```
program sample4;
var i,n,r,s,t:integer;
    x:array[1..9000] of
                    integer;
begin
  randomize;
  write('s,n=');  read(s,n);
  for i:=1 to n do x[i]:=i;
  for i:=1 to s do
  begin
    r:=i+random(n-i+1);
    t:=x[i];
    x[i]:=x[r];
    x[r]:=t;
    write(x[i],'  ')
  end
end.
```

Fig. 23.5

c) Program **sample4** is very easy to understand. First we choose at random a number r from 1 to n and we exchange $x[r]$ with $x[1]$. Then we choose at random a number r from 2 to n and we exchange $x[r]$ with $x[2],\ldots$, and finally we choose a number r at random from s to n and exchange $x[r]$ with $x[s]$. Then $x[1]$ to $x[s]$ are printed. This corresponds exactly to drawing s from n numbers without replacement.

d) The program **resample** is a recursive version of **sample1**. It prints the sample in descending order. By exchanging the two statements in line 7 it prints them in ascending order. It is easy to comprehend without comment. For $s = 100$, $n = 10000$ **sample1** to **sample4** require less than a second on my AT. With **resample** I had to go down to about $n = 6500$ to prevent overflow. But then it took less than a second to choose a 100-subset. None of these programs is suitable for the following problem: From the phone book of New York with $n = 2000000$ entries you are to select $s = 2000$ entries for a phone interview. See the references for section 23 for a state of the art solution of this problem.

```
program resample;
var s,n:integer;
procedure choose(s,n:integer);
begin
  if s>0 then if random(n)<s then
  begin
    write(n,' '); choose(s-1,n-1)
  end
  else choose(s,n-1)
end;

begin randomize;
  writeln('s,n'); readln(s,n);
  choose(s,n)
end.
```

Fig. 23.6

Exercises:

1. **Generation of a random permutation.** In $x[1..n]$ numbers are stored. These numbers are to be permuted so that all permutations have the same probability. Write the corresponding program. We will need this program later.

2. Generate a random permutation $x[1..m]$ of $1..m$, and find the length k of the longest "saw-tooth" subsequence $x[p+1] < x[p+2] > x[p+3] < \ldots > x[p+k]$ (with teeth upwards and a last downward segment).

24. Coin Tosses for Poor People

In this and the next two sections we treat some methods to generate random digits, which lead to interesting programs and interesting insights. We do not treat the most popular linear congruential RNG since the corresponding programs are straightforward and the topic is dealt with extensively elsewhere. See for instance Knuth's "bible", vol. 2.

a) We compute $x = \log_b a$ with $1 < a < b$. By definition this is the solution x of

$$b^x = a , \quad 0 < x < 1.$$

Let

$$x = 0.d_1 d_2 d_3 \ldots = d_1/2 + d_2/4 + d_3/8 + \ldots$$

be the binary representation of x. Then

(1) $$b^{d_1/2+d_2/4+d_3/8+\cdots} = a.$$

We square both sides and set at once $a := a * a$:

$$b^{d_1+d_2/2+d_3/4+\cdots} = a.$$

If $a < b$ then $d_1 = 0$. Otherwise $d_1 = 1$ and we set $a := a/b$. In both cases we have

(2) $$b^{d_2/2+d_3/4+\cdots} = a.$$

This is again (1) with the first digit shaved off. Thus we get a simple algorithm for printing successive binary digits of x:

1. $a := 2$; $b := 10$.
2. $a := a * a$.
3. If $a < b$ then print 0 and go to 2.
4. If $a > b$ then print 1, set $a := a/b$ and go to 2.

Instead of 2 and 10 we could use any two suitable numbers for a and b. We are computing the binary digits of an irrational number. Only about 40 digits will be correct. The remaining digits will be corrupted by rounding errors. The idea is to use these digits as a cheap source of tosses of a good coin.

In the following recursive program toss we can use any a, b with $1 < a < b$. Of course, we should also have $b^p \neq a^q$ for integer p, q but this last condition is practically always satisfied. Fig. 24.3 shows the output of the program toss.

Now we write a program which generates n bits. After each 1000 bits it prints on separate lines the number d of "ones" and the number $i - d$ of "zeros". With $a = 2$ and $b = 10$ we get after 10000 bits $d = 5069$ "ones" and $i - d = 4931$ "zeros". The intermediate results are also what one expects from a coin. The first 1000 bits are the same as those in Fig. 24.3.

b) There are primes p whose reciprocals have decimal expansions with maximal period $p - 1$. If $p = 2q + 1$ with q also a prime, then the possible periods are $1, 2, q, 2q = p - 1$. These candidates are easy to check. In this way we find that $p = 2063, 10463, 20087, 100667, 2040287$ have maximal period. The programs period and half_period print a full and a half period, respectively. The half period for numerator num=1 and prime=2063 is shown in Fig. 24.6.

```
program toss;
procedure goodcoin
   (a,b:real;i:integer);
begin
   a:=a*a;
   if a<b then write(0)
   else
   begin
      a:=a/b; write(1)
   end;
   if i>1 then
      goodcoin(a,b,i-1)
end;

begin
   goodcoin(2,10,10000)
end.
```

Fig. 24.1

```
program randcoin;
var i,d,n: integer;
    a,b: real;
begin
write('a,b,n='); read(a,b,n);
i:=0; d:=0;
while i<n do
begin
   a:=a*a;
   if a<b then write(0)
   else begin
      write(1); a:=a/b; d:=d+1
   end;
   i:=i+1;
   if i mod 1000=0 then
   begin
      writeln; writeln;
      writeln(d,' ',i-d);
      writeln
   end
end
end.
```

Fig. 24.2

```
0100110100010000010011010100001001111101111100001100100110010010100000
0001111110110011110000000011011010001111000110111100110010010010111011
1110000110100110011100011111001001001100010010101100111000011010011001
1011110100000000110001011000010011000100001111010101101110000100011011
0011111110011110000110100001100100000101111100001110100000000010010111
0111100001100110101101010110011010111 010011010001110000110110011111111
111011001 0111100111101011001110100000011001100011 0000111110111111110
1010011011110001001011 0101100100011011110101011011111101101011110100011
1110101001100001111100000101001001111001011001011000010111110101001100
1001001101000010011000011011000011101010100100111000000111110001010110
0001000010000111100001100001101110100101011010001100110101110011001100
0111011010101010010010011011101000001110010100101110010111111110110110
0110101010011001010111101111010011110100011010001010111110110110111001110
1001000111110111001110101111000001100110011010000011100110110101011000
0001000000010101000000000110101100100000001000
```

Fig. 24.3

```
program period;
var num,prime,rem,digit: integer;
begin
  write('num,prime'); readln(num, prime); rem:=num;
  repeat
    rem:=10*rem;
    digit:=rem div prime; write(digit);
    rem:=rem mod prime
  until rem=num
end.
```

Fig. 24.4

```
program half_period;
var i, num, prime, rem, digit: integer;
begin
  write('num, prime='); readln(num, prime); rem:=num;
  for i:=1 to (prime-1) div 2 do
  begin
    rem:=10*rem; digit:=rem div prime;
    write(digit); rem:=rem mod prime
  end
end.
```

Fig. 24.5

```
00048473097430925836160930683470673777605428986912263693650024236548
71
51808046534173533688802714493456131846825012118274357731459040232670
8
67668444724672806592341250605913717886572952011633543383422200678623
3
64032961706253058943286476005816771691711100339311682016480853126514
7
84294716432380029083858555016965584100824042656325739214735821619001
4
54192922927775084827920504120262869607367910809500727096461463887542
4
13960252060106640814348036839554047548230731943771206980126030053320
4
07174018419777023751817744115365971885603490150266602035870092098885
11
87590887057682985942801745031507513330101793504605593795443528841492
9
71400872515753756665050896752302472127968977217644207464043625787687
8
33252544837615123606398448860882210373242850218128938439166262707561
8
03199224430441105186621425109064469219583131362094037809015996122152
2
05310712554532234609791565681047018904507998061076102762966553562772
6
61173048957828405235094522539990305380513814832767813863305865244789
1
420261754726126
```

Fig. 24.6

The next table shows the frequency of digit d in a full period and its first half.

digit d	0	1	2	3	4	5	6	7	8	9
period	206	206	206	207	206	206	207	206	206	206
half period	130	105	112	101	101	105	106	94	101	76

The digits of a full period are too evenly distributed to be qualified as "random". But in a half period we observe fluctuations around the expected value 103, which seem to be of the correct size. A closer look at the half period reveals hidden regularities. Let F_i be the frequency of the digit i. Then $F_i + F_{9-i} = 206$, except $F_3 + F_6 = 207$. The same phenomenon occurs for other primes with maximal period. For $p = 20663$ we get the following frequency table:

digit d	0	1	2	3	4	5	6	7	8	9
period	2066	2066	2066	2067	2066	2066	2067	2066	2066	2066
half period	1044	1037	1029	1041	1019	1047	1026	1037	1029	1022

This regularity could matter in applications.

Exercise:

The tables above show that the frequencies of the digits in a full period differ at most by 1. Prove that if division by p has period $p - 1$ then this will always be so. Similarly, the numbers of occurrences of pairs, triples etc. of digits differ at most by 1. Write programs for checking this for pairs and triples.

25. Shift Registers: Another Source of Cheap Coin Tosses

Fig. 25.1 to 25.6 show six shift registers. Each is filled by an initial vector of binary digits (bits). The registers generate a stream of bits as follows: First the modulo 2 sum s of the tapped cells (indicated by arrows leading from these cells) is formed. Then the initial vector moves one place left and the resulting empty cell is filled by s.

Take for instance Fig.25.3. The register can assume 2^7 or 128 states, but 0000000 is a special state. It cannot be reached from a non-zero vector. Starting from any non-zero vector we can reach only 127 states. We have tapped the cells so that the vector will indeed traverse all these states before the initial vector recurs. In the program **shift** we have printed 134 bits, just to show you that the first and last 7 bits are identical. Thus 127 is a period, and, since it is a prime, there is no smaller period. (A smaller period would divide 127.)

In this program we have not used

$$s := (a + b) \bmod 2$$

because for bits Turbo Pascal has a far more efficient operation

```
s:=a xor b;
```

"xor" stands for "exclusive or" and is equivalent to addition mod 2. There is also an operation **xor** for integers, which we will not use.

These registers can also be described by linear difference equations. Hence they are called *linear shift registers* (LSR). For instance, the registers in Fig. 25.1 to 25.6 can be described by

$$x_n = x_{n-1} + x_{n-4}, \qquad x_n = x_{n-5} + x_{n-6}, \qquad x_n = x_{n-6} + x_{n-7},$$

$$x_n = x_{n-5} + x_{n-9}, \qquad x_n = x_{n-7} + x_{10}, \qquad x_n = x_{n-9} + x_{n-11},$$

respectively, all with addition mod 2.

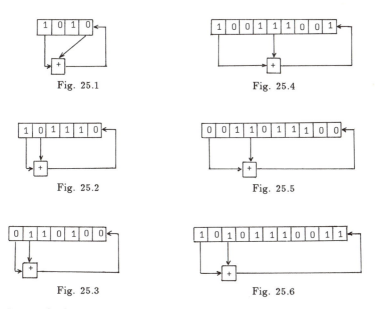

Fig. 25.1

Fig. 25.4

Fig. 25.2

Fig. 25.5

Fig. 25.3

Fig. 25.6

Let us look again at the program **shift**. Here the variables a to g and s are not really integers, but bits. They are from the restricted range 0..1. In Turbo Pascal only predefined types are allowed inside a procedure name. I must first predefine a new type **range=0..1** as in **shift1**. This program is considerably faster than **shift**, and, even more importantly, it uses half as much memory. The reason is that Turbo Pascal reserves two bytes for an integer but only one byte for a variable which can take only the values 0 and 1. For such a variable and for variables of type

```
program shift;                      program shift1;
procedure                           type range=0..1;
   x(a,b,c,d,e,f,g,n:integer);      procedure x(a,b,c,d,e,f,g:
var s:integer;                                    range;n:integer);
begin                               var s:range;
  if n>0 then                       begin
  begin                               if n>0 then
    s:=a xor b; write(s);             begin
    x(b,c,d,e,f,g,s,n-1)                s:=a xor b; write(s);
  end                                   x(b,c,d,e,f,g,s,n-1)
end;                                    end
begin                               end;
  x(0,1,1,0,1,0,0,134)              begin x(0,1,1,0,1,0,0,134)
end.                                end.
```

Fig. 25.7   Fig. 25.8

Boolean a single bit would suffice, but that would be fairly complicated
to implement and Turbo Pascal uses a whole byte.

Exercises:

1. Show that all shift registers in Fig. 25.1 to 25.6 generate sequences with
   maximal period.

2. Show that binary LSR-sequences are purely periodic as in Fig. 6.13.

3. In the sequence 1983113835952... each digit from the fifth on is the mod
   10 sum of the preceding four digits. Does the sequence contain the word
           a) 1234;   b) 3269;   c) 5198;   d) 1983 a second time;   e) 1357?
   The questions a) to d) should be answered without a computer, just by
   plain thinking. For e) a computer should be used. Experiment with
   different initial values. How does the length of the period depend on the
   initial values?

4. The two LSR in Fig. 25.9 and 25.10 generate streams of binary and base
   5 digits of maximum period. Check this with the computer. Initially the
   cells are loaded by a non-zero vector. We call the two streams x and y.
   We use x to transform y into a sequence of decimal digits by means of
   the following algorithm: If $x_n = 0$, leave y_n unchanged. If $x_n = 1$, set
   $y_n := y_n + 5$. What is the period of the resulting sequence of decimal
   digits?

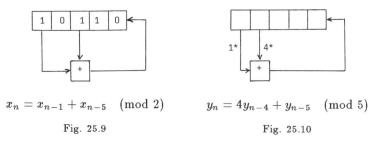

$$x_n = x_{n-1} + x_{n-5} \quad (\text{mod } 2) \qquad y_n = 4y_{n-4} + y_{n-5} \quad (\text{mod } 5)$$

Fig. 25.9   Fig. 25.10

5. The LSR defined by the recurrences $x_n = x_{n-14} + x_{n-17}$ (mod 2) and $y_n = 4y_{n-1} + 3y_{n-7}$ (mod 5) generate sequences with periods $2^{17} - 1$ and $5^7 - 1$, respectively. These sequences are combined as in Exercise 4. What is the period of the resulting sequence? Write programs to print the combined sequences in 4. and 5. How do you check the periods of these sequences?

Additional Remarks about LSR. The quality of random digits generated by LSR goes up (they satisfy more standard tests) as the order of the corresponding difference equation goes up. The recursion given for $n > 55$ by

$$x_n = (x_{n-24} + x_{n-55}) \bmod m,$$

is recommended by Knuth as a very good generator for random digits in even base. Initially $x_1, \ldots, x_{55}$ are not all even. For $m = 2^e$ the length of the period is

$$2^f (2^{55} - 1), \qquad 0 \le f < e.$$

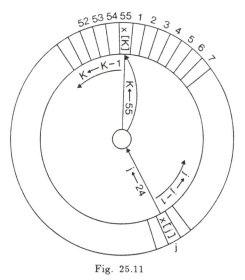

Fig. 25.11

For $m = 10$, we get the program `randig10`, which generates n random decimal digits and counts the frequencies $f[i]$ of the digits i. We first fill $x[1], \ldots, x[55]$ with (low quality) random digits of the computer. Initially $j = 24$ and $k = 55$. Then we go around the circle in Fig. 25.11 counterclockwise, i.e. diminishing j and k by 1, and add up pairs mod 10. From 1 j or k jumps to 55. The results d are the (high quality) decimal random digits.

```
program randig10;                  for i:=1 to n do
var d,i,j,k,n:integer;             begin
    f:array[0..9] of integer;        d:=x[k]+x[j];
    x:array[1..55] of 0..9;          if d>9 then d:=d-10;
begin randomize;                     f[d]:=f[d]+1; x[k]:=d;
write('n=');readln(n);               j:=j-1; if j=0 then j:=55;
for i:=0 to 9 do f[i]:=0;            k:=k-1; if k=0 then k:=55
for i:=1 to 55                     end;
  do x[i]:=random(10);             for i:=0 to 9 do
j:=24; k:=55;                        writeln(i,f[i]:10)
                                   end.
```

Fig. 25.12

26. Random Sequence Generation by Cellular Automata

In an LSR-sequence a small chunk of the sequence suffices to restore the whole sequence. Since randomness is tantamount to total unpredictability, we must turn to nonlinear shift registers (NLSR) or "cellular automata". It is well known that simple quadratic difference equations can lead to unpredictable, chaotic behavior. But first we introduce two new Turbo Pascal operations for bits:

a **xor** b = a+b **mod** 2 (exclusive or, was used already once)

a **or** b = a+b+ab **mod** 2 (inclusive or)

That is, a or $b = 1$ if at least one of a, b is equal to 1. The operations **xor** and **or** are considerably more efficient than their modular counterparts. We test this by means of the program **test**. In this program writeln(chr(7)) tells the computer to ring the bell. We first generate $n + 1$ (10001) random bits. Now the bell rings. Then we perform n additions mod 2. The bell rings again. Now we perform the **xor** operation on the same n pairs of bits. This time the bell rings six times faster. The next two rings tell us that the **or** operation is about nine times faster than the corresponding modular operation.

A simple, well studied and very good generator of good coin tosses is

$$(1) \qquad y_n = x_{n-1} + x_n + x_n x_{n+1} \quad (\text{mod } 2).$$

This can be written more simply and efficiently as

$$(2) \qquad y_n = x_{n-1} \text{ xor } (x_n \text{ or } x_{n+1}).$$

It transforms an infinite sequence (x_n) into another sequence (y_n). Since we cannot handle infinite sequences we will use a circular array as in Fig. 26.2. For $i = 2, 3, ..., n-1$ we can use the recurrence

$$(3) \qquad y[i] := x[i-1] \ \textbf{xor} \ (x[i] \ \textbf{or} \ x[i+1]),$$

but

$$y[1] := x[n] \ \textbf{xor} \ (x[1] \ \textbf{or} \ x[2]), \qquad y[n] := x[n-1] \ \textbf{xor} \ (x[n] \ \textbf{or} \ x[1]).$$

If we set $x[0] := x[n]$ and $x[n+1] := x[1]$, then we can use the recurrence (3) for $i \leftarrow 1$ to n. The program orxor1 is based on this idea. First the array $x[1..n]$ is filled by n random bits. Then the sequence $y[1..n]$ is computed from $x[1..n]$. Finally we set $x[i] := y[i]$ for $i = 1, 2, ..., n$, and we repeat the operation max times.

```
program test;
const n=10000;
var i:integer; a:byte;
    x:array[0..n] of byte;
begin
for i:=0 to n do x[i]:=random(2);
write(chr(7));
for i:=0 to n-1 do
  a:=(x[i]+x[i+1]) mod 2;
write(chr(7));
for i:=0 to n-1 do
  a:=x[i] xor x[i+1];
writeln(chr(7));
for i:=0 to n-1 do
  a:=(x[i]+x[i+1]+x[i]*x[i+1])
      mod 2; write(chr(7));
for i:=0 to n-1 do
  a:=x[i] or x[i+1];
write(chr(7))
end.
```

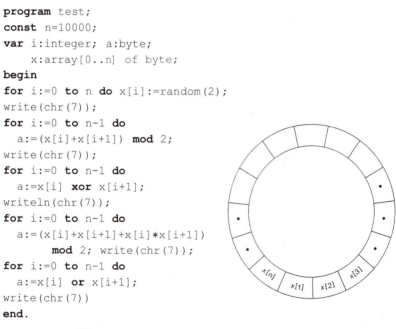

Fig. 26.1   Fig. 26.2

Exercise:

Explore the similar rule $y_i = x_{i-1} \ \textbf{xor} \ (x_i \ \textbf{or} \ (1 - x_{i+1}))$, or

$$y_i = (1 + x_{i-1} + x_{i+1} + x_i x_{i+1}) \pmod 2.$$

```
program orxor1;
var i,max,n,count:integer;
    x,y:array[0..1000] of byte;
begin randomize;
  write('n,max=');readln(n,max);count:=0;
  for i:=1 to n do x[i]:=random(2);
  repeat count:=count+1;
    x[0]:=x[n]; x[n+1]:=x[1];
    for i:=1 to n do
    begin
      y[i]:=x[i-1] xor (x[i] or x[i+1]); write(y[i])
    end;
    write('   ':3);
    for i:=1 to n do x[i]:=y[i]
  until count=max
end.
```

Fig. 26.3

27. The Period Finding Problem

Given a function f which maps a finite domain D of integers into itself, and an arbitrary starting point $x \in D$, the sequence $x_0 = x$, $x_1 = f(x_0)$, $x_2 = f(x_1)$, $x_3 = f(x_2),\ldots$ is ultimately periodic by the pigeonhole principle. That is, for some t and c we have $t + c$ distinct values $x_0, x_1, \ldots, x_{t+c-1}$, but $x_{t+c} = x_t$. This implies, in turn, that $x_{i+c} = x_i$ for all $i > t$.

The problem of finding the unique pair (t, c) will be called the *cycle problem* for f and x. The integer c is the *cycle length* and t is the *tail*. This problem arises when analysing random number generators that produce successive "random" numbers by applying some function to the preceding value of the sequence. A simple solution of the cycle problem is due due to R. W. Floyd. We work with two chips y and z on Fig. 27.1. The slow chip y (the turtle) proceeds in unit steps $x_0 \mapsto x_1 \mapsto x_2 \mapsto \ldots$. The fast chip z (the hare) proceeds twice as fast: $x_0 \mapsto x_2 \mapsto x_4 \mapsto \ldots$. That is:

```
y:=x; z:=x; c:=0;
repeat
  y:=f(y); z:=f(f(z)); c:=c+1
until y=z;
```

When $y = z$ the hare has caught up with the turtle somewhere on the cycle. Now we check if $x = y$. In this case the tail $t = 0$ and we have a pure cycle of length c. If not, the turtle is sent once more around the cycle:

```
                    c:=0;
                    repeat
                       y:=f(y);  c:=c+1
                    until y=z;
```

Now c is the cycle length. To find t the hare is sent to the start, both the hare and the turtle move at the same unit speed. They meet at x_t! (Ex. 1.) Thus we can compute t as follows.

```
                    z:=x;  t:=0;
                    repeat
                       z:=f(z);  y:=f(y);  t:=t+1
                    until z=y;
```

Now t is the length of the tail. What remains to be done is to define some function f, for instance

```
               function f(u:integer);integer;
               begin f:=(u*u+a) mod m end;
```

Here two new variables a and m are introduced. We will enter them as global variables into the main program. By collecting the fragments we get the program FINDCYCLE.

```
program FINDCYCLE;
var a,m,x,y,z,c,t:integer;

function f(u:integer):integer;
begin f:=(u*u+a) mod m end;

begin
write('x,a,m=');  read(x,a,m);
y:=x;  z:=x;  c:=0;
repeat
   y:=f(y);  z:=f(f(z));  c:=c+1
until y=z;
if x=y then
   writeln('pure cycle, length=',c)
else begin c:=0;
   repeat y:=f(y);  c:=c+1
   until y=z;
   writeln('cycle length=',c);
   z:=x;  t:=0;
   repeat z:=f(z);  y:=f(y);  t:=t+1
   until z=y;
   writeln('tail length=',t)
   end
end.
```

Fig. 27.2   Fig. 27.1

Exercises for Sections 1-27:

1. Play with two chips on Fig. 27.1 and similar figures with different c and t until you comprehend and can prove all of the facts we mentioned, it is not difficult.

2. Take the binary digit generator from Section 24 , starting with $a = 2$, $b = 10$. Incorporate this algorithm into the program FINDCYCLE, find cycle length c and tail length t. Count all zeros and all ones generated by all function calls. How many times are the bits from the tail counted? **Answer:** $c = 173469$, $t = 1362641$, number of zeros $= 3532883$, number of ones $= 3529924$. The bits of the tail are counted exactly three times, those of the cycle are counted many times.

3. We define an infinite binary sequence as follows: Start with 0 and repeatedly replace each 0 by 001 and each 1 by 0.
   a) Write a program which prints this sequence.
   b) What is the 10000th bit of the sequence?
   c) What is the proportion of zeros among the first 10000 bits of the sequence?
   d) Can you show that the sequence is not periodic?
   e) Try to find empirically a formula for the place numbers of the 1's, that is, for the sequence 3, 6, 10, 13, 17, ... Find also a formula for the place numbers of the 0's.

4. Start with a finite sequence $a[0], a[1], \ldots, a[n]$ and extend it successively by defining $a[n + 1] = \max_{0 \le i < n}(a[i] + a[n - i])$. You will find that the first differences are ultimately periodic. For instance, you could start with $a[i]:=\texttt{random(10)}$ for $i = 1, 2, ..., 9$.

5. *Unsolved problem of Dickson.* Given k integers $a[1] < a[2] < \ldots < a[k]$, define $a[n + 1]$ for $n \ge k$ as the least integer greater than $a[n]$ which is not of the form $a[i] + a[j]$, $i, j \le n$. Is the sequence of differences $a[n + 1] - a[n]$ eventually periodic? Take $k = 2$, $a[1] = 1$, $a[2] = 6$. Can you detect a periodicity in the sequence of differences? What about the set $\{1, 4, 9, 16, 25\}$? Is the cycle finding algorithm of any use here?

6. Start with the digit 1 and use repeatedly the replacement rule T:
$$0 \to 0000, \quad 1 \to 1321, \quad 2 \to 0021, \quad 3 \to 1300.$$
Thus we get $T(1) = 1321$, $T^2(1) = T(1321) = T(1)T(3)T(2)T(1) = 1321\ 1300\ 0021\ 1321, \ldots$. The sequence $T^{n+1}(1)$ has $T^n(1)$ as an initial segment. Thus, if we start with 1 and apply T repeatedly, we get longer and longer initial segments of an infinite sequence.
   a) Show that this sequence does not change when T is applied.
   b) Check if there is a period in the first 30000 digits.
   c) Show that this sequence is not periodic.

Remark. By placing a dot somewhere in the infinite sequence, for example $t = 1.321130000211321\ldots$ we get the infinite expansion in some base ≥ 4. Then one can prove by advanced techniques that t is either rational (if periodic) or transcendental. We cannot get an algebraic irrational like $\sqrt{2}$ by fixed replacement rules. So t is transcendental.

7. *Numbers with the equisum property.* Write a program, which generates at random words with $2n$ decimal digits and checks if the sum of the first n digits is equal to the sum of the last n digits. Estimate the corresponding probability $p(n)$, and also $\sqrt{n}\,p(n)$. Guess a good approximation for $p(n)$.

8. We return for the last time to the sequence defined by $a_0 = 3$, $a_1 = 0$, $a_2 = 2$, $a_n = a_{n-2} + a_{n-3}$, $n > 2$. One can show that n prime $\Rightarrow n \mid a_n$ is always true. Show that $n \mid a_n$ for $n = 271441 = 521^2$. This may be the smallest counterexample to the conjecture $n \mid a_n \Rightarrow n$ prime. There is no counterexample up to $n = 140000$. With a modification of the primitive program in the solutions it will take my AT over a week to decide if $n = 271441$ is indeed the smallest counterexample. With matrices we can speed up the program from $O(n^2)$ to $O(n \log n)$ with a huge saving in computation time.

9. Find the smallest n with the following property: In the binary representation of $1/n$ the bit patterns of the binary representations of all the numbers 1 to 1990 do occur somewhere. You are supposed to find this number by means of a program which conducts a systematic search. The number can also be found by Olympiad type hard thinking plus some elementary number theory.

10. A die is rolled repeatedly until the sum $X_1 + X_2 + \cdots + X_N$ of the points surpasses 100. Find by simulation
    a) the most probable value of N b) the most probable sum.

CHAPTER 4

Statistics

28. Matched Pairs

We will study in depth the simple but important problem of matched pairs, which can be treated in a new way by means of the computer.

a) The data in Table 1 are from the first controlled marijuana study. They show for 9 subjects the changes X and Y in their mental performance 15 minutes after smoking an ordinary cigarette and a marijuana cigarette, respectively. Positive X's and Y's represent improvements. To deal with a possible effect of smoking the different substances in a certain order, the flip of a coin decided for each subject which type of cigarette was smoked first; this is one of several methods called *randomization*. We have also tabulated the difference $D = Y - X$ and $|D|$ for each subject.

X	-1	-1	-3	3	-3	-3	2	4	10	$\Sigma X = 8$
Y	1	-3	-7	-3	-9	5	-6	-7	-17	$\Sigma Y = -46$
$D = Y - X$	2	-2	-4	-6	-6	8	-8	-11	-27	$\Sigma D = -54$
$|D| = d_k$	2	2	4	6	6	8	8	11	27	$\Sigma |D| = 74$

Table 1. Source: SCIENCE 162, 1234–1242.

The experiment was made to investigate the effect of marijuana. The table shows that different subjects experience different effects; can we nevertheless infer something about the effects on the group, for example on most members or on the average? Could it be that the fluctuations in mental performance shown in the table are just random? To deal with such a possibility, we formulate two hypotheses. The first is called

The null hypothesis H: The fluctuations shown in Table 1 are just random. There is no difference in the effects of the two types of cigarettes.

Since the experimenters, on the basis of anectodal evidence, suspected that marijuana has a more detrimental effect than ordinary cigarettes, they wanted to test the following alternative which the many negative differences in the D-row make credible:

The alternative A: Mental performance is lower after smoking marijuana than after smoking tobacco.

To the hypothesis H we give the precise meaning that the signs of the differences were decided by a fair coin. With n differences we then have 2^n equiprobable cases.

Now we choose a *test statistic* T, which expresses quantitatively the amount of evidence for A contained in the set of the differences. We can always define T in such a way that the smaller T is, the more likely it is that A is true. Suppose the observed value of T is t. Next we compute

$$P = P(T \leq t|H) = \text{ probability that } T \leq t \text{ if } H \text{ is true.}$$

This is the *observed significance level* or *P-value* of the test. The smaller the P-value the stronger is the evidence for A contained in the set of the differences. For instance, we could choose $T = \#$ of positive signs of the differences. The observed value of T is $t = 2$. There are obviously $\binom{9}{0} + \binom{9}{1} + \binom{9}{2} = 1 + 9 + 36 = 46$ cases, which are favorable to the event $T \leq 2$. Thus

$$P = P(T \leq 2|H) = 46/2^9 = 9\%.$$

We have not enough evidence for A to get excited about. But this so called sign test is not very "powerful". It does not make good use of the data. There are not only few positive differences, but they are also significantly smaller on the average than the negative differences.

A far better statistic would be the sum of the differences $S = \sum D$. As we see in Table 1, the value S obtained from our data is -54. If the null hypothesis is true, the signs of the differences are as likely to be positive as negative. We want to know the probability that $S \leq -54$ if the signs of the differences are assigned by fair coin tosses. We observe that if T is the sum of the positive differences then $S = -74 + 2T$, so it suffices to find the probability that $T \leq 10$ which in our problem can be found very easily as follows.

We have 2^9 or 512 possible and equiprobable cases. The favorable cases are simply all 24 subsets of the last row in Table 1 with sum 10 or less, i.e.:

8, 8, 8+2, 8+2, 8+2, 8+2, 6, 6, 6+4, 6+4, 6+2, 6+2, 6+2,
6+2, 6+2+2, 6+2+2, 4, 4+2, 4+2, 4+2+2, 2, 2, 2+2, 0.

Thus

$$P(T \leq 10|H) = \frac{24}{512} = \frac{3}{64} = 4.6875\%.$$

If the observed significance level is 5% or less most journals are willing to publish the result. This 5%-barrier was erected to stem the flood of pseudo-discoveries.

b) Let us consider another slightly larger example: Does maternal mal-
nutrition retard the mental development of a child?

A simpler study which may shed some light on this is to relate the
mental development of identical twins to their weights at birth. The IQ's
of 12 pairs of identical twins was measured years later and compared with
the weights at birth. Table 2 shows the IQ X of the heavier twin and Y
of the lighter twin.

X	100	124	108	91	100	91	79	80	95	104	100	119
Y	101	123	106	97	106	84	70	70	84	92	85	104
$\lvert Y - X \rvert = d_k$	1	1	2	6	6	7	9	10	11	12	15	15

Table 2. Source: Child Development, vol. 38, 623-629.

H: The heavy and the light twins develop the same IQ.
A: The heavy twin usually develops a higher IQ.

The sum of the positive differences $Y - X$ is $T = 1 + 6 + 6 = 13$. Under
H all 2^{12} or 4096 subsets of the differences have the same probability.
The favorable cases are those subsets of the d_k with sum $T \leq 13$. We find
these subsets by brute force. Sort the subsets by the maximum element.
By teamwork we quickly get all solutions:

12, 12+1, 12+1,
11, 11+1, 11+1, 11+2, 11+1+1,
10, 10+2, 10+1, 10+1, 10+2+1, 10+2+1, 10+1+1,
9, 9+2, 9+1, 9+1, 9+2+1, 9+2+1, 9+1+1, 9+2+1+1,
7, 7+6, 7+6, 7+2, 7+1, 7+1, 7+2+1, 7+2+1, 7+1+1, 7+2+1+1,
6, 6, 6+6, 6+2, 6+2, 6+1, 6+1, 6+1, 6+1, 6+6+1, 6+6+1, 6+2+1,
6+2+1, 6+2+1, 6+2+1, 6+1+1, 6+1+1, 6+2+1+1, 6+2+1+1,
2, 2+1, 2+1, 2+1+1,
1, 1, 1+1, 0.

There are 4096 possible cases, and those 60 cases listed above are favorable for $T \leq 13$. Thus

$$P = P(T \leq 13 \mid H) = 60/4096 = 15/1024 = 1.465\%.$$

This is strong evidence for the alternative A that the better nourished
twin develops a higher IQ. But we had to pay a stiff computational price
for the answer.

c) Next we consider a famous experiment by Charles Darwin. He took
15 pairs of seeds of the same plant and planted them into 15 pots. One
seed of each pair was produced by cross fertilization, the other by self
fertilization. For pot # i he measured the height x_i of the cross fertilized
plant and the height y_i of the self fertilized plant. For the difference
$z_i = x_i - y_i$ he got (in 1/8-ths of an inch):

6, 8, 14, 16, 23, 24, 28, 29, 41, −48, 49, 56, 60, −67, 75.

The sum of the negative differences is $t = 115$.

H: There is no difference between the two kinds of seed.

A: Cross fertilized seeds develop stronger plants.

If H is true we have 2^{15} possible and equiprobable cases. The favorable cases are those with $T \leq t$. Let us generalize the problem slightly. What is the number $q(t, n)$ of all subsets whose sum T is $\leq t$ from the set $D = \{d_1, d_2, \ldots, d_n\}$?

For $q(t, n)$ we have the recursion

(1) $$q(t, n) = q(t, n - 1) + q(t - d_n, n - 1)$$

with the boundary conditions

(2) $$q(t, n) = 0 \text{ for } t < 0 \quad \text{and} \quad q(0, n) = q(t, 0) = 1.$$

Indeed, there are $q(t, n-1)$ subsets without d_n and $q(t-d_n, n-1)$ subsets with d_n. This recursion can be immediately translated into the Pascal program in Fig. 28.1.

```
program rematch;
const d:array[1..15] of integer=
      (6,8,14,16,23,24,28,29,
       41,48,49,56,60,67,75);
function q(t,n:integer):integer;
begin
  if t<0 then q:=0
  else if (t=0) or (n=0)
    then q:=1
  else q:=q(t,n-1)+q(t-d[n],n-1)
end;
begin
  writeln(q(115,15))
end.
```

Fig. 28.1

Running the program we get $q(115, 15) = 863$ and

$$P = \frac{863}{2^{15}} = \frac{863}{32768} = 2.63\%.$$

The program **rematch1** in Fig. 28.2 is more flexible and can solve a wide variety of problems. Yet we are dealing with a tree recursive process and the program is too slow for large problems (which frequently occur in practice). Fig. 28.3 is a straightforward translation of the recursion into an iterative program. It is easy to write, but it is space consuming. Space and time complexity are proportional to $(t + 1)(n + 1)$. This program uses for the first time a *matrix* $q[i, j]$, which is a table of numbers with $t + 1$ rows and $n + 1$ columns.

Finally we will construct the most efficient program for computing $q(t, n)$, which will run on the cheapest programmable pocket calculator. We compute $q(t, n)$ rowwise, and we denote the current row by $r[0], r[1], \ldots, r[t]$. To find the next row we use the recurrence

$$r1[t] = r[t] + r[t - d], \text{ where } d \text{ belongs to the current row.}$$

```
program rematch1;
var i,t,n: integer;
   d:array[1..50] of integer;
function q(t,n:integer):
                      integer;
begin
  if t<0 then q:=0
  else if (t=0) or (n=0)
                 then q:=1
  else
    q:=q(t,n-1)+q(t-d[n],n-1)
end;
begin
write('t,n='); readln(t,n);
for i:=1 to n do begin
  write('?'); read(d[i]) end;
writeln;writeln('q=',q(t,n))
end.
```

Fig. 28.2

```
program matchit;
var d,t,n,i,j:integer;
    q:array[0..150, 0..30]
                  of integer;
begin
write('t,n='); readln(t,n);
for i:=0 to t do q[i,0]:=1;
for i:=1 to n do
begin
  write('d['i']='); read(d);
  for j:=0 to d-1 do
    q[j,i]:=q[j,i-1];
  for j:=d to t do
    q[j,i]:=q[j,i-1]+q[j-d,i-1]
end;
writeln;
writeln('q(',t,',',n,')=',q[t,n])
end.
```

Fig. 28.3

d	i\j	0 1 2 ...	j−d	...	j	...	t
	0	1 1 1 ...	1	...	1	...	t
6	1						
8	2						
14	3						
⋮	⋮						
d	i−1		r[j−d]	...	r[j]		
	i				r1[j]		
	⋮						
	n						r[t]

$r1[j] := r[j] + r[j - d]$ is stored in $r[j]$.

Fig. 28.4.

If we start at the end of the row, then we can store $r1[t]$ into $r[t]$, and so we need just one array $r[t]$. Fig. 28.4 shows the details of the computation, and Fig. 28.5 shows the corresponding program. For a cheap calculator it must be translated into BASIC. In Fig. 28.5 it is advisable to declare `r:array[0..200] of real` and to change the output into `writeln('r(',t,')=',r[t]:0:0)`. In this way we can avoid overflow with big numbers beyond `maxint`, or even `MaxLongInt`.

The time required by matchit1 is, as we mentioned, of the order tn which means we could easily deal with even 1000 plants. However, there is something strange about the method. While the most natural way to represent results of measurement is by real numbers, the method makes essential use of the data being integers. In practice the result of a measurement will be expressed as an integer multiple of a unit or some subunit of it, so the fact that the method does not work with a set of arbitrary real numbers is of no practical significance. However, mathematicians would like to devise a good algorithm for finding the number of subsets of a given set

```
program matchit1;
var d,i,j,n,t:integer;
    r:array[0..200] of
                  integer;
begin
    write('t,n=');readln(t,n);
    for i:=0 to t do r[i]:=1;
    for i:=1 to n do
    begin
      write('?');readln(d);
      for j:=t downto d do
      r[j]:=r[j]+r[j-d]
    end;
    writeln('r(',t,')=',r[t])
end.
```

Fig. 28.5

of integers with sum $\leq t$ which will be practicable even if the integers do not have the limited range one has in problems coming from statistics.

One way very large integers could come up in a seemingly innocuous problem is if the numbers originally given are not integers but fractions with not necessarily the same denominators. For instance, we could generate a set of 100 fractions with randomly picked 3-digit integers as numerators and denominators and ask how many subsets of this set have sum ≤ 50. While all the numbers in our original problem are between 0.1 and 10, if we bring all the fractions over a common denominator in order to be able to work with integers, we get very large integers and the method of matchit1 can not be used. We cannot check all 2^{100} subsets.

D. B. Johnson and S. D. Kashdan proved: *Given a set of n integers, counting the number of subsets with sum $\leq t$ is NP-hard.* For a full explanation of the meaning of this theorem we refer to the book *Computers and Intractability* by M. R. Garey and D. S. Johnson, 1979, p. 225. (The reader should ignore the "log" in their account of Lawler's work.) Johnson's and Kashdan's result implies that a polynomial-time algorithm for this problem would yield polynomial-time algorithms for the hundreds of computational problems in the class NP, but is unlikely to exist. See also Section 49.

29. Permutation Test, P by Simulation

Let us now go back to the Marijuana example. By plotting Y versus X (Fig. 29.1) we see that there seems to be little if any relation be-

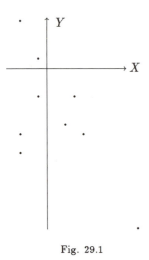

Fig. 29.1

tween the performances after smoking of the two types of cigarettes. So there is no advantage in considering pairs. Instead we treat all 18 data $(-3, 5, 10, -17, -3, -7, 3, -3, 4, -7, -3, -9, 2, -6, -1, 1, -1, -3)$ as spins of the same Random Number Generator (RNG), which is the effect of any smoking on performance. The observed subset X due to tobacco smoking has sum $S = \Sigma\, X = 8$. The subset Y due to Marijuana smoking has sum $T = \Sigma\, Y = -46$.

H: The large difference is due to chance.

A: The RNG for Marijuana gives lower performance than the RNG for tobacco.

If H is true, X is just a random 9-subset of the 18 data. How likely is it that a randomly chosen 9-subset of the data has sum $S \geq 8$? We draw **rep** random 9-samples of the data, and we count with the variable **count** how often $S \geq 8$. The drawing of the random samples is done by means of the program **sample1** in section 23.

```
program RandTest;
const n=18; k=9; x:array[1..n] of integer=
(-3,5,10,-17,-3,-7,3,-3,4,-7,-3,-9,2,-6,-1,1,-1,-3);
var i,j,sum,rep,count,n1,s: integer;
begin randomize;
  write('rep='); readln(rep); count:=0;
  for j:=1 to rep do
  begin sum:=0; i:=1; n1:=n; s:=k;
    repeat
      if random(n1)<s then begin s:=s-1; sum:=sum+x[i] end;
      i:=i+1; n1:=n1-1
    until n1=0;
    if sum>=8 then count:=count+1
  end;
  writeln('count=',count,' P=',count/rep:0:6)
end.
```

Fig. 29.2

Three runs of the program **RandTest** with **rep**=32000 gave the **count**-values 478, 497, 490, 545, 494 with the mean **count**=501. Hence we have

$$P = P(S \geq 8 \mid H) \approx 501/32000 = 1.5656\% \,.$$

This method is only feasible with a computer. It gives us an estimate $\hat{P} = 1.5656\%$ for an unknown probability P. It also gives an estimate of the error in P. Let us modify the program RandTest into RandTst1 (see Exercise 2 following Section 30). The new program also counts the number of occurrences of $S \geq 8$ in each thousand. We get 32 numbers,

0	9
1	0012233344
1	5555567788889
2	01111233

Fig. 29.3

which we summarize in the stem-and-leaf plot in Fig. 29.3. Here the tens' digits are placed on the left of the vertical line. From these numbers we extract an error estimate in P. We throw away the lower and upper quarter of these data. The range of the remaining data is the so called *interquartile range* $R = 19 - 13 = 6$. Let $\tilde{C} = (15 + 16)/2 = 15.5$ be the median of the data. Tukey gives a robust rule of thumb for an estimate of the true median $\tilde{m}$. A conservative 95% confidence interval for $\tilde{m}$ is

$$\tilde{C} \pm \frac{1.58R}{\sqrt{N}} = 15.5 \pm 1.586/\sqrt{32} \approx 15.5 \pm 1.68.$$

Dividing by 1000 and halving the result we get $P \in [0.0138, 0.0172]$ as a 95% confidence interval for P.

30. Permutation Test, P by Computation

We will take up the Marijuana example for the last time. The null hypothesis is that it makes no difference what kind of cigarette the subject smoked. If this were so then all 18 numbers could be regarded as values of the same random variable and the subset of 9 values corresponding to tobacco would be just a randomly picked subset of 9 out of the 18 values. However, the sum of these 9 values is 8, which seems rather large for a 9-subset of this set. To decide how likely it is that this happened just by chance, we step with the computer through all $\binom{18}{9}$ or 48620 subsets of the data

$$(-3, 5, 10, -17, -3, -7, 3, -3, 4, -7, -3, -9, 2, -6, -1, 1, -1, -3).$$

It turns out that exactly 766 subsets have the sum 8 or more, as in Table 1 in 28. Thus we have again $P = 766/48620 = 1.5755\%$. We will call this value the "exact" significance level. The P-value of the program RandTest will converge to this value for rep $\to \infty$. Fig. 30.1 shows an algorithm for this traversal of the 9-subsets which has two drawbacks: it is hard to comprehend, and it is inefficient. My AT requires merely 5 seconds, which means the program is good enough for small problems. There is little hope for medium sized and large problems. The computation time grows exponentially with the size of the problem. The number of 18-subsets of a 36-set is already 9075135300. In general, a $2n$-set has $\sim 2^{2n}/\sqrt{\pi n}$ n-subsets. In Section 32 we will deal with the problem more efficiently with an easy to understand program. For very large problems we could use the program RandTest in the preceding section. It gives practically the same significance level.

The program in Fig. 30.1 traverses the 9-subsets in lexicographic order and counts those with sum 8 or more. Inputs are $n = 18$, $k = 9$, sum=8 and the data array above. The result $P = 1.5755\%$ gives us confidence in the method of Section 29.

The program is not comprehensible without comment. So we will give a short comment. How do you generate lexicographically all k-subsets of an n-set $\{1, 2, \ldots, n\}$? You start with $(1, 2, \ldots, k)$: for i:=1 to k do c[i]:=i;. The next subset is found by scanning the current subset from right to left for i:=1 to k do c[i]:=i; so as to locate the rightmost index that has not yet attained its maximum value while c[j] = n-k+j do j:=j-1;. This element is incremented by one: c[j]:=c[j]+1;. All positions to its right are reset to the lowest values possible: for i:=j+1 to k do c[i]:=c[i-1]+1;

```
program permTest;
const d:array[1..18] of integer=
      (-3,5,10,-17,-3,-7,3,-3,4,-7,-3,-9,2,-6,-1,1,-1,-3);
var n,k,sum,i,s,j,count:integer; c:array[0..10] of integer;
begin
   write('n,k,sum=');readln(n,k,sum); c[0]:=-1;count:=0;s:=0;
   for i:=1 to k do c[i]:=i;
   repeat
     for i:=1 to k do s:=s+d[c[i]];
     if s>=sum then count:=count+1;
     j:=k; s:=0;
     while c[j]=n-k+j do j:=j-1;
     c[j]:=c[j]+1;
     for i:=j+1 to k do c[i]:=c[i-1]+1
   until j=0;
   writeln('count=',count)
end.
```

Fig. 30.1

Exercises for sections 28-30:

1. Should hospitals allow parents of sick children to live in the hospital together with their children (rooming-in model) to shorten their stay? Of

50 children 25 pairs were formed. In each pair the relevant variables were nearly the same. The parents of one child in each pair were allowed to stay. The following table shows the number of days each child stayed in the hospital, X without, Y with rooming-in.

X	29	44	18	15	15	7	15	11	12	12	7	11	7	22	7	14	28	31	18	16	10	11	29	8	23
Y	29	32	16	8	8	7	14	8	7	12	7	8	7	17	6	19	12	8	8	6	10	8	21	17	15

Table 3. Source: Fernandez-Jung, F., Dissertation FU Berlin, 1983.

Find $D = X - Y$ and compute the P-value with the programs Matchit and Rematch.

2. Adapt program RandTst1 to the data in Table 3 and find your 95% confidence interval for the P-value associated with this test.

3. The table below shows the frequency of one letter words per 1000 words in 8 essays of A. Hamilton and 7 essays of J. Madison.

Hamilton	24	21	23	24	33	28	28	37
Madison	20	27	19	30	11	17	27	

Source: F. Mosteller and D. L. Wallace, *Inference and Disputed Authorship: The Federalist.* 1964, p. 248.

H: These are samples from the same distribution.

A : These are samples from different distributions.

Find the P-value with the program RandTest.

4. Apply the program PermTest to the table in Exercise 3. Find the P-value.

5. Rewrite RandTest in Section 29 by using the procedure choose in Section 23.

6. Let us look again at the Marijuana data in Table 1. Write a program which chooses a random subset of the fourth row with sum T and checks if $T \leq 10$ or $T \geq 64$. If so a counter should be increased by 1. Let rep be the number of repetitions. Then estimate P by means of count/rep/2 and compare with the exact P-value.

31. The Two Sample Problem of Wilcoxon

We use a concrete example to derive an efficient algorithm which solves a large number of two sample problems. Fig. 31.1 shows the stem-and-leaf plot of the ages at death of the 31 US presidents who died of natural causes.

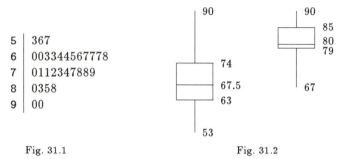

5	367
6	003344567778
7	0112347889
8	0358
9	00

Fig. 31.1   Fig. 31.2

Let us divide the presidents according to height into two categories: *short* (<5' 8") and *tall* (>5' 8"). Then we get
short: 67, 79, 80, 85, 90
tall: 53, 56, 57, 60, 60, 63, 63, 64, 64, 65, 66, 67, 67, 68, 70, 71, 71, 72, 73, 74, 77, 78, 78, 83, 88, 90.
In a **box plot** we enter just 5 data: the minimum, the lower quartile, the median, the upper quartile, and the maximum.
If we compare the box-plots of the two sets we get the impression that we are dealing with samples from two quite different populations (Fig. 31.2).
Let us sort the data increasingly by age at death. But what about ties? These can be broken by a look into the Encyclopedia Americana, which reveals the exact ages. Thus we find for the tie 67:
W. Wilson (tall) < B. Harrison (short) < G. Washington (tall)
and for the tie 90: H. Hoover (tall) < J. Adams (short) .
Here $a < b$ means that a died at a younger age than b.
By writing 0 for tall and 1 for short presidents, we get the binary word

$$W = 00000\ 00000\ 00100\ 00000\ 00001\ 10100\ 1$$

This looks quite significant. Most of the short presidents are concentrated in the right tail. So we consider the two hypotheses

H : It is just a random fluctuation.

A : Short people on the average live longer.

Our test statistic will be the rank sum RS of the 1's in W, which is the sum of their positions in the sequence. For the actual data $RS = 13+25+26+28+31 = 123$. We find an approximate value of $P = P(RS \geq 123 \mid H)$ by simulation.

The program simupres chooses 5 of the ranks 1 to 31 at random and adds them up. This experiment is repeated rep times, and for each occurrence of the event $RS \geq 123$ a counter count is increased by 1. Fig. 31.3 shows the program. With input rep=32000, $n = 31$, $m = 5$, max=123 we get count=299 and

$$P = \text{count/rep} = 0.00934375.$$

Exercise: Finding small probabilities by simulation requires a large number of trials. Modify the program **simupres** to count the words for which the word or its reflexion has $RS \geq 123$, which raises the probability of success to $2P$. Compare the estimates so obtained with those from the original program.

```
program simupres;
var rep,i,k,m,n,max,r,rs,count:
                           integer;
    b:array[1..31] of byte;
begin randomize;
write('rep,n,m,max=');
readln(rep,n,m,max); count:=0;
for k:=1 to rep do
begin rs:=0;
    for i:=1 to n do b[i]:=0;
    for i:=1 to m do
    begin
        repeat r:=1+random(n)
        until b[r]=0;
        rs:=rs+r;  b[r]:=1
    end;
    if rs>=max then count:=count+1
end;
writeln('count=',count,'
                P=',count/rep)
end.
```

Fig. 31.3

We could also find a confidence interval for P as we did in Section 29. But in this particular case we will find the exact P-value by finding the number of binary words with 26 zeros and 5 ones with rank sum ≥ 123 by means of a recursive formula.

The number of all words with 26 zeros and 5 ones is $\binom{31}{5} = 169,911$. Such a word can be encoded by a nonincreasing sequence $z_1, z_2, \ldots$ of 5 nonnegative integers $z_i \leq 26$ which give the number of zeros to the right of the first, $\ldots$, fifth 1 in the sequence assigned to it. Every nonincreasing sequence $z_1, z_2, \ldots$ of the above kind corresponds to a word of 26 zeros and 5 ones.

The rank sum of a word can be easily computed from our encoding. The rank of the first 1 in the word is $1 + \#$ of zeros to its left $= 1 + (26 - z_1)$. Similar formulas for the ranks of the other 1's give

$$RS = 1 + 2 + 3 + 4 + 5 + 5 * 26 - z_1 - z_2 \ldots..$$

Thus the words with $RS \geq 123$ are the ones encoded by nonincreasing sequences of 5 nonnegative integers with sum $U \leq 22$. (The number of such sequences is $1 +$ the number of *partitions* into at most 5 positive parts of the numbers $1, \ldots, 22$.) The integers in the sequences, or summands in the partitions representing our words must be ≤ 26 since there are at most 26 0's to the right of a 1. In our example the condition $U \leq 22$

automatically excludes summands > 22.

Let $w(u, m, n)$ be the number of nonincreasing sequences of m non-negative integers with sum $\leq u$, and largest term $\leq n$. We are going to compute $w(22, 5, 22)$ recursively. We have

$$w(u, m, n) = 0 \text{ for } u < 0, \quad \text{and} \quad w(0, m, n) = w(u, m, 0) = 1,$$

for positive values of u, m and n. A sequence counted by $w(u, m, n)$ either has largest term $\leq n - 1$ or it consists of n, followed by $m - 1$ terms $\leq n$ with sum $u - n$. This gives us the recursion

$$w(u, m, n) = w(u, m, n - 1) + w(u - n, m - 1, n);$$

and this will be true even for $m = 1$ if we set $w(u, 0, n) = 1$ for nonnegative u and n. The straightforward translation of this recursion yields Fig. 31.4.

```
program wilcorec;
var u,m,n:integer;
function w(u,m,n:integer):
                    integer;
begin
  if u<0 then w:=0
  else if (u=0) or
               (m=0) or (n=0)
    then w:=1
  else w:=w(u,m,n-1)+
              w(u-n,m-1,n)
end;
begin
write('u,m,n=');readln(u,m,n);
writeln('w=',w(u,m,n))
end.
```

Fig. 31.4

```
program wiliter1;
var u,m,n,x,y,z:integer;
    w:array[0..25,0..6,0..27]
                    of integer;
begin
write('u,m,n=');read(u,m,n);
for z:=0 to u do
  for x:=0 to m do w[z,x,0]:=1;
for z:=0 to u do
  for y:=0 to n do w[z,0,y]:=1;
for x:=0 to m do
  for y:=0 to n do w[0,x,y]:=1;
for x:=1 to m do
  for y:=1 to n do
    for z:=1 to u do
    if z<y then
        w[z,x,y]:=w[z,x,y-1]
      else w[z,x,y]:=
      w[z,x,y-1]+w[z-y,x-1,y];
writeln('w=',w[u,m,n])
end.
```

Fig. 31.5

With input $u = 22$, $m = 5$, $n = 22$ we get in 0.3 seconds $w(22, 5, 22) = 1601$, and $P = 1601/169911 = 0.00942$. By simulation we got $P = 0.00934$.

The program wiliter1 shows a straightforward translation into an iterative program. With $u = 22$, $m = 5$, $n = 22$ the run time is again 0.3 seconds. But for more extensive problems it is considerably faster than the recursive program. Its time complexity is proportional to $u * m * n$.

Unfortunately it requires $(u+1)*(m+1)*(n+1)$ storage cells for the matrix $w[u, m, n]$. These storage requirements can be reduced considerably.

We compute $w(z, x, y)$ for $y = 0, 1, 2, \ldots$. We set up the layer $y = 0$ by assigning the value $w = 1$ to each lattice point (z, x) with $0 \leq z \leq u$, $0 \leq x \leq m$ (Fig. 31.6). Using the recurrence for w we now compute successively the w-values at the lattice points for layers $y = 1, 2, \ldots, n$. Each computed value is immediately stored into the preceding layer, so that we need to remember just the latest layer. In this way the recursion

$$w(z, x, y) = w(z, x, y-1) + w(z-y, x-1, y)$$

reduces to setting

$$w(z, x) := w(z, x) + w(z-y, x-1) \quad \text{for} \quad z \geq y;$$

for $z < y$ the values of $w(z, x)$ need not be changed. We get the program wiliter2 in Fig. 31.7. With $u = 22$, $m = 5$, $n = 22$, we now get $w = 1601$ in just one second. This program runs on the cheapest programmable pocket calculators, since it requires only $23 * 6 = 138$ memory locations for the matrix $w[z, x]$. Fortunately we have $w(u, m, n) = w(u, n, m)$ (exercise 1); so we can always put the larger of m and n into the last place. This reduces memory space.

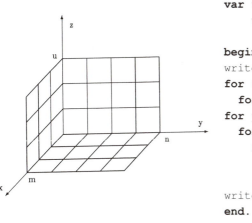

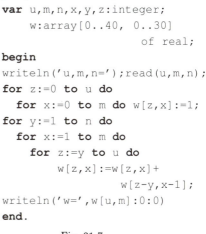

```
program wiliter2;
var u,m,n,x,y,z:integer;
        w:array[0..40, 0..30]
                       of real;
begin
writeln('u,m,n=');read(u,m,n);
for z:=0 to u do
   for x:=0 to m do w[z,x]:=1;
for y:=1 to n do
   for x:=1 to m do
      for z:=y to u do
         w[z,x]:=w[z,x]+
                   w[z-y,x-1];
writeln('w=',w[u,m]:0:0)
end.
```

Fig. 31.6   Fig. 31.7

32. The General Two Sample Test

In the presidents example we lost some information by replacing ages by ranks. It would be better to use the ages. Suppose we take a random

5-subset of the 31 ages. What is the probability that the sum S of these ages is $\geq 67 + 79 + 80 + 85 + 90 = 401$? Let $q(n, k, s)$ be the number of k-subsets of an n-set $\{d_1, d_2, \ldots, d_n\}$ of nonnegative integers, whose sum is $\leq s$. Then

(1)　　　　　　　$q(n, 0, s) = 1$　　for $s \geq 0$,

(2)　　　　　　　$q(n, k, s) = 0$　　for $n < k$ or $s < 0$,

and

(3)　　　　$q(n, k, s) = q(n - 1, k, s) + q(n - 1, k - 1, s - d_n)$.

Indeed, there are $q(n-1, k, s)$ subsets without d_n and $q(n-1, k-1, s-d_n)$ subsets with d_n. To use this recursion we could transform the ages by means of the reflection $x \to 90 - x$. This would make the observed ages of the short presidents small. Instead of $s \geq 67 + 79 + 80 + 85 + 90 = 401$ we would get $s \leq 0 + 5 + 10 + 11 + 23 = 49$. The formulas (1) to (3) can be translated into the program TwoSample in Fig. 32.1. It gives $q(31, 5, 49) = 2260$ and $P = 1.33\%$. So we see that ranks gave us a P-value which is too good. They overestimate tiny differences, which occur several times. All differences get the same weight. The Two Sample Test rates them correctly. Ties do not matter.

```
program TwoSample;
const d:array[1..31] of integer=(0,0,2,5,7,10,11,12,12,13,16,
    17,18,19,19,20,22,23,23,23,24,25,26,26,27,27,30,30,33,34,37);
var n,k,s:integer;

function q(n,k,s:integer): integer;
begin
   if (k=0) and (s>=0) then q:=1
   else if (n<k) or (s<0) then q:=0
        else q:=q(n-1,k,s)+q(n-1,k-1,s-d[n])
end;

begin
   write('n,k,s='); readln(n,k,s);
   writeln; writeln('P=', q(n,k,s)/169911.0:7:4')
end.
```

Fig. 32.1

Exercises for Sections 31-32:

1. Prove that always $w(u, m, n) = w(u, n, m)$.

2. Write iterative programs for the Two Sample Test imitating wiliter1 and wiliter2 in Fig. 31.5. and Fig. 31.7.

3. Find the P-value of the hypothesis in Exercise 3 at the end of Section 30 by means of the programs wiliter2 and TwoSample.

4. Apply the program RandTest to the presidents example. Now permutes the data at random before running the program. Originally the data are sorted increasingly. I found that the Applesoft BASIC RNG was quite sensitive to this and gave $P \approx 0.65\%$, but by one random permutation the P-value became correct. Try to find if the Turbo Pascal RNG has a similar bug.

5. Apply the permutation test to the presidents data. With the program PermTest in Fig. 30.1 you should step through all the 5-subsets of the 31 data and count those with sum$\geq$ 401 and compare with the output of Fig. 32.1.

6. Apply the Two Sample Test to the Marijuana Example. It should give the same result as the program PermTest, and faster. But you should first turn the data into nonnegative integers by addition of 17 to each of the 17 numbers. The input is still $n = 18$, $k = 9$, but $s = 9 * 17 - 46 = 107$. Indeed, in the original problem $s \geq 8$ is equivalent to $s \leq -46$. By adding 17 to each number we get $s \leq 107$. See Table 1 in Section 28. To gain even more speed you should use the iterative programs suggested in Exercise 2 above and given in the solutions section.

33. Kendall's Rank Correlation

Since World War II plutonium for US atomic weapons has been produced in Hanford, Washington. Until 1965 radioactive wastes were stored in open pits, and they leaked into the Columbia river, which flows through Oregon into the Pacific. For nine Oregon districts an index of exposure X was defined and compared with cancer mortality Y (per 100000 man years for 1959-1964). Fig. 33.1 shows the data sorted by the index of exposure.

X	1.25	1.62	2.49	2.57	3.41	3.83	6.41	8.34	11.64
Y	113.5	137.5	147.5	130.1	129.9	162.3	177.9	210.3	207.5

Fig. 33.1. Source: Journal of Environmental Health, v. 27, 1965, 883–897.

Replace each cancer mortality by its rank. You get the permutation

$$p = 145326798$$

since the first tabulated has rank 1, the second has rank 4, the third rank 5, etc.

Is this a "random permutation"? The extreme cases 123456789 and 987654321 are certainly not random. Consider the 36 pairs that can be picked from such a sequence in the order in which they occur; i.e. (i, j) with i occurring before j. A pair with the first entry $<$ the second is called *rising* or *concordant*, a pair with the second entry $<$ the first is called *falling* or *discordant*, or an *inversion*. The first extreme case above has no inversions, the second has 36. By symmetry in a random permutation we expect 18 rising and 18 falling pairs. But p has $\text{inv}(p)=6$ and its reflection $p' = 897623541$ has $\text{inv}(p')=30$.

By symmetry

$$P(\text{inv}(p) \leq 6 \mid H)$$
$$= P(\text{inv}(p) \geq 30 \mid H),$$

where

H: It is just a random fluctuation.

A: p has too few (too many) inversions.

```
program cancer;
var rep,copy,i,j,k,m,n,inv,
    min,max,count:integer;
    x:array[0..15] of byte;
begin randomize;
write('rep,n,min,max=');
readln(rep,n,min,max);
count:=0;
for m:=1 to rep do
begin inv:=0;
    for i:=1 to n do x[i]:=i;
    for i:=n downto 2 do
    begin
        k:=1+random(i); copy:=x[i];
        x[i]:=x[k]; x[k]:=copy
    end;
    for i:=1 to n-1 do
        for j:=i+1 to n do
            if x[i]>x[j] then
                inv:=inv+1;
    if (inv<=min) or (inv>=max)
                then count:=count+1
end;
writeln('count=',count,'
                P=',count/rep/2)
end.
```

Fig. 33.2

First we want to find this probability by simulation. We will generate rep random 9-permutations, count the inversions, and test if $\text{inv}(p) \leq 6$ or $\text{inv}(p) \geq 30$. For each occurrence of this event a counter count is increased by 1. Fig. 33.2 shows the program. The input rep=32000, $n = 9$, min $= 6$, max $= 30$ results in count $= 400$. Thus

$$P \approx \frac{1}{2}(\text{count}/\text{rep}) = 0.00625.$$

To find the exact value of this probability we must find the number of 9-permutations with at most 6 inversions and divide this number by 9!.

Let $p(n, k)$ be the number of n-permutations with at most k inversions. Our aim is to derive a computer friendly recursion formula for $p(n, k)$. Suppose I have an $(n-1)$-permutation $X_1 X_2 \ldots X_{n-1}$ of $\{1, 2, \ldots, n-1\}$. I can make of it an n-permutation by inserting element n into one of the n places numbered 0 to $n - 1$, as indicated by boxes:

$\boxed{\text{n-1}}\; X_1 \;\boxed{\text{n-2}}\; X_2 \ldots \boxed{\text{i}}\; X_{n-i} \ldots \boxed{2}\; X_{n-2} \;\boxed{1}\; X_{n-1} \;\boxed{0}.$

If I place element n into place i, it contributes i inversions. To get at most k inversions altogether the $(n-1)$-permutation must contribute at most $k-i$ inversions. The number of $(n-1)$-permutations with this property is $p(n-1, k-i)$. By inserting element n successively into places 0 to $n-1$ we get

$$p(n, k) = p(n-1, k) + p(n-1, k-1) + \cdots + p(n-1, k-n+1).$$

This recursion is not yet computer friendly. If we substitute $k-1$ in place of k we get

$$p(n, k-1) = p(n-1, k-1) + p(n-1, k-2) + \cdots + p(n-1, k-n).$$

Using this to replace most of the sum in the previous formula we get

$$p(n, k) = p(n, k-1) + p(n-1, k) - p(n-1, k-n).$$

If we take into account $p(n, 0) = p(1, k) = 1$ for $n \geq 1$, $k \geq 0$, and $p(n, j) = 0$ for $j < 0$, we get

$$k < n \rightarrow p(n, k) = p(n, k-1) + p(n-1, k),$$
$$k \geq n \rightarrow p(n, k) = p(n, k-1) + p(n-1, k) - p(n-1, k-n),$$
$$p(n, 0) = p(1, k) = 1.$$

The programs **kendallr** and **kendallit** show the recursive and iterative programs, respectively.

```
program kendallr;
var n,k:integer;
function p(n,k:integer):real;
begin
  if (k=0) or (n=1) then p:=1
  else if k<n then p:=p(n-1,k)
                        +p(n,k-1)
  else p:=p(n-1,k)+p(n,k-1)
                        -p(n-1,k-n)
end;
begin
  write('n,k='); readln(n,k);
  writeln('p=',p(n,k):0:0)
end.
```

Fig. 33.3

```
program kendallit;
var n,k,i,j:integer;
    p:array[0..20, 0..50]
                       of real;
begin
  write('n,k='); readln(n,k);
  for i:=1 to n do p[i,0]:=1;
  for i:=0 to k do p[1,i]:=1;
  for i:=2 to n do
    for j:=1 to k do
    if j<i then p[i,j]:=
             p[i,j-1]+p[i-1,j]
    else
       p[i,j]:=p[i,j-1]+p[i-1,j]
                      -p[i-1,j-i];
  writeln(p[n,k]:0:0)
end.
```

Fig. 33.4

By entering $n = 9$, $k = 6$, we get $p(9, 6) = 2298$ and $P = P(inv \leq 6) = 2298/9! = 0.0063$.

For the computation of $p(9, 6)$ my AT requires less than 1 second. But the recursion is quite complicated and for $p(12, 11) = 431886$ my AT needed 1.3 minutes. The iterative program in Fig. 33.4 is very fast and solves all problems arising in practice.

The Permutation Test and Correlation. Let us look again at the table in Fig. 33.1. With the help of a computer we can take a test statistic of our own invention. For instance, let us take the *dot product*

$$T = X \cdot Y = \sum X[i]Y[i].$$

of X, Y. Note that the components of X are sorted in ascending order. It is easy to see that T is a maximum if also the components of Y are sorted increasingly. Because of some inversions the observed value Cor of T is smaller than the maximum. We first find Cor $= 7440.3660$. Now we permute the components of Y at random and call the permuted vector Z; this is repeated 10000 times. We set count:=count+1 each time

```
program cancer1;
const n=9;
type vector=array[1..n] of real;
const X:vector=(1.25,1.62,2.49,2.57,3.41,3.83,6.41,8.34,11.64);
      Y:vector=(113.5,137.5,147.5,130.1,129.9,162.3,177.9,
                210.3,207.5);
var count,i,j,r:integer; Cor,copy:real; Z:vector;
function dot(S,T:vector):real;
var i:integer; sum:real;
begin
  sum:=0.0; for i:=1 to n do sum:=sum+S[i]*T[i]; dot:=sum
end;
begin Cor:=dot(X,Y); writeln(Cor:0:4); count:=0; randomize;
for j:=1 to 10000 do
begin Z:=Y;
  for i:=n downto 2 do
    begin r:=1+random(i); copy:=Z[i]; Z[i]:=Z[r]; Z[r]:=copy
  end;
  if dot(X,Z)>=Cor then count:=count+1
end;
writeln(count)
end.
```

Fig. 33.5

dot$(X, Z) \geq$ Cor. Since it is difficult to estimate very small probabilities, we have repeated the program cancer1 ten times, getting the count values 2, 3, 3, 3, 4, 4, 9, 5, 5, 4 and the estimate $\widehat{P} = 0.00042$.

Next we present a program, cancer2, which uses the old established but more complicated *Correlation Coefficient* as test statistic. This topic can be skipped as far as the rest of this book is concerned. The correlation coefficient has the advantage that significance tables can be precomputed.

```
program cancer2;
            ⟨declarations as in program cancer1⟩
procedure unitize( var A:vector);
var i:integer;sum, sum2:real;
begin sum:=0.0;  sum2:=0.0;
for i:=1 to n do sum:=sum+A[i];  sum:=sum/n;
for i:=1 to n do A[i]:=A[i]-sum;
for i:=1 to n do sum2:=sum2+sqr(A[i]);  sum2:=sqrt(sum2);
for i:=1 to n do A[i]:=A[i]/sum2
end;
            ⟨ function dot as in cancer1⟩
begin
unitize(X);  unitize(Y);  Cor:=dot(X,Y);  writeln(Cor:0:4);
randomize;  count:=0;
for j:=1 to 10000 do
begin Z:=Y;
    ⟨ random permutation of the components of Z as in cancer1⟩
unitize(Z);  if dot(X,Z)>=Cor then count:=count+1 end;
writeln(count);  readln;  end.
```

Fig. 33.6

The correlation coefficient of two vectors, which have to have the same number of components, can be read off from the way cancer2 computes this quantity. The procedure unitize takes a vector as input. First it subtracts from each component the arithmetic mean of all components, which makes the sum of the components = 0. Then unitize transforms the modified vector into a *unit* vector (the sum of the squares of the components is 1) by dividing its components by its norm (length).

Now look at the main program. The *procedure calls* (uses of the procedure) unitize(X) and unitize(Y) transform X and Y into unit vectors and return them back to the main program. The dot product of X and Y is the *correlation coefficient, Cor*. It turns out that $Cor = 0.9259$, which is almost the maximum 1. Now we permute the components of Y at random and call the permuted vector Z. If $dot(X, Z) \geq Cor$ we set count:=count+1. This is repeated 10000 times and the variable count is printed. 10 repetitions of the program cancer2 resulted in the count-values 0, 3, 3, 6, 4, 5, 4, 5, 5, 6. The significance level 0.00041 is almost the same as with `cancer1` above. It would be the same if we had used the same random permutations.

`cancer2` shows a new feature of Pascal. In the list of input variables of **procedure** `unitize` we put **var** in front of the identifier A. This causes the procedure `unitize` to work on the variable A itself. Otherwise it would work on a copy of A which would be discarded at the exit from the procedure.

For $n = 9$ and $Cor = 0.898$, tables give the P-value $P=0.0005$. We have $Cor = 0.9259$. So our P-value should be smaller, which it is. But the tables are valid for normally distributed quantities. We make no assumptions about the distributions of X and Y, and program `cancer1` is much simpler. *Statistical tables are no longer needed!*

Exercises:

1. Fig. 33.7 shows the draft lottery for 1970. (The table is also on the disk with the programs.) The days of the year were assigned the ranks 1 to 366, presumably at random. Thus each birthday had its own rank. Now men with ranks 1,2,3,... were drafted until the army had enough recruits. Men with low ranks were almost certainly drafted, those with high ranks almost never. Our aim is to show that Fig. 33.7 is not a random calendar.

a) For each month we find the median of the ranks of its days. Check the table below.

month	1	2	3	4	5	6	7	8	9	10	11	12
median of ranks	211	210	256	225	226	207.50	188	145	168	201	131.5	100

b) Assign a rank to each median in the table above. You will get a permutation of the numbers 1 to 12.

c) Determine the number C of rising (concordant) pairs. You will find $C = 11$, instead of the expected value 33.

d) H: It is a random calendar. Find $P = P(C \leq 11 \mid H)$ and draw your own conclusions.

2. The following table gives the cholesterol level of males of two age groups.

X (20-30 years)	135	222	251	260	269	235	386	252	173	156	352
Y (40-50 years)	294	311	286	264	277	336	208	346	173	254	346

Source: Dixon and Massey, Introduction to Statistical analysis.

H : The cholesterol of both groups is the same.

A : The older group has higher cholesterol level.

a) Find the P-value using the program **TwoSample**.

b) Find the P-value using the program **RandTest**.

	Jan	Feb	Mar	Apr	May	Jun	Jul	Aug	Sep	Oct	Nov	Dec
1	305	086	108	032	330	249	093	111	225	359	019	129
2	159	144	029	271	298	228	350	045	161	125	034	328
3	251	297	267	083	040	301	115	261	049	244	348	157
4	215	210	275	081	276	020	279	145	232	202	266	165
5	101	214	293	269	364	028	188	054	082	024	310	056
6	224	347	139	253	155	110	327	114	006	087	076	010
7	306	091	122	147	035	085	050	168	008	234	051	012
8	199	181	213	312	321	366	013	048	184	283	097	105
9	194	338	317	219	197	335	277	106	263	342	080	043
10	325	216	323	218	065	206	284	021	071	220	282	041
11	329	150	136	014	037	134	248	324	158	237	046	039
12	221	068	300	346	133	272	015	142	242	072	066	314
13	318	152	259	124	295	069	042	307	175	138	126	163
14	238	004	354	231	178	356	331	198	001	294	127	026
15	017	089	169	273	130	180	322	102	113	171	131	320
16	121	212	166	148	055	274	120	044	207	254	107	096
17	235	189	033	260	112	073	098	154	255	288	143	304
18	140	292	332	090	278	341	190	141	246	005	146	128
19	058	025	200	336	075	104	227	311	177	241	203	240
20	280	302	239	345	183	360	187	344	063	192	185	135
21	186	363	334	062	250	060	027	291	204	243	156	070
22	337	290	265	316	326	247	153	339	160	117	009	053
23	118	057	256	252	319	109	172	116	119	201	182	162
24	059	236	258	002	031	358	023	036	195	196	230	095
25	052	179	343	351	361	137	067	286	149	176	132	084
26	092	365	170	340	357	022	303	245	018	007	309	173
27	355	205	268	074	296	064	289	352	233	264	047	078
28	077	299	223	262	308	222	088	167	257	094	281	123
29	349	285	362	191	226	353	270	061	151	229	099	016
30	164		217	208	103	209	287	333	315	038	174	003
31	211		030		313		193	011		079		100

Fig. 33.7 Draft Lottery results for 1970.
Source: Selective Service System, Office of the Director, Washington, D. C.

34. The Binomial Distribution

Fig. 34.1 shows a spinner. Let p and $q = 1 - p$ be the probabilities of the outcomes 1 (success) and 0 (failure), respectively. Any binary word like 100110...01 with x ones and $n - x$ zeros has probability $p^x q^{n-x}$. There are altogether $\binom{n}{x}$ such words, and so the probability of exactly x successes in n spins (trials) is

$$b(x) = \binom{n}{x} p^x q^{n-x}$$

or, more precisely,

(1) $$b(n, p, x) = \binom{n}{x} p^x q^{n-x}, \qquad x = 0, 1, 2, \ldots, n.$$

Here we are dealing with a function of 3 variables and so tabulation is hopeless; $b(n, p, x)$ is difficult to evaluate in the form (1), even with a computer. So we change (1) into a recursion

(2) $$b(0) = q^n, \quad b(x) = b(x-1) * r * \frac{n - x + 1}{x}, \quad r = \frac{p}{q}, \quad x = 1, 2, \ldots, n.$$

With (2) there is another difficulty. We must first evaluate $b(0)$. For $n = 126$ and $p = q = 0.5$ we get, correct to 11 significant digits,

$$0.5^n = 1.1754943508E - 38,$$

but for $n = 127$ we get a run time error, because 0.5^{127} lies outside the permissible range of Turbo Pascal, which is

$$2^{-127} < x \le 2^{127}.$$

If $x \le 2^{-127}$ we have *underflow*, for $x > 2^{127}$ we have *overflow*. Both are run-time errors because they are discovered during the running of the program. Some versions of Pascal round underflowed results to 0; still others let you set what should happen in case of an underflow. Further we need an individual probability $b(n, p, x)$ only quite rarely and then only for small n. What we really need is a sum

$$s = b(c) + b(c + 1) + \cdots + b(d - 1) + b(d).$$

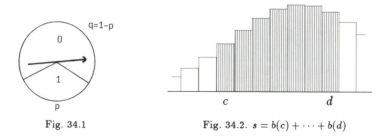

Fig. 34.1　　　　　　　Fig. 34.2. $s = b(c) + \cdots + b(d)$

To avoid underflow we must take logarithms in (2):

$$\ln b(0) = n \ln q , \quad \ln b(x) = \ln b(x-1) + \ln r + \ln \frac{n-x+1}{x} .$$

Then we can find the sum s in Fig. 34.2 by means of the program in Fig. 34.3, in which we have set $m = n + 1$.

```
program bin1;
var m,n,c,d,x:integer;
    p,q,r,L,s:real;
begin
  write('n,p,c,d=');
  readln(n,p,c,d);
  m:=n+1;q:=1-p;
  r:=ln(p/q); L:=n*ln(q);
  for x:=1 to c do
    L:=L+r+ln(m/x-1);
  s:=exp(L);
  for x:=c+1 to d do
  begin
    L:=L+r+ln(m/x-1);
    s:=s+exp(L)
  end;
  writeln(s)
end.
```

Fig. 34.3

```
program bin;
var m,n,c,d,x:integer;
    p,q,r,L,s:real;
begin
  write('n,p,c,d=');
  readln(n,p,c,d);
  m:=n+1;q:=1-p;
  r:=ln(p/q); L:=n*ln(q);
  for x:=1 to c do
    L:=L+r+ln(m/x-1);
  if L<-50 then s:=0
  else s:=exp(L);
  for x:=c+1 to d do
  begin
    L:=L+r+ln(m/x-1);
    if L>-50 then s:=s+exp(L)
  end;
  writeln(s:10:8)
end.
```

Fig. 34.4

If we run this program we usually get a run-time error. The culprit is the expression **exp(L)**. It turns out that if the absolute value of L is too large, **exp** cannot handle it. Taking this into account we arrive at our final program **bin** in Fig. 34.4, which is very robust.

A Confidence Interval for p.

A random process is controlled by the spinner in Fig. 34.5 with unknown parameter p. I make n spins and get X successes. Then

$$\widehat{p} = X/n$$

is a *point estimate* for p. We show now how one can find an *interval* such that we can say with a specified level of confidence that p lies in the interval.

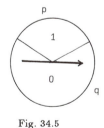

Fig. 34.5

Suppose we get $x = 80$ successes in $n = 200$ trials. Then $80/200=0.4$ is the point estimate for p. Of course we can get 80 or fewer successes even if p is larger than 0.4, but the larger p is the less likely this is. Let us find the value $\widehat{p}_2$ of p for which $P(X \leq 80) = 2.5\%$ exactly. The equation we must solve is

$$B(p) = P(X \leq 80) = \sum_{x=0}^{80} \binom{200}{x} p^x (1-p)^{200-x} = 0.025 \ .$$

We do this by intelligent search with the program bin in Fig. 34.4. The inputs $n = 200$, $c = 0$, $d = 80$ and different p-values give $\widehat{p}_2 = 0.47147$. The trial steps are given in the following table:

p	0.45	0.47	0.471	0.472	0.4714	0.47147
$B(p)$	0.088	0.0275	0.02577	0.0241	0.025106	0.0249906

So if $p \geq 0.47147$ the probability of getting only 80 successes in 200 trials is 2.5% or less. We say we can assert with 97.5% confidence that $p < 0.47147$. This means that if we repeatedly make assertions to which a confidence level of 97.5% can be attached in this way, we shall be right, in the long run, at least 97.5% of the time.

It may seem we could have expressed the same thing more simply by saying that the probability that $p \leq 0.47147$ is at least 97.5%. However, strictly speaking, this type of statement is appropriate only when p is a random variable which may take various values. In the present situation p is not a random variable. A random variable can take a different value each time we perform the experiment which defines it; p has a single well defined value, but we do not know what it is. The probability refers not to the value of p, but to the correctness of the statement we are making about it. That is why we are using the word "confidence" here.

If p were much *smaller* than 0.4, it is unlikely that we would have got 80 or more successes. The solution of the equation

$$B(p) = P(X > 80) = \sum_{x=80}^{200} \binom{200}{x} p^x (1-p)^{200-x} = 0.025$$

is $\widehat{p}_1 \approx 0.33155$. If $p < \widehat{p}_1$ then the probability of obtaining 80 or more successes is $< 2.5\%$. In other words, $\widehat{p}_1 < p$ is also a 97.5% interval for p. Therefore the intersection

$$0.33155 \leq p \leq 0.47147$$

of the two intervals is a 95% confidence interval.

Cutting off equal probabilities at the two ends is somewhat arbitrary. It usually does not give the shortest possible confidence interval. One could find very slightly shorter confidence intervals by varying the probabilities cut off on the left and right.

Exercises:

1. For the example in the text with $X = 80$, $n = 200$ find a confidence interval for p of a) 98%, b) 99%.

2. For $X = 50$ and $n = 100$ find a confidence interval for p of a) 95%, b) 98%.

3. An unknown proportion p of a population is smoking. In a random sample of 1000 persons 600 were smoking. Find a confidence interval for p of a) 95%, b) 98%.

4. We want to estimate the proportion p of left handers among all students. In a random sample of $n = 100$ there were $X = 10$ left handers. Find 95% and 98% confidence intervals for p.

5. 100 coin tosses resulted in $X = 60$ successes. To see how likely this is if the coin is fair, find the probability of the event $|X - 50| \geq 10$ for a fair coin.

6. 900 coin tosses resulted in $X = 486$ successes. This time test the likelihood of this by computing the probability of getting $X \geq 486$ for a fair coin.

7. 600 rolls of a die resulted in $X = 120$ sixes. How likely is this if the die is fair. What is the probability of $|X - 100| \geq 20$?

8. **The Taxi Problem: an exploratory study.** Taxis in cities are usually numbered from 1 to T. I want to find a 95% confidence interval (CI) for the unknown number T by observing n taxi numbers, which I order according to size, getting $X_1 \leq X_2 \leq \ldots \leq X_n = M$ = maximum observed taxi number. We have $\leq$ since the same taxi could show up again. $M \leq T$ is known, so we take M as the lower end of the CI. We find the upper end of the CI by using that $P(\text{all } X \leq M) = (M/T)^n$. We want to exclude T-values for which this probability is $< 5\%$ so we set the upper end of the CI for T at $T = M\sqrt[n]{20}$. Thus the statement $M \leq T \leq M\sqrt[n]{20}$ will be wrong only in 5% of the cities we visit. Write a program with inputs n, T. It should choose n taxi numbers at random by means of $X \leftarrow 1 + \text{random}(T)$, find the confidence interval, and check if it covers T or not. This should be repeated 1000 times. Does it cover T in about 95% of the cases? How does the confidence interval depend on n?

    It is easy to find an almost "best point estimate" $\widehat{T}$ of T. (The definition of "best point estimate" can be found in statistics books and is not essential here.) We consider all n observed gaps $X_1 - 0, X_2 - X_1, X_3 - X_2, \ldots, X_n - X_{n-1}$. One can show that the lengths of all the gaps, including the first and the last, have the same probability distribution. So we use as an estimate for the last unknown gap $T - M$ the mean of the observed gaps, which is M/n. So $\widehat{T} = M(1 + 1/n)$. This is the *gap estimate* of T. Each time find also $\widehat{T}$ and compare with T.

35. Hypergeometric Distribution. Fisher's Exact Test

After many bitter disappointments the medical community has decided that the results of an investigation or a new treatment (e.g. a new drug) can only be recognized as valid if the experiment was **controlled, randomized, and double blind**. This means

a) The effects of a treatment can be judged only by comparison. We need an *experimental group* E which receives the treatment, and a *control group* C, which does not receive the treatment (*controlled experiment.*)

b) The only way to exclude extraneous factors is to assign subjects to groups E and C by a random device (*randomization*).

c) A subject must not know if he/she belongs to E or C. Hence each person from C must receive a placebo treatment, which has no effect, but cannot be distinguished from the new treatment. This assures that the effect of the treatment is due to the treatment itself and not to psychological factors. Belief can heal! (*blind experiment*).

d) The doctor who measures the effect of the treatment must not know if a subject belongs to C or E. This assures that the diagnosis is not influenced subconsciously (*double-blind experiment*).

Does vitamin C help in preventing the common cold? This question was studied in the past repeatedly with quite contradictory results. I found the most extensive and most reliable data in the *Canadian Medical Association Journal*, September 1972, 503-508 (Toronto Study).

The 2×2-table in Fig. 35.1 comes from that source. It was a controlled, randomized, and double-blind experiment.

	cold	no cold	sum
Group E	302	105	407
Group C	335	76	411
sum	637	181	818

Fig. 35.1

The 407 subjects in E received daily a vitamin C pill and the 411 subjects in the control group C got a placebo pill instead, which tasted exactly the same as a vitamin C pill. To simplify computations we concentrate on the smallest of the 4 numbers in the 2×2-table which is 76. How many should we expect, if vitamin C has no effect? A person picked at random will fall into the second row with probability $\widehat{p}_1 = 411/818$ (estimated value) and into the second column with probability $\widehat{p}_2 = 181/818$. The probability of falling into the lower right cell is $\widehat{p}_1 \widehat{p}_2$ So we should expect $n\widehat{p}_1\widehat{p}_2$ in that cell, i.e.

$$n\widehat{p}_1\widehat{p}_2 = 818 * \frac{411}{818} * \frac{181}{818} = \frac{411 * 181}{818} \approx 91 \,.$$

It looks as though the difference $91 - 76 = 15$ is just a random fluctuation. But appearances sometimes deceive. We really cannot tell without computation.

H : Vitamin C does not help.

A : Vitamin C helps.

Suppose H is true. Then we can think of the experiment as follows: $s = 181$ of 818 people remained uninfected for reasons unrelated to vitamin C. The actual result is one possible outcome of the following random experiment. We give placebos to $r = 411$ people, selected at random from among the 818, and vitamin C to the 407 others. Let x be the number of placebo receivers among the 181 who happened to remain uninfected. The numbers of people in the other three categories can then be expressed in terms of x as shown in Fig. 35.2.

	cold	no cold	sum
Group E	$226 + x$	$181 - x$	407
Group C	$411 - x$	x	$411 \leftarrow r$
sum	637	181	$818 \leftarrow n$

$\uparrow$
s

Fig. 35.2

To test our hypothesis, we observe that if vitamin C were effective, we would expect fewer individuals in the untreated yet uninfected category than we would on a random basis; the uninfected people would tend to be in the treated group. Therefore we choose our P-value to be the probability that x should be only 76, as it was in the actual experiment, or even smaller, by chance alone.

Let $h(x)$ be the probability of getting the table in Fig. 35.2. Then

$$P = P\big(0 \leq x \leq 76 \,|\, H\big) = \sum_{x=0}^{76} h(x) \ .$$

The probability $h(x)$ can be computed as follows. The total number of ways in which the s people who did not get the cold can be selected from among the n persons is $\binom{n}{s}$. The number of ways s people can be selected to form the noninfected group with x persons from among the r who did not get vitamin C and $s - x$ from among the $n - r$ who did get vitamin C is $\binom{r}{x}\binom{n-r}{s-x}$. Therefore

$$h(x) = \frac{\binom{r}{x}\binom{n-r}{s-x}}{\binom{n}{s}} \ , \quad h(0) = \frac{\binom{n-r}{s}}{\binom{n}{s}} = \frac{(n-r)\cdots(n-r-s+1)}{n\cdots(n-s+1)} \ .$$

A random variable with probabilities given by this type of formula is said to have a *hypergeometric* distribution; we shall not need the word, but since we came across an example of it we might as well give its name. For computing successive values of $h(x)$ we shall use the recursion formula

$$h(x) = h(x-1) * \frac{(r - x + 1)(s - x + 1)}{x(n - s - r + x)}$$

which follows from our formula for $h(x)$.

When we use these formulas we must take precautions to avoid overflow and underflow, which we do as in Fig. 34.4.

```
program left_tail;
var i,n,r,s,x:integer; L,p:real;
begin
  write('n,r,s,x='); readln(n,r,s,x); L:=0; p:=0;
  for i:=1 to s do L:=L+ln((n-r-i+1)/(n-i+1));
  if L>=-50 then p:=exp(L);
  for i:=1 to x do
  begin
    L:=L+ln((r-i+1)/i*(s-i+1)/(n-r-s+i));
    if L>=-50 then p:=p+exp(L)
  end;
  writeln('P=',p);
  readln; end.
```

Fig. 35.3

In $h(x)$, and hence in the input, we can switch r and s. The input $n = 818$, $s = 181$, $r = 411$, $x = 76$ yields the output $P = 7.4 * 10^{-3} < 1\%$.

Vitamin C definitely helps. We have a huge data base. This enabled us to assert with a high degree of confidence that vitamin C offers some protection against cold, but, for most people if not all, the protection is very small.

When using the program left_tail in other problems, observe the following: We are considering the probabilities of various distributions of individuals in the cells, given that the row and column sums are kept at the values they had in the experiment under discussion. Under these conditions the number of individuals can be as low as 0 only in the cell or cells which had the lowest number of individuals in the actual experiment. For this reason we must be careful that the x to which we apply the program left_tail should be the number of individuals in such a cell, since the program assumes that the possible numbers of individuals in the selected cell go all the way down to 0; r and s should be the sums of the row and column of that cell.

Exercises:

1. Until almost the end of the 19th century, mortality after surgery was extremely high. Then surgeons switched from dirty to clean bandages, which did not help much. Finally the surgeon Joseph Lister started to use sterilized bandages.

	steril.	unsteril.
survived	34	19
died	6	16

Fig. 35.4. From Ch. Winslow, The Conquest of Epidemic Disease. Princeton 1943, p. 303.

Fig. 35.4 shows the results of 75 amputations he performed, 35 without and 40 with sterilized bandages.

H : Sterilization does not help.

A : Sterilization helps.

Find the P-value.

2. Write a program `right_tail` to compute $P = h(x) + h(x + 1) + \cdots + h(n)$. With the programs `left_tail` and `right_tail` solve the following problem: To estimate the number of trout in a pond, 1000 trout were caught, marked, and released again. A few days later again 1000 trout were caught and 100 turned out to be marked. Find a 95% confidence interval for the number N of trout in the pond.

Hint: Proceed by trial and error in finding confidence intervals for p as in Section 34.

3. During 1970–1974 exactly 194 patients with hip fractures were brought to a municipal hospital. Of the 73 patients delivered Mon.-Wed. nobody died. Of the 121 patients delivered Thu.-Sun. ten died. The table in Fig. 35.5 summarizes the data.

arrival	survived	died	sum
Mon–Wed.	73	0	73
Thu-Sun.	111	10	121
sum	184	10	194

Fig. 35.5. Source: D. McNeil: Hip Fractures —Influence of Delay in Surgery on Mortality. Wisconsin Medical Journal, 74, Dec. 1975, 129–130.

H : Death does not depend on the day of the week the patient was admitted.

A : Patients arriving on weekends have a greater mortality rate (probably because treatment is delayed).

Find the P-value.

Why did we count Thursday as a weekend day? Surgery is only a last resort. A patient arriving on a Thursday is first observed for one to two days before a decision on surgery is made.

4. **Quiet Don.** In 1965 M. Sholokhov received the Nobel Prize for literature for the novel *Quiet Don*. Since 1928 there were rumors both in the USSR and abroad that the novel is not his work. In an anonymous study by a Soviet critic the novel is attributed to the Cossak writer F. Kryukov, who died of typhoid in 1920.

1000 words were chosen at random from each of *Marking Time* (Kryukov), *The Way and the Road* (Sholokhov) and *Quiet Don* (?). Each time the number of lexemes was counted. [A lexeme is a dictionary entry. For instance: *write, writes, wrote, written, writing* are the same lexeme.] Fig. 35.6 shows the result.

	Words	Lexemes
Marking Time (Kryukov)	1000	589
The Way and the Road (Sholokhov)	1000	656
The Quiet Don (?)	1000	646

Fig. 35.6. Source: G. Kjetsaa, *The Battle of the Quiet Don: Another Pilot Study.*
Computers and the Humanities, Vol. 11, pp. 341–346, 1977.

	L	$1000 - L$	Sum
Kryukov	589	411	1000
Quiet Don	646	354	1000
Sum	1235	765	2000

Fig. 35.7

Since there is almost no difference between Sholokhov and the author of the *Quiet Don* we test Kryukov versus the author of the *Quiet Don*, Fig. 35.7

H : Kryukov could have written the Quiet Don.

A : Kryukov did not write it.

Find the P-value.

36. Fishing Laws

Three countries, Anchuria, Celophania, and Sikinia, lie on the shore of the same lake. The weights of the fish in the lake are uniformly distributed between 0 and 1. That is, catching the next fish is equivalent to a call of the function random in Pascal. The inhabitants of these countries are passionate anglers. To conserve the fishing population strict fishing laws were enforced. Fishing was allowed just once a week, and each nation had to observe its national stopping rules A, C, S defined below.

Let $W_1, W_2, W_3, \ldots$ be the weights of successive fish caught by an angler. The stopping rules were:

A: Stop as soon as $W_{n-1} < W_n$.
C: Stop as soon as $W_1 + W_2 + \cdots + W_n > 1$.
S: Stop as soon as $W_n > W_1$.

The number X of fish caught is a random variable assuming the values 2,3,4,.... Let
$$p_n = P(X = n),$$
$$q_n = P(X > n).$$
These are the probabilities that a catch has *exactly* n or *more than* n fish, respectively. Let $E = E(X)$ be the expected number of fish in a catch. We first simulate the catches of $m = 10000$ anglers under each of these rules. Then we try to estimate p_n, q_n, and $E(X)$. Finally, we will try to prove our observations. With two rules we will have total success, in one case we will be partially successful.

In program Anchuria, prec and next stand for the preceding and next fish and $c[x]$ is the frequency of a catch of size x. sum accumulates the number of all the fish caught by the m anglers. With these comments the program is easy to comprehend.

```
program Anchuria;
const m=10000;
var sum,x,n:integer;
    c:array[2..20] of integer;
    prec, next, p, q: real;
begin
writeln('n':5,'p':9,'q':9);
writeln; sum:=0;
for x:=2 to 20 do c[x]:=0;
for n:=1 to m do
begin x:=2;
   prec:=random; next:=random;
   while prec>=next do
   begin prec:=next;
      next:=random; x:=x+1
   end;
   c[x]:=c[x]+1; sum:=sum+x
end
q:=1;
for n:=2 to 20 do
   if c[n]>0 then
   begin p:=c[n]/m; q:=q-p;
      writeln(n:9,p:9:4,q:9:4)
   end;
writeln;writeln('E=',sum/m:0:4)
end.
```

Fig. 36.2

n	2	3	4	5	6	7	8
p	0.4992	0.3326	0.1258	0.0343	0.0067	0.0012	0.0002
q	0.5008	0.1682	0.0424	0.0081	0.0014	0.0002	0.0000

$$E = 2.7211$$

Output of program Anchuria

Looking at the output of Anchuria, we guess that $q_n = 1/n!$. For E we get the estimate 2.7211. This reminds us of the number $e = 2.71828182845\ldots$, which is computed to 1000 decimals in Section 59.

Exercises:

1. Write the corresponding program for the rule C.
2. Write the corresponding program for the rule S.

It turns out that the output for rule C is similar to that for rule A. For rule S the output is puzzling. It is easy to guess q_n and p_n, but not the expectation. With 10 repetitions we get the estimates 10.84, 46.56, 90.92, 10.60, 214.85, 132.89, 22.06, 14.44, 47.59, 26.27 for the expectation.

Let us first treat rule A. Clearly $X > n$ precisely when $W_1 > W_2 > \cdots > W_n$. The probability of such a decreasing sequence of length n is $1/n!$. Thus

$$q_n = \frac{1}{n!}, \qquad p_n = q_{n-1} - q_n = \frac{n-1}{n!}, \qquad np_n = 1/(n-2)!,$$

$$E(X) = \sum_{n \geq 2} \frac{1}{(n-2)!} = 1 + \frac{1}{1} + \frac{1}{2} + \frac{1}{6} + \frac{1}{24} + \cdots = e.$$

Let us now tackle rule S. Clearly $X > n$ exactly when $W_1 = \max(W_1, \ldots, W_n)$. The probability of this is $q_n = 1/n$. Thus $X = n$ with probability

$$p_n = q_{n-1} - q_n = \frac{1}{n(n-1)}, \qquad np_n = \frac{1}{n-1}, \qquad n = 2, 3, 4, \ldots$$

and

$$E(X) = \sum_{n \geq 2} n\, p_n = 1 + \frac{1}{2} + \frac{1}{3} + \cdots,$$

which is infinite. No wonder that we got such erratic results for the expectation. We were trying to approximate infinity.

Finally we discuss the rule C. The probability q_n is the volume of the region $W_1 + \cdots + W_n \leq 1$ in n-dimensional space. This is a simplex (n-dimensional tetrahedron) whose vertices are the origin and the endpoints of the unit coordinate vectors. The following reasoning shows that the volume V_n of this simplex is $1/n!$.

The n-dimensional "Cavalieri's Principle" implies that the volume of an n-dimensional pyramid is C_n(base)(height), where the constant C_n depends only on n. We determine C_n by decomposing a cube with edges of unit length into $2n$ pyramids with the $2n$ faces as the bases and the center of the cube as the vertex of all the pyramids. This gives $2nC_n\frac{1}{2}1 = 1$ and hence $C_n = \frac{1}{n}$.

The points with nonnegative coordinates satisfying $W_1 + \cdots + W_n \leq 1$ can be regarded as a pyramid with the unit vector in the W_n-direction as altitude and the $n-1$-dimensional pyramid consisting of the points with nonnegative coordinates satisfying $W_1 + \cdots + W_{n-1} \leq 1$, $W_n = 0$ as base. Thus $V_n = \frac{1}{n}V_{n-1}$. This and $V_1 = 1$ imply $V_n = 1/n!$.

Exercise:

Antithetic variables. If U_1 is uniformly distributed in (0,1), then $U_2 = 1 - U_1$ has the same distribution. Suppose the count X in rule C is based on U_1 and Y is based on U_2. If X is below its expected value then Y will be above. The average $(X + Y)/2$ is also an estimate of e with a lower spread (in technical language *variance*). Write a program which uses this important idea to get a better estimate of e. In writing the program it helps to realize that always $\min(X, Y) = 2$.

37. Estimation of a Probability

In statistics we usually try to estimate small probabilities. This is quite hard to do by simulation and requires a great many repetitions of the experiment. In probability theory we usually estimate probabilities of very complicated events, but the probabilities are usually not small. So fewer repetitions suffice and we can also get more precise estimates if we want.

Here is an artificial probability problem, which leads to an instructive simulation program. Let $M = 0, 1, 2, 3, 4, 5, 6, 7, 8, 9$ be a set of 10 persons. Each person chooses at random two of the others as "friends". A person chosen by nobody will be called "lonesome". Find the expected number of lonesome people and the probability that nobody remains lonesome.

In the program `lonesome` (Fig. 37.1) the variable sum accumulates all lonesome people in n repetitions of the experiment. The variable count counts all instances with no lonesome people in n repetitions. L counts the lonesome people in one experiment. If at the end of an experiment $L = 0$ then count is increased by 1.

Running the program with $n = 10000$ yields count=2746, sum=10385. Thus we estimate the probability that nobody is lonesome to be $P = 0.2746$ and the expected number of lonesome people to be $E = 1.0385$.

```
program lonesome;
var i,j,e,n,r,s,count,sum:integer; x:array[0..9] of integer;
begin randomize; write('n='); readln(n); count:=0; sum:=0;
  for j:=1 to n do
  begin e:=0;
    for i:=0 to 9 do x[i]:=0;
    for i:=0 to 9 do
    begin
      repeat r:=random(10) until r<>i;
      repeat s:=random(10) until (s<>i) and (r<>s);
      x[r]:=1; x[s]:=1
    end;
    for i:=0 to 9 do if x[i]=0 then e:=e+1;
    sum:=sum+e; if e=0 then count:=count+1
  end;
  writeln('count=',count,'  sum=',sum)
end.
```

Fig. 37.1

We can easily find the expected value, but the probability is messy. Consider any person, say #0. Any other person does not choose #0 with probability $(8/9)(7/8)=7/9$. None of the 9 others chooses #0 with probability $(7/9)^9$. The expected number E of lonesome people is np with $n = 10$ and $p = (7/9)^9$. Thus $E = 10 * (7/9) \approx 1.041597$.

The number of choices which leave nobody lonesome can be expressed by means of the "inclusion-exclusion principle" (see, e.g., Mathematics of Choice by Ivan Niven, NML15, Ch. 5). One gets a sum of 8 terms. Some exceed 10^{15} and half are negative. One cannot rely on single-precision arithmetic to compute this sum but if one is careful one gets $P(\text{nobody is lonesome}) = 0.274293751\ldots$.

38. Runs

a) *Runtest.* The binary word

$$W = 101011010010110001011001010110001010100110010100100$$

consists of 36 runs (blocks of successive equal digits), 18 runs of ones and 18 runs of zeros. W was the result of a student trying to write down 50 coin tosses without using a coin. W looks good, it is quite irregular, and it has 23 ones and 27 zeros. Yet the student is a miserable "forger", his word has too many runs. How many runs do we expect in an n-word? The first digit starts the first run. The next bit is different from the preceding one with probability $1/2$. So we can expect $(n-1)/2$ additional runs. For the number R of runs in an n-word we get

$$E(R) = 1 + (n-1)/2 = (n+1)/2.$$

For $n = 50$ we have $E(R) = 25.5$. But we have observed $R = 36$ runs. What is the probability to get $R \geq 36$ runs with $n = 50$ tosses? Let us find the answer by simulation. We repeat the experiment $m = 10000$ times, each time generating $n = 50$ bits and counting the changes. If the preceding bit P is different from the succeeding bit S the run counter R is increased by 1. Initially $R = 1$. Each time a 50-word is completed and $R \geq 36$, the variable count is increased by 1. We ran the program four times with $m = 10000$ and $n = 50$, and we got for count: 1, 2, 0, 3. As an estimate for the probability $P(R \geq 36)$ we get 0.00015. It is extremely difficult to estimate such very small probabilities. So we try to get the result by reasoning. In a way this is an easy problem, equivalent to "What is the probability of getting 35 or more successes in 49 tosses of a fair coin?" We recall our program bin (Fig. 34.4). By entering $n = 49$, $p = 0.5$, $c = 0$, $d = 14$ we get

$$P(R \geq 35) = P(R \leq 14) = 1.9 * 10^{-3}.$$

This is a big surprise! Our estimate is about 13 times smaller than the correct value. We have disovered a defect in the Random Number Generator (RNG) of Turbo Pascal. Quite a number of RNG's have this defect. Variability around the expectations is smaller than with truly random numbers. In particular, very large deviations from expectations are much rarer. These are personal observations based on countless simulations with various RNG's.

Runtest was run with Turbo 3.0. Turbo 4.0 gave vastly better results. Ten runs of the program gave 13, 17, 23, 23, 15, 23, 23, 18, 18, 14 with a probability estimate $1.87 * 10^{-3}$. So the defect seems to be eliminated by now. This is strange since the RNG of Turbo4 is in one respect worse than that of Turbo3. It has shorter period (see next section). But length of period is only one of many quality indicators.

b) *Maximum Run Test.* Let us now generate n random bits one by one and keep track of maxrun, the length of the longest success run. Theory predicts for its expected length

$$E(n) \approx \log_2 n - 0.66725 \ldots.$$

For $n = 1024$ we have $E(n) \approx 9.33$. It makes sense to test a conjecture, but why test something that has been proved? Testing a theorem is a test of the RNG of Turbo Pascal. This is important since it gives you confidence in the RNG, or it reveals its defects.

```
program runtest;
var i,j,r,p,s,m,n,count:
                  integer;
begin     randomize;
write('m,n='); readln(m,n);
count:=0;
for i:=1 to m do
begin r:=1; p:=random(2);
  for j:=1 to n-1 do
  begin s:=random(2);
    if s<>p then r:=r+1;
    p:=s
  end;
  if r>=36 then
    count:=count+1
end;
writeln('count=',count)
end.
```

Fig. 38.1

```
program maxruntest;
var i,j,n,count,maxrun,sum:
                    integer;
begin randomize;
write('n=');readln(n);sum:=0;
for j:=1 to 100 do
begin maxrun:=0; count:=0;
  while count<n do
  begin i:=0;
    while (random(2)=1)
          and (count<n) do
      begin i:=i+1;
      count:=count+1 end;
    if i>maxrun then maxrun:=i
  end;
  write(maxrun,' ');
  sum:=sum+maxrun
end;
writeln; writeln(
  'average maxrun=',sum/100)
end.
```

Fig. 38.2

9	8	9	10	9	10	9	10	10	8	10	8	10	9	9	10	9	9	9	9
9	9	9	9	10	8	8	9	10	9	10	9	10	10	8	10	8	10	9	9
10	9	9	9	9	9	9	9	9	9	8	8	9	9	9	10	9	9	10	12
10	8	10	9	7	10	9	9	9	9	9	14	9	9	9	9	9	9	9	9
10	10	9	10	9	11	10	8	9	8	10	9	9	9	9	9	14	9	9	9

average maxrun = 9.26

Fig. 38.3

The program maxruntest generates n random bits and finds the length of the longest run of ones. It repeats the experiment 100 times and prints maxrun for each repetition. The output in Fig. 38.3 shows that the average 9.27 is a very good estimate of the expected value 9.33. But there seems to be too little variation in the data. But appearances can deceive. We really do not know how much variability we should expect. In fact theory shows that little variation is to be expected, and it is independent of n. For instance, if one could generate $n = 2^{1000}$ random bits, it would be very probable that the length of the longest run would be in $997..1002$. So the maximum run is highly predictable. Maxruntest was run with Turbo 3.0. Turbo 4.0 gave averages which were consistently too high by about 1. The RNG of Turbo 4.0 is definitely different from that of Turbo

3.0, and its bit generator `random(2)` is inferior to that of Turbo 3.0 as far as maximum runs are concerned (problem 18).

c) *Waiting for a Success Run.* A fair coin with faces "0" and "1" is flipped until the word 1111 occurs for the first time. Let X be the number of flips until stop.

We want to find the probability p_n that exactly n flips will be required until stop and the waiting time m to reach our goal:

$$\text{(i)} \quad p_n = P(X = n), \qquad \text{(ii)} \quad m = E(X) = \sum_{n \geq 1} n * p_n.$$

Fig. 38.4 shows that

$$p_i = 0 \ \text{ for } \ i = 1, 2, 3, \quad p_4 = \frac{1}{16}, \quad p_5 = \frac{1}{32}.$$

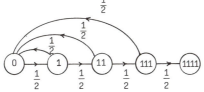

Fig. 38.4

```
program mean;
var i,n:integer; p:array[0..2000] of real;
begin write('n='); readln(n);
m:=0; p[0]:=1; p[1]:=0; p[2]:=0; p[3]:=0; p[4]:=1/16;
for i:=5 to n do p[i]:=p[i-1]-p[i-5]/32;
for i:=4 to n do m:=m+i*p[i];
writeln('m=',m) end.
```

Fig. 38.5

If more than four steps are needed from 0 to 1111 then we must traverse one of the loops leading back to the start. The total probability rule gives for this case

$$p_n = \frac{1}{2}p_{n-1} + \frac{1}{4}p_{n-2} + \frac{1}{8}p_{n-3} + \frac{1}{16}p_{n-4}, \qquad n \geq 5.$$

Replacing n by $n - 1$ and subtracting we get

$$p_n = p_{n-1} - \frac{p_{n-5}}{32}, \qquad n \geq 6.$$

This recurrence is also valid for $n = 5$ if we define $p_0 = 1$. Then

$$p_n = p_{n-1} - \frac{p_{n-5}}{32}, \qquad n \geq 5.$$

Fig. 38.5 shows an iterative program to find $m = E(X)$. For $n = 1000$ we get $m = 29.999999999$. This should be exactly $m = 30$, but we do not get it more precisely, even if we increase n considerably. This is due to rounding errors which can be eliminated by summing the series backwards.

Exercises:

1. Write a program with inputs n, r which computes the mean waiting time $m(r)$ for a word with r ones in a row. Try to guess a formula for $m(r)$.

2. A fair coin with faces 0 and 1 is flipped until the word 1001 occurs for the first time. Let X be the number of flips until stop. Draw a graph similar to Fig. 38.4 and find

$$p_n = P(X = n) \quad \text{and} \quad E(X) = \sum_{n \geq 1} n * p_n.$$

39. The Period of Random(2)

In the last Section we have studied the output of **maxruntest** for $n = 2^{10} = 1024$. For $n = 2^{11}, \ldots, 2^{15}$ the output becomes increasingly pathological. We consider first Turbo 4.0. With $n = 2^{15}$ the output has period 4: 10, 14, 14, 10,.... For $n = 2^{16}$ the period drops to 2: 14, 10, 14, 10,.... Finally, for $n = 2^{17} = 131072$ the sequence becomes constant: 14,14,14,.... Is 2^{17} the period of **random(2)**? The program in Fig. 39.1 tests this.

```
program period_test;
var i,n:longint; x:integer;
begin write('n='); read(n);
for i:=1 to n+40 do
  begin x:=random(2);
  if i<=40 then write(x:2);
  if i=41 then writeln;
  if i>n then write(x:2)
end
end.
```

```
program period_real;
var i,n:real; x:integer;
begin write('n='); read(n);
i:=0;
repeat i:=i+1; x:=random(2);
  if i<=40 then write(x:2);
  if i=41 then writeln;
  if i>n then write(x:2)
until i>=n+40
end.
```

Fig. 39.1   Fig. 39.2

For input $n = 2^{17} = 131072$, Fig. 39.1 has output

0 0 1 0 0 0 1 1 1 1 0 0 0 0 0 0 1 1 1 0 0 0 1 1 0 1 1 1 0 1 0 1 0 0 0 1 1 0 1 0
0 0 1 0 0 0 1 1 1 1 0 0 0 0 0 0 1 1 1 0 0 0 1 1 0 1 1 1 0 1 0 1 0 0 0 1 1 0 1 0.

We see that the 40 digits after the 2^{17}-th are the same as the first 40 digits. This is quite convincing evidence for a period 2^{17}, and it can be easily extended. But is it the smallest period? To find out we check if 2^{16} is also a period. We get

0 0 1 0 0 1 1 1 1 0 0 0 0 0 0 1 1 1 0 0 0 1 1 0 1 1 1 0 1 0 1 0 0 0 1 1 0 1 0 0
1 1 0 1 1 0 0 0 0 1 1 1 1 1 1 0 0 0 1 1 1 0 0 1 0 0 0 1 0 1 0 1 1 1 0 0 1 0 1 1.

After 2^{16} bits we get the complementary sequence.

Let us now check the period of random(2) for Turbo 3.0 (Fig. 39.2). For $n = 2^{17}$ we get the output

1 1 0 0 0 0 1 0 1 1 0 0 0 1 1 1 1 0 1 0 1 1 1 1 1 1 1 0 0 0 1 1 1 1 1 0 1 1 0
0 0 1 1 1 1 0 1 0 0 1 1 1 0 0 0 0 1 0 1 0 0 0 0 0 0 0 0 1 1 1 0 0 0 0 0 1 0 0 1.

After 2^{17} bits we get the complementary sequence. So the period is exactly 2^{18}. Longer period means better quality. So Turbo 3.0 has a better bit generator than Turbo 4.0. Both periods are far too short for a good random number generator. We get a vastly better bit generator from x:=trunc(2*random).

The RNG used in Turbo Pascal for IBM-compatibles, versions 4 and later, and Turbo Pascal for the Macintosh works as follows. There is a predefined variable RandSeed of type longint. (In the Macintosh version the variable is called Seed.) The randomize statement sets it equal to a number obtained from the clock but one can assign it any longinteger starting value. Each time the RNG is called it performs the substitution

$$\text{Randseed}:=a*\text{RandSeed}+1 \quad (\text{mod } 2^{32}),$$

where $a = 134775813 = 3 * 17 * 131 * 20173$. Bits beyond the 32nd are discarded in longint arithmetic. So the mod-operation is automatic.

The value of the function random of type real (it is called randomR on the Macintosh) is just RandSeed with a binary point in front of bit 32. The value of the integer function random(n) is computed by omitting the 16 lower bits (RandSeed shr 16) and taking the result mod n. in particular, random(2) is just the 17th bit (from the right) of RandSeed.

If we print out the value of a longint x which has a 1 in position 32, we get a negative number y. Its absolute value is the one obtained by complementing the digits of x. It is important to note that mod 2^{32}, x and y are equal; y is the number in the interval $-2^{31} \le y < 2^{31}$ which is congruent to x mod 2^{32}. Representing the negative number y by the corresponding x is called the 2's *complement representation* of negative numbers. Note however that when the RNG uses RandSeed to produce a random integer or a random variable of type real with values in $[0,1)$,

RandSeed is treated as a nonnegative number and the bit in the 32nd position is treated as just another binary digit.

The computation of random(2), which is the 17th bit of RandSeed, can be regarded as a mod 2^{17} computation, since the lowest 17 digits in the result of a multiplication or addition do not depend on higher digits of the factors or summands. The factor a above is congruent to 33767 mod 2^{17}. (This number is prime.) So we can think of the next value of RandSeed as resulting from the substitution

$$\text{RandSeed} := 33797 * \text{RandSeed} + 1 \text{ mod } 2^{17}.$$

We see now that the bit generator trunc(2*random) can be computed faster, without converting to real, by means of the formula

$$\text{random(65536) shr 15.}$$

In Section 3.6 of vol. 2 of *The Art of Computer Programming* Knuth summarizes rules for a good linear congruential RNG. Check if these rules are satisfied by the RNG of Turbo Pascal versions 4+.

Exercise:

Observe what proportion of the integers generated by Turbo Pascal's random(26000) function is < 13000. Explain your observation.

40. Waiting for a Complete Set: An Intractable Problem?

Spin the spinner in Fig. 40.1 until all numbers from 1 to n have occurred. Let S be the number of spins to get such a "complete set". We want to estimate the expected number $E(S)$ by simulation. Fig. 40.2 shows the program for the equiprobable case. The variable **spins** counts the spins and the variable **count** counts the different elements already collected. The program repeats the experiment once and counts the spins needed for one complete set. Ten runs of the program with $n = 10$ gave for the number of spins: 25, 33, 20, 28, 26, 24, 25, 43, 28, 40, with the mean 29.2.

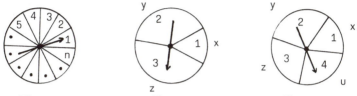

Fig. 40.1   Fig. 40.3   Fig. 40.4

```
program complete_set;
const n=10;
var i, spins, count,r:integer; x:array[1..n] of integer;
begin randomize; spins:=0;
  for i:=1 to n do x[i]:=0;
  for count:=1 to n do
  begin
    repeat r:=1+random(n); spins:=spins+1 until x[r]=0;
    x[r]:=1
  end;
  writeln('spins=', spins)
end.
```

Fig. 40.2

The equiprobable case is very easy to handle exactly. In the Appendix it is shown that

$$(1) \qquad E(S) = n \left(1 + \frac{1}{2} + \frac{1}{3} + \cdots + \frac{1}{n} \right) \sim n \ln n \,.$$

For $n = 6$ and $n = 10$ we get for $E(S)$ 14.7 and 29.29, respectively.

The case of unequal probabilities is more complicated. It is still possible to give a closed formula similar to the one in (1). But the size of the formula grows like $n!$. To show you what this means I have calculated for you the exact formulas for the spinners in Fig. 40.3 and 40.4. The relations $x + y + \ldots = 1$ could be used to eliminate one of the variables but the formulas are simpler if we do not do that.

In Fig. 40.3, I get for $E(S)$

$$1 + \frac{x}{1-x}\left(1 + \frac{y}{z} + \frac{z}{y}\right) + \frac{y}{1-y}\left(1 + \frac{x}{z} + \frac{z}{x}\right) + \frac{z}{1-z}\left(1 + \frac{x}{y} + \frac{y}{x}\right),$$

and for Fig. 40.4, with probabilities x, y, z and u, I get for $E(S)$

$$1 + \frac{x}{1-x}\left[1 + \frac{y}{z+u}(1 + \frac{u}{z} + \frac{z}{u}) + \frac{z}{y+u}(1 + \frac{y}{u} + \frac{u}{y}) + \frac{u}{y+z}(1 + \frac{y}{z} + \frac{z}{y})\right]$$

$$+ \frac{y}{1-y}\left[1 + \frac{z}{u+x}(1 + \frac{x}{u} + \frac{u}{x}) + \frac{u}{z+x}(1 + \frac{z}{x} + \frac{x}{z}) + \frac{x}{z+u}(1 + \frac{z}{u} + \frac{u}{z})\right]$$

$$+ \frac{z}{1-z}\left[1 + \frac{u}{x+y}(1 + \frac{y}{x} + \frac{x}{y}) + \frac{x}{u+y}(1 + \frac{u}{y} + \frac{y}{u}) + \frac{y}{u+x}(1 + \frac{u}{x} + \frac{x}{u})\right]$$

$$+ \frac{u}{1-u}\left[1 + \frac{x}{y+z}(1 + \frac{z}{y} + \frac{y}{z}) + \frac{y}{x+z}(1 + \frac{x}{z} + \frac{z}{x}) + \frac{z}{x+y}(1 + \frac{x}{y} + \frac{y}{x})\right].$$

For $n = 10$ the exact formula would fill hundreds of gigantic volumes. It would be completely useless. In this case simulation is the only way to get numerical results.

Exact solutions to most problems are intractable. Mathematicians have solved exactly just a few problems with high symmetry resulting in slick formulas.

Exercises:

1. *Waiting for two complete sets.* A spinner has outcomes $1, 2, \ldots, n$ with equal probabilities. What is the expected number of spins until all numbers from 1 to n have occured at least twice? Write a simulation program and run it for $n = 6$ and $n = 10$.

2. Extend the program in Fig. 40.2 so that it generates 100 complete sets and finds the mean number of spins. Run the program for $n = 6$ and $n = 10$ and compare the results with the exact values $14.7\ldots$ and $29.29\ldots$.

3. Write a simulation program for the complete set problem with outcome i having probability p_i. Run it for $n = 10$ and $p_i = \frac{1}{i} - \frac{1}{i+1}$, $i = 1, 2, \ldots, 9$ and $p_{10} = \frac{1}{10}$.

4. On an $n \times n$ chessboard rooks are placed on cells sequentially and independently with the same probability $1/n^2$, until the board is covered, i.e. all cells are threatened. Find by simulation the expected number of rooks if several rooks on the same cell are allowed. (Multinomial case.) Run the program for $n = 8$. Theory predicts that $E_M \approx 17.12865$.

5. (Continuation.) Suppose that each successive occupancy is permitted only in one of the currently unoccupied cells (hypergeometric case). Find

by simulation the expected number of rooks needed to cover the board. Theory predicts $E_H \approx 15.0029$.

6. *Total Tour of a regular n-gon.* A beetle starts at some vertex of a regular n-gon and performs a random walk on the edges. Find by simulation the expected time for such a Total Tour of the n-gon. (The beetle covers one edge in one time unit.) Hint. Take $n = 3, 4, 5, \ldots$ and try to guess a formula.

7. Start with n coins. Toss all coins and eliminate those showing "1". Find by simulation the expected number $E(n)$ of tosses until all coins are eliminated. As a check we give $E(3) = 22/7$, $E(4) = 368/105$, $E(15) = 5.287$. See Problem 2 of Section 45.

Additional Exercises for Sections 21 to 40:

1. Write a program which performs n crap games and counts the number of wins. Here are the rules:

   Roll two dice and find the total of points showing. If the total is 7 or 11 you win. If it is 2,3, or 12 you lose. On any other total, call that total the "point" and continue rolling dice until either the total 7 comes up (a loss) or the point comes up (a win).

   Estimate the probability of a win by repeating the experiment 10000 times. Simple theory shows that the winning probability is $w = 244/495 = 0.492492\ldots$. How many experiments should you perform to be 95% sure that $w < 0.5$? (The last question is for readers familiar with the rudiments of statistics.)

2. A fair coin is flipped until either the word 0011 or the word 1111 comes up. In the first case Abby wins and in the second case Carl wins. Write a simulation program which finds an estimate of Abby's winning probability.

3. *Symmetric Random Walk on the Line.* A particle starts at the origin and jumps each second one step to the left or to the right, each with probability 0.5. Let D_n be its distance from the origin after n steps. Find $E(D_n^2)$.

4. A particle starts at the origin and performs a random walk in the plane. At each step it goes from its current position (x, y) to one of its four neighbors $(x+1, y)$, $((x-1, y)$, $(x, y+1)$, $(x, y-1)$ with probability $1/4$. Let D_n be its distance from 0 after n steps. Find $E(D_n^2)$.

5. A particle starts at 0 and makes unit steps. The direction of each step is uniformly distributed between 0 and 2π. Let D_n be its distance from 0 after n steps. Find by simulation $E(D_n^2)$.

6. *Neighbors in Lotto.* Six numbers are drawn at random without replacement from the numbers $1 .. 49$. Find by simulation the probability that

the sample drawn contains at least two consecutive numbers, such as 18 and 19.

7. *Random Walk in Space.* A particle starts at the origin 0 and makes unit steps from its current position (x, y, z) to one of its six neighbors $(x \pm 1, y, z)$, $(x, y \pm 1, z)$, $(x, y, z \pm 1)$, each with probability $1/6$. Let D_n be its distance from 0 after n steps. Guess by simulation $E(D_n^2)$.

8. *Random Walk in Space in Random Directions.* A particle in space starts at the origin 0 and makes unit steps. Around its current position P a unit sphere is drawn and a random point R is chosen on this sphere. The next step goes from P to R. Let D_n be its distance from 0 after n steps. Guess by simulation $E(D_n^2)$.

9. *Maximum Test.* What is the probability that among any three successive base ten random digits the middle digit is strictly larger than any of its neighbors? Generate 10000 successive triples of base ten digits and find the proportion of maxima. Use

   a) The RNG of the computer. b) The sophisticated RNG `random10` in Fig. 25.12.

10. *Poker-Test.* Generate 1000 5-words of base ten random digits, find the proportions of the 7 types defined in the table and compare with the given probabilities.

type	example	probability
all distinct	30862	0.3024
one pair	32082	0.5040
two pairs	02772	0.1080
one triple	96066	0.0720
triple and pair	39399	0.0090
quadruple	80888	0.0045
quintuple	55555	0.0001

11. Read again the rules of the crap game in Exercise 1. Draw a graph for this game and find the winning probability exactly (up to 11 digits).

12. *It all depends on the information.* If you toss a thumbtack, it can point up (U) or down (D) as in Fig. 40.5. Let $p = P(D)$. We are given the outcome $DDDUDUDDDDDU$, and we want to test $H : p = \frac{1}{2}$ versus $A : p > \frac{1}{2}$. Let X be the number of D's. To compute the P-value we need more information.

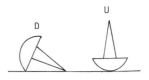

Fig. 40.5

a) I am informed that it was decided in advance to make 12 tosses. Then

$$P = P(9 \le X \le 12 \mid H) = 2^{-12} \left(\binom{12}{3} + \binom{12}{2} + \binom{12}{1} + \binom{12}{0} \right) = \frac{299}{4096} \approx 7.3\%$$

b) I am informed that the experiment was halted as soon as three U's were scored. Now the results $X = 9, 10, 11, 12, 13, \ldots$ are as extreme or more extreme than what we had, and we must find

$$P = P(X \geq 9 \mid H) = \sum_{x=9}^{\infty} p(x), \quad p(x) = \binom{x+2}{2} 2^{-x-2} \frac{1}{2} = \binom{x+2}{2} 2^{-x-3} .$$

(i) Find the P-value by simulation.

(ii) Find the exact P-value by avoiding an infinite sum.

13. The sequence $u_1, \ldots, u_n$ is unimodal if there exists an s with

$$u_1 < u_2 < \cdots < u_s < u_{s+1} > \ldots > u_n \quad (s = 1 \text{ or } s = n \text{ are allowed})$$

a) How do you test for unimodality?

b) Let p_n be the probability that a sequence of n randomly chosen real numbers is unimodal. (p_n does not depend on the probability distribution as long as the chance of getting the same number twice is 0.) Find p_1 to p_9 by simulation and try to guess a formula for p_n.

c) Find the exact value of p_n.

14. *Palindromes.* Given is an array $A[1..n]$ of integers, $n < 1000$.

a) Find a maximal section $A[i, \ldots, j]$ such that $A[i] = A[j]$, $A[i+1] = A[j-1], \ldots$. Print the length of this section. The algorithm should have complexity $O(n^2)$.

b) Let $A[i] := \text{random(10)}$. What is the expected length of the longest palindrome? This answer is to be found by simulation.

15. *Palindromes again.* We generate a sequence consisting of digits from the alphabet $\{0, 1, 2\}$ each occurring with equal probability. Estimate the probability P that this word starts with a palindrome (of length ≥ 2).
Hints:

a) Look at sequences of length n. They start with a palindrome with probability $P_n < P$. We can estimate P_n by simulation. For large n we have $P_n \approx P$.

b) For small n you can find P_n by brute force. This is handy for testing whether your program is correct.

16. *A Total Beetle Tour.* A beetle starts at some vertex of a cube and performs a random walk on the edges of the cube until it has visited all vertices. What is the expected time for such a Total Beetle Tour? (The beetle covers one edge in one time unit.)

17. Suppose the random walk in Exercise 16 stops only when all the edges have been traversed. Find by simulation the expected time until stop. (Answer: ≈ 48.5.)

18. Try to find the periods of $\text{random(3)}, \text{random(4)}, \text{random(5)}$.

41. The Birthday Problem. An $O(n^{1/4})$-Method of Factoring

Let n be the number of days in a year and let $q(n, s)$ be the probability that s persons chosen at random all have different birthdays. Then $p(n, s) = 1 - q(n, s)$ is the probability of at least one multiple birthday. Assume for simplicity that all birthdays are equally probable. Multiplying along the path in Fig. 41.1 we get

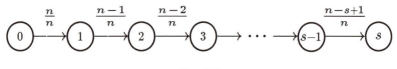

Fig. 41.1

$$(1) \qquad q(n, s) = \frac{n}{n} \, \frac{n-1}{n} \, \frac{n-2}{n} \cdots \frac{n-s+1}{n}.$$

In Fig. 41.1 we are in state i, if we have collected i different birthdays in a row.

Let $E(n)$ be the expected waiting time until we get a double birthday. A useful and easily proved variant of the defining formula for the expected waiting time of an event is the following: If $q(n, s)$ is the probability that the event will not have taken place after s trials, then

$$(2) \qquad E(n) = \sum_{s=0}^{n} q(n, s).$$

In our case, there is an asymptotic formula for $E(n)$:

$$(3) \qquad E(n) \sim \sqrt{\frac{\pi n}{2}} - \frac{1}{3}.$$

```
program birthday;
var i,n,s:integer; q:real;
begin
write('n,s=');readln(n,s);
q:=1;
for i:=0 to s-1 do
   q:=q*(1-i/n);
writeln('p(n,s)=',1-q)
end.
```

Fig. 41.2
Computes $p(n, s)$.

```
program birthexp;
var i,n:integer; q,e:real;
begin write('n='); read(n);
q:=1; e:=q;
for i:=1 to n-1 do
   begin q:=q*(1-i/n);
   e:=e+q end;
writeln(e,'   ',
            sqrt(pi*n/2)-1/3)
end.
```

Fig. 41.3.
Computes $E(n)$ and $\sqrt{\pi n/2} - 1/3$.

The birthday problem has important applications in Computer Science, for instance in hashing, an arrangement of data so as to facilitate storage and retrieval. We will use the birthday problem to devise a factoring algorithm, which for large n is considerably more efficient than trial division. For numbers which are accessible to us (up to 11 digits) it cannot compete with trial division. The method is due to H. Pollard, and it sometimes fails. Then you must start with new starting values.

We want to factor a large number n with an unknown prime factor p. If we can find a pair (a, b) of integers, so that $p \mid (a - b)$, but $a - b$ is not a multiple of n, then we can find a proper factor of n by computing $\gcd(a - b, n)$. We find sets which are good sources from which to pick a and b as follows.

Consider a mapping $x := f(x) \bmod p$. If f is a many-to-one mapping then the range of f consits of fewer than p residue classes and iterates of f may have an even smaller range. Thus if we select a and b from these sets, our chances of finding a and b which are congruent mod p are improved.

Pollard suggested to use the function $f(x) = x^2 + c$ with any integer $c \notin \{0, -2\}$, for instance $c = 1$. The ranges of the iterates of this function modulo any prime p seem to shrink until a fairly small set is reached on which the mapping is one-to-one. For example, for $p = 4903$ the limit set consists of 112 residue classes; for $p = 6947$ it consists of only 48.

More relevant to H. Pollard's method of factoring n is how long it is likely to take if we start with some number, e.g. $x_0 = 101$, and compute $x_{i+1} = f(x_i)$ until we get two numbers congruent mod p, where p is a factor of n. The computation is practicable because it suffices to compute mod n; since the unknown prime p is a factor of n, $p \mid (x_i - x_j)$ if and only if $p \mid ((x_i - x_j) \bmod n)$. Experience suggests that for a prime modulus p the sequence $x_0, x_1, \dots \bmod p$ has the features of a random sequence until a residue class is encountered a second time. From there on the sequence is periodic.

The birthday formula gives us the estimate $\sqrt{\pi p / 2}$ for the number of steps until a residue class mod p is repeated. Assuming randomness until the repetition starts, one could even guess that the expected length of the repeating cycle is half that much. Since we do not know p, we have no easy way of observing exactly when the repetition starts, short of testing all previously obtained numbers each time we compute a new x_i. However, it is wasteful to aim at detecting the first pair whose difference is 0 mod p. Once the repetitions mod p start, the sequence will have a rapidly increasing number of such pairs. A good way to find one of these is to use Floyd's cycle-finding algorithm.

In Algorithm 1, t will usually be a proper factor of n, i.e. $t < n$. A number n can have two prime factors that are discovered in the same step of Algorithm 1. The product of the factors may be the number n itself.

```
                                    t ← x ← y ← 1;
                                    while t=1 do
                                    begin p← 1;
t ← x ← y ← 1;                        for i← 1 to 20 do
while t=1 do                          begin x← x² +1 mod n;
begin x← x² +1 mod n;                   y ← y²+1 mod n;
  y← y² +1 mod n;                       y ← y²+1 mod n;
  y← y² +1 mod n;                       p ← p*abs(x-y) mod n
  z← abs(x-y) mod n;                  end;
  t←gcd(z,n)                          t ← gcd(p,n)
end;                                end;
writeln(t);                         writeln(t);
```

Fig. 41.4. Algorithm 1.   Fig. 41.5. Algorithm 2.

In that case Algorithm 1 fails and we should try an initial value other than 1, or use $x^2 - 1$ instead of $x^2 + 1$, or try $x^2 + c$ with some other c.

A more serious defect of Algorithm 1 is that the gcd must be computed in each step of the algorithm. To speed up the algorithm we could accumulate the product mod n of say 20 consecutive z-values and find the gcd of this product with n. Indeed, if p is a common factor of n and one or more z-values, it is also a common factor of n and the product of the z's. Thus we get Algorithm 2. Now it will be even more probable that two factors of n will be collected in the same 20 consecutive steps. In this case we will have to go back and "single step" through the previous 20 steps in order to separate them. Algorithm 2 fails when two factors are discovered in the same step. However, this rarely occurs when the small factors have first been removed by trial division.

Remarks: a) We avoided $c = 0$ and $c = -2$ in $f(x) = x^2 + c$ since in that case there are closed formulas for the n-th iterate of f of the form $x_m = (x_0)^{2^m}$ or $x_m = r^{2^m} + r^{-2^m}$, respectively. This is a very nonrandom feature.

b) Any 11-digit integer can be factored in a short time by means of the program `factor1` in Fig. 17.15. We need at least double precision arithmetic for Algorithms 1 and 2 to pay off.

Example. Factor $n = 91643$ using $x_{i+1} = x_i^2 - 1$, $x_0 = 3$.

$$x_1 = 8, \qquad x_2 = 63, \qquad z_1 = 63 - 8 = 55, \qquad \gcd(55, n) = 1$$
$$x_3 = 3968, \quad x_4 \equiv 74070, \quad z_2 \equiv x_4 - x_2 \equiv 74007, \quad \gcd(74007, n) = 1$$
$$x_5 \equiv 65061, \quad x_6 \equiv 35293, \quad z_3 \equiv x_6 - x_3 \equiv 31225, \quad \gcd(31225, n) = 1$$
$$x_7 \equiv 83746, \quad x_8 \equiv 45368, \quad z_4 \equiv x_8 - x_4 \equiv 62941, \quad \gcd(62941, n) = 113.$$

$$n = 113 * 811.$$

Exercises:

1. Factor a) $n = 136891$, b) $n = 164009$, c) $n = 176399$ with Pollard's method. 11-digit accuracy is sufficient here. Use Algorithm 1. Small primes have been removed already.

2. Collect birthdays at random, until you get three birthdays on the same day. Repeat this experiment 10000 times and estimate the expected waiting time.

Additional exercises for Sections 1-41:

1. *Two-Pile Game.* We have two piles with a and b chips, respectively. Each second a pile X is chosen at random and a chip is moved from X to the other pile Y. Let T be the waiting time until one pile becomes empty. Find by simulation a formula for the expected time $E(T) = f(a, b)$.

2. *Three-Pile Game.* We have three piles with a, b, c chips, respectively. Each second a pile X is chosen at random, then another pile Y is chosen at random, and a chip is moved from X to Y. Let T be the waiting time until one pile becomes empty. Write a simulation program for $E(T) = f(a, b, c)$. By choosing small values for a, b, c try to guess $f(a, b, c)$. The formula was found as late as 1992 by computer experimentation. (See the Crash Course in Probability.)

3. Write a program which computes $f(m, n) = \sqrt{m\sqrt{(m+1)\ldots\sqrt{n}}}$. Explore $f(2, m)$ for ever larger values of n and study convergence speed. Find $f(n, 2n)$, $f(n, 3n)$, $f(n, \infty)$ for large n.

CHAPTER 5

Combinatorial Algorithms

42. Sorting

Sorting is one of the major tasks occupying commercial computers. We present a tiny sample of the existing methods; the reader will find it rewarding to study the topic in more detail in computer science textbooks or in vol. 3 of Knuth's classic work.

42.1 Sorting by Selection. In this simplest sorting method we proceed as follows: Find the smallest element in $a[1..n]$ and exchange it with $a[1]$. Then find the smallest element in $a[2..n]$ and exchange it with $a[2]$, etc. until the whole array is sorted.

Initially we will store in the array $a[1..n]$ random integers, which are to be sorted.

For $n = 1000$ and $n = 10000$ selection_sort, see Fig. 42.1, requires 4.2 and 430 seconds, respectively, on my AT.

```
program selection_sort;
const n=1000;
var i,j,min,copy:integer;
    a:array[0..n] of integer;
begin
for i:=1 to n do
  a[i]:= random(n);
for i:=1 to n do
begin
  min:=i;
  for j:=i+1 to n do
    if a[j]<a[min] then min:=j;
  copy:=a[min]; a[min]:=a[i];
  a[i]:=copy
end;
for i:=1 to n do
  write(a[i],' ')
end.
```

Fig. 42.1

The length of time an algorithm takes can be measured in two ways: the time it takes in the *worst case*, and the *expected* length of time. The notion of expected length of time presupposes that the data are produced by a random process with a certain probability distribution. We will use the uniform distribution. The computation time of selection_sort is proportional to n^2 in both the worst case and the expected time. Indeed, the **if** statement is executed $(n-1) + (n-2) + \cdots + 1$ times, and the other statements may be performed fewer times.

42.2 Sorting by Insertion. Insertion sort is almost as simple as selection_sort, but more flexible. This method is used by bridge players to sort their hands. They look at the cards one at a time, inserting each in its proper place among those already sorted.

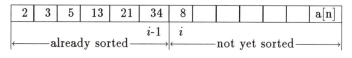

Fig. 42.2

1. The first element of the not yet sorted array is copied: v:=a[i].
2. We go backwards among the elements already sorted and move each one place to the right as long as they are larger than v which is then stored into the gap.

while a[j-1]>v do begin a[j]:=a[j-1]; j:=j-1 end; v:=a[j];

3. To prevent the while-loop from running past the left end of the array if s is the smallest element in the array, we must put a "sentinel" into $a[0]$ with $a[0] \leq s$. We use $a[0] := -1$.

```
program insertion_sort;
const n=1000;
var i,j,v:integer;
    a:array[0..n] of integer;
begin
for i:=1 to n do
  a[i]:=random(n);
a[0]:=-1;
for i:=2 to n do
begin
  v:=a[i]; j:=i;
  while a[j-1]>v do
    begin a[j]:=a[j-1];j:=j-1 end;
  a[j]:=v
end;
for i:=1 to n do write(a[i]:4)
end.
```

Fig. 42.3

For $n = 1000$ and $n = 10000$ insertion_sort takes 2.5 and 260 seconds, respectively, on my AT. It is better than selection_sort, but its worst case and expected time complexities are still proportional to n^2. The computation will be longest if the elements are in decreasing order to start with. The n elements of the array form $n(n-1)/2$ pairs. Each pair is out of sort with probability 0.5. Thus the expected number of out of sort pairs is $n(n-1)/4$. To bring them into order requires $n(n-1)/4$ exchanges.

42.3 Shellsort. Insertion_sort takes less time if the elements are close to their correct order to start with. Also, in general the time required is proportional to n^2 and hence it takes much less time to sort h arrays of about n/h elements than to sort single arrays of n elements. Shellsort speeds up sorting by taking advantage of these facts.

```
program shellsort;
label 0;
const n=1000;
var i,j,h,v:integer;
    a:array[0..n] of integer;
begin h:=1;
for i:=1 to n do a[i]:=random(n);
repeat h:=3*h+1 until h>n;
repeat h:=h div 3;
  for i:=h+1 to n do
  begin v:=a[i];j:=i;
    while a[j-h]>v do
    begin a[j]:=a[j-h];j:=j-h;
      if j<=h then goto 0
    end;
  0:a[j]:=v
  end
until h=1;
for i:=1 to n do write(a[i]:4)
end.
```

Fig. 42.4

For a given value of h we sort separately the subarrays consisting of elements which are multiples of h apart. We first do this for large values of h. This will move elements long distances and in general put the array much closer to its natural order. We then use smaller h-values and finally we use $h = 1$ which is just insertion_sort, to get each element into the right place. Some sequences of h-values are better than others. Knuth recommends the sequence ..., 1093, 364, 121, 40, 13, 4, 1; here each number is obtained from the next one by the substitution $h := 3 * h + 1$.

In the program shellsort we stop the search for the proper place of v from going off the left end of the array by means of a test and a goto statement to the end of the loop. Doing it by means of sentinels as in insertion_sort would require a large extension of the array.

The worst case and average case complexity of shell_sort depends very much on the h-sequence. Little is known about this dependence. For $n = 10000, 20000, 30000$ it requires about 3, 6, 12 seconds on my AT.

Pratt (1979) has shown that for many sequences the worst case complexity is proportional to $n^{3/2}$. Sedgewick (1986) gave an $O(n^{4/3})$ upper bound. There is a possibility that even better sequences may be found.

42.4 Quicksort.   The most popular and the fastest general sorting method is Quicksort, due to C. A. R. Hoare (1960). It is a very instructive example of the Divide-and-Conquer paradigm and recursion.

Partitioning of an Array. We want to partition an array a[1..r] so that all elements $< v = a[(l + r)$ div $2]$ are to the left of all elements $> v$. We scan the array from the left until we find an element a[i]$\geq$v, and from the right until an element a[j]$\leq$v is found. Now the elements a[i] and a[j] are swapped, and the process of scanning and swapping is repeated until we meet somewhere. As a result the array is partitioned into a left part with numbers $\leq v$ and a right part with numbers $\geq v$.

The partitioning is performed by the **repeat** ... **until** loop in Fig. 42.5. As an example, let us partition

$$56\ 15\ 61\ 50\ 99\ 5\ 22\ 50.$$

50 15 22 5 99 50 61 56

Here $v = 50$. After partitioning we have the permutation shown in the figure, with the numbers < 50 to the left of the numbers > 50. At the end we will have $i = 5$, $j = 4$. Occasionally we will do unnecessary work. In the extreme case of n equal numbers we will do $\lceil \frac{n}{2} \rceil$ unnecessary interchanges.

5 5 5 5 5 5

```
program quicksort;
const n=10000;
var k:integer;
    a:array[0..n] of integer;
procedure sort (l,r:integer);
var v,t,i,j:integer;
begin v:=a[(l+r) div 2];
i:=l; j:=r;
repeat
  while a[i]<v do i:=i+1;
  while v<a[j] do j:=j-1;
  if i<=j then
    begin
      t:=a[i]; a[i]:=a[j];
      a[j]:=t; i:=i+1; j:=j-1
    end
until i>j;
if l<j then sort(l,j);
if i<r then sort(i,r)
end;

begin randomize;
for k:=1 to n do
  a[k]:= random(n);
sort(1,n);
for k:=1 to n do
  write(a[k], ' ')
end.
```

Fig. 42.5

It looks as though it is easy to avoid the interchange of equal numbers in replacing "<" by "<=" in both while-loops. But in this case the element v of the array is no longer a sentinel for both scans. The array with identical entries would cause the while-loops to run off the array boundaries. To avoid this would cause a complication of the program. For simplicity we put up with some additional swaps which occur rarely. After each partition we call up the same procedure **sort** recursively for the left part **l..j** and the right part **i..r**, checking each time if $l < j$ and $i < r$. See Fig. 42.5. My AT requires about 1.5, 3, 5 seconds, for $n = 10000, 20000, 30000$, respectively. **Quicksort** has the excellent average case complexity proportional to $n \log_2 n$. Its worst case complexity is proportional to n^2, but that occurs very rarely.

Exercises:

1. Suppose $n = 2^m$ equal numbers are sorted by **quicksort**. Show that $0.5 n \log_2 n$ interchanges will be made.
2. Could we replace **if** i<=j by **if** i<j in Fig. 42.5? Analyze the array $2\ 2\ 2\ 5\ 2\ 2\ 2$.

42.5 Bucket Sort.   We often have to sort a list of n integers with
each integer lying between min and max. If the difference $max - min$
is not too large, there will be relatively few different numbers and so we
introduce for each i, $bucket[i]$ to hold the frequency of i. By scanning
the list once we place the numbers into their corresponding buckets. A
second scan could print the numbers in ascending order. It is much better
to print each number and how often it occurs. This sort is so quick that
the generation of the random integers takes more time than the sorting.
So we would like to start counting the sorting time after the generation
of the data is over. The statement writeln(chr(7)) generates a sound
after the data are generated. The sorting method has time complexity
$O(n + max - min)$ and requires space $n + max - min$ instead of n.

```
program bucket_sort;
const n=8000; m=16; min=0; max=15;
var i:integer; a:array[1..n] of integer;
    bucket:array[min..max] of integer;

begin
    for i:=1 to n do a[i]:=random(m); writeln(chr(7));
    for i:=min to max do bucket[i]:=0;
    for i:=1 to n do bucket[a[i]]:=bucket[a[i]]+1;
    for i:=min to max do write(i:11,bucket[i]:5);
readln; end.
```

0	487	1	486	2	495	3	525	4	517	5	500
6	491	7	465	8	507	9	483	10	513	11	514
12	487	13	513	14	503	15	514				

Fig.42.6

43. The kth Smallest Element of an n-Set

We want to find the kth smallest element of an n-set, e.g. the middle element (the median) or some percentile. We could proceed as in
selection_sort and find the smallest, second smallest, ..., until we reach
the kth smallest element. This is a good method if k is small. We could
also sort the set and print the element $a[k]$. Generally, we can reach our
goal faster if we use a modification of the partitioning process, which we
have constructed for quicksort. Our new process rearranges the array
$a[1], \ldots, a[n]$ so that ultimately

(1) $a[1], \ldots, a[k-1] \le a[k] \le a[k+1], \ldots, a[n]$

but it avoids most of the work of ordering the entries on either side of the
k-th element among themselves.

We begin with the assignment $v := a[k]$. Starting from the left, we look for the first i such that $a[i] \geq v$, and starting from the right, we look for the first j such that $a[j] \leq v$ and we interchange them. We continue doing this until one of the indexes i, j has moved past k. The entry $a[k] = v$ will be a stopping place for whichever index gets there first, or possibly for both at the same time. This pair i, j will be used in the last interchange of the current partitioning step. (The values produced by the final assignments $i := i + 1$, $j := j - 1$ are not used in an interchange.)

Consider the case when the index i is the first to reach k or reaches it at the same time as j. In the latter case (1) has been attained.

If $k < j$, then the entries $a[j], \ldots, a[n]$ need not be moved any more because they are $\geq$ than the first k entries of the array. So we set $l = 1$, $r = j - 1$ and repeat our partition procedure on the subarray $a[l], \ldots, a[r]$ with the aim of rearranging it so that it satisfies

(1') $$a[l], \ldots, a[k - 1] \leq a[k] \leq a[k + 1], \ldots, a[r].$$

The case when the right-hand index j is the first one to coincide with k is similar. In this situation we have rearranged the array so that $v \leq a[k], \ldots, a[n]$. This implies that the k-th largest element is at least v. Moreover, $a[1], \ldots, a[i] \leq v$. In this case we need not move $a[1], \ldots, a[i]$ any more. We set $l = i + 1$, $r = n$ and perform a partition of this subarray to achieve (1') or to reduce the problem to the partitioning of a still shorter subarray. The computation must stop because the length of the subarray which remains to be rearranged gets smaller in each step. Fig. 43.1 is an iterative implementation of this algorithm.

```
program kselect;
const n=20;
var l,r,k,v,t,i,j:integer;
    a:array[0..n] of integer;
begin write('k='); readln(k);
for i:=1 to n do a[i]:=random(n);
for i:=1 to n do write(a[i],' ');
writeln; l:=1; r:=n;
while  l<r do
begin v:=a[k]; i:=1; j:=r;
  while (i<=k) and (k<=j) do
  begin
    while a[i]<v do i:=i+1;
    while v<a[j] do j:=j-1;
    t:=a[i]; a[i]:=a[j]; a[j]:=t;
    i:=i+1; j:=j-1;
  end;
  if k<i then r:=j;
  if j<k then l:=i;
end;
writeln(a[k]);
readln; end.
```

Fig. 43.1

Exercise

Let $(a_1, a_2, \ldots, a_m)$ and $(b_1, b_2, \ldots, b_n)$ be increasingly sorted sequences of real numbers. Also set $a_{m+1} = b_{n+1} = max$, where max is larger than any of the a_i's or b_i's. Find a merging algorithm which forms an increasingly sorted sequence $(c_1, c_2, \ldots, c_{m+n})$ containing all the a_i and b_j, followed by $c_{m+n+1} = max$.

44. Binary Search

Fig. 44.1 shows a program which creates an increasing array $a[1..n]$ with random differences and for input v returns the location x for which a[x]=v. If v does not occur in the array it returns the number 0.

```
program Bin_Search;
var i,n,v:integer; a:array[0..10000] of integer;
function BinSearch(v:integer):integer;
var x,l,r:integer;
begin l:=1; r:=n;
  repeat x:=(l+r) div 2;
     if v<a[x] then r:=x−1 else l:=x+1
  until (v=a[x]) or (l>r);
  if v=a[x] then BinSearch:=x
  else BinSearch:=0
end;
begin write('n,v='); readln(n,v); a[0]:=0;
  for i:=1 to n do a[i]:=a[i−1]+random(3);
  writeln('x=',BinSearch(v))
end.
```
Fig. 44.1

Exercises:

1. Modify **Bin_Search** so that it also finds in the array the longest
     a) strictly increasing interval   b) interval of constancy
   If there is more than one such interval it should give the leftmost such interval (and the number of such intervals).

2. Modify **Bin_Search** so that it searches for given v in a[i]:=trunc(t*i), $t = (\sqrt{5} + 1)/2$.

45. Binary Guessing (20 Questions)

Carl thinks of a number $x \in [1..n]$. We instruct the computer to guess this number with a minimum of guesses. Carl always answers "yes" and "no". A program for the guessing game is shown in Fig. 45.1.

```
program binguess;
var l,r,m,n,x,guesses,c:integer;
begin
  write('n='); readln(n); write('x='); readln(x);
  l:=1; r:=n; guesses:=0;
  repeat
    m:=(l+r) div 2; guesses:=guesses+1;
    writeln('is x<=',m,'? (answer 1/0 for yes/no)');
    readln(c);
    if c=1 then r:=m else l:=m+1
  until r=l;
  writeln('x=' ,r,' guesses=', guesses)
readln; end.
```

Fig. 45.1

Exercises:

1. If $n = 2^{10}$ and Carl thinks of the number $x = 777$ then the sequence of zeros and ones is 0011110111. How is this sequence related to the binary expansion of 777? Generally, what is the relation between the number x and the (0,1)-sequence of answers, which guesses x?

2. Random Guessing. Carl thinks of a number $G \in 1..n$. Abby tries to guess G as follows: She chooses a random subset $R \subset 1..n$ and Carl replies "yes" or "no" if $G \in R$ or $G \notin R$, respectively. Let $E(n)$ be the expected number of guesses to find G. Estimate $E(n)$ by simulation for different values of n. To check your simulation program you can use the following exact expectations: $E(2) = 2$, $E(3) = 8/3$, $E(4) = 22/7$, $E(5) = 368/105$, $E(16) = 5.287$. You lose surprisingly little compared to deterministic guessing. To write the program it helps to play the game with a coin several times. Then describe by a program what you did. How is this related to Ex. 7 of Section 40?

3. Simulation in 2. gives unsatisfactory results, if the coin tosser **random(2)** is used. Replace **random(2)** by the equivalent **trunc(2*random)**, and the results will improve considerably, although a simulation run will take 6 to 7 times longer. **random** seems to be far better than **random(2)** for Turbo 3.0, 4.0, 5.0 for IBM-compatibles.

46. The n-Queens Problem

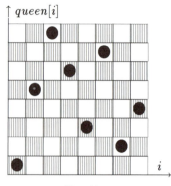

Fig. 46.1

Place 8 queens on an 8x8-chessboard peacefully, i.e. with no two queens in the same row, column, or diagonal. The solution in Fig. 46.1 can be written as a permutation 1 5 8 6 3 7 2 4. We want the computer to print *all* solutions in this form. Let us generalize the problem to the $n \times n$-chessboard. For $n > 8$ we will not print the solutions, we just count them with a variable count.

We introduce three boolean variables:

row[i]:=true: Row i is not attacked

upl[k]:=true: Diagonal with $i + j = k$ is not attacked; k takes values from 2 to $2n$.

upr[k]:=true: Diagonal with $i - j = k$ is not attacked; k takes values from $1 - n$ to $n - 1$.

Here upl and upr stand for upleft (from right to left) and upright (from left to right). Finally we introduce the assignment

queen[i]:=j: Queen in column i is placed in row j.

In each row and each column we must have one queen. We place queens columnwise, beginning with the first column, by setting queen[1]:=1. Then we tentatively move the second queen in the second column up until we find the first square not attacked. On this square we place the second queen. Then we move the third queen in the third column up until we find a square not attacked by the first two queens, etc. If we cannot place a queen because each square of its column is attacked, then we backtrack and try to move up the preceding queen if possible, otherwise we backtrack one more step. As soon as we have a solution, we count or print it, remove the last queen from the board and try to move the preceding queen up. In this way we find all solutions in increasing lexographic order. Even the first queen will get to the n-th row eventually.

In Fig.46.2 the precedure trycol(i) tries to place the queen in column i. The program counts the number of solutions. To print them we must replace count:=count+1; by

```
begin for k:=1 to n do write(queen[k]:3); writeln end;
```

The program in Fig. 46.2 is written for $n = 9$ and the printout is for $n = 8$. The otherwise redundant local variable k is declared for the purpose of replacing counting by printing.

```pascal
program nqueens;
const n=9; p=18; q=8;
var i, count:integer;
    queen:array[1..n] of 1..n; upr:array[-q..q] of boolean;
    row:array[1..n] of boolean; upl:array[2..p] of boolean;
procedure trycol(i:integer);
var j,k:integer;
begin
  for j:=1 to n do
    if row[j] and upl[i+j] and upr[i-j] then
      begin
        queen[i]:=j; row[j]:=false;
        upl[i+j]:=false; upr[i-j]:=false;
        if i<n then trycol(i+1) else count:=count+1;
        row[j]:=true; upl[i+j]:=true; upr[i-j]:=true
      end
end;

begin
  for i:=1 to n do row[i]:=true;
  for i:=2 to p do upl[i]:=true;
  for i:=-q to q do upr[i]:=true;
  count:=0; trycol(1); writeln('count=',count)
end.
```

```
1 5 8 6 3 7 2 4    3 6 8 1 5 7 2 4    5 2 4 6 8 3 1 7    6 3 1 8 5 2 4 7
1 6 8 3 7 4 2 5    3 6 8 1 4 7 5 2    5 2 4 7 3 8 6 1    6 3 1 8 4 2 7 5
1 7 4 6 8 2 5 3    3 7 2 8 5 1 4 6    5 2 6 1 7 4 8 3    6 3 7 4 1 8 2 5
1 7 5 8 2 4 6 3    3 7 2 8 6 4 1 5    5 2 8 1 4 7 3 6    6 3 7 2 4 8 1 5
2 4 6 8 3 1 7 5    3 8 4 7 1 6 2 5    5 3 1 6 8 2 4 7    6 3 7 2 8 5 1 4
2 5 7 4 1 8 6 3    4 2 5 8 6 1 3 7    5 3 1 7 2 8 6 4    6 4 2 8 5 7 1 3
2 5 7 1 3 8 6 4    4 2 7 5 1 8 6 3    5 3 8 4 7 1 6 2    6 4 1 5 8 2 7 3
2 6 1 7 4 8 3 5    4 2 7 3 6 8 1 5    5 1 4 6 8 2 7 3    6 4 7 1 3 5 2 8
2 6 8 3 1 4 7 5    4 2 7 3 6 8 5 1    5 1 8 4 2 7 3 6    6 4 7 1 8 2 5 3
2 7 3 6 8 5 1 4    4 2 8 5 7 1 3 6    5 1 8 6 3 7 2 4    6 1 5 2 8 3 7 4
2 7 5 8 1 4 6 3    4 2 8 6 1 3 5 7    5 7 4 1 3 8 6 2    6 8 2 4 1 7 5 3
2 8 6 1 3 5 7 4    4 1 5 8 6 3 7 2    5 7 1 4 2 8 6 3    7 2 4 1 8 5 3 6
3 1 7 5 8 2 4 6    4 1 5 8 2 7 3 6    5 7 1 3 8 6 4 2    7 2 6 3 1 4 8 5
3 5 2 8 1 7 4 6    4 6 1 5 2 8 3 7    5 7 2 4 8 1 3 6    7 3 1 6 8 5 2 4
3 5 2 8 6 4 7 1    4 6 8 2 7 1 3 5    5 7 2 6 3 1 4 8    7 3 8 2 5 1 6 4
3 5 7 1 4 2 8 6    4 6 8 3 1 7 5 2    5 7 2 6 3 1 8 4    7 4 2 5 8 1 3 6
3 5 8 4 1 7 2 6    4 7 3 8 2 5 1 6    5 8 4 1 3 6 2 7    7 4 2 8 6 1 3 5
3 6 4 1 8 5 7 2    4 7 1 8 5 2 6 3    5 8 4 1 7 2 6 3    7 5 3 1 6 8 2 4
3 6 4 2 8 5 7 1    4 7 5 3 1 6 8 2    6 2 7 1 4 8 5 3    7 1 3 8 6 4 2 5
3 6 2 5 8 1 7 4    4 7 5 2 6 1 3 8    6 2 7 1 3 5 8 4    8 2 4 1 7 5 3 6
3 6 2 7 5 1 8 4    4 8 1 3 6 2 7 5    6 3 5 7 1 4 2 8    8 2 5 3 1 7 4 6
3 6 2 7 1 4 8 5    4 8 1 5 7 2 6 3    6 3 5 8 1 4 2 7    8 3 1 6 2 5 4 7
3 6 8 2 4 1 7 5    4 8 5 3 1 7 2 6    6 3 1 7 5 8 2 4    8 4 1 3 6 2 7 5
```

Fig. 46.2

In Fig. 46.3 q_n is the number of peaceful arrangements of n queens on an $n \times n$-board; t_n is the computation time on an IBM PC AT (12 Mhz).

n	1	2	3	4	5	6	7	8	9	10
q_n	1	0	0	2	10	4	40	92	352	724

n	11	12	13	14	15	16
q_n	2680	14200	73712	365596	2279184	14772512

n	8	9	10	11	12	13	14
t_n	1 sec	4 sec	10 sec	1 min	6 min	35 min	3.5 hours

Fig. 46.3

The computation time can easily be cut in half. How? For q_n no closed formula is known and none is to be expected. The n-queens problem is a paradigmatic example for the BACKTRACK method.

History of the problem: It was first posed in 1848 by Max Bezzel in a chess magazine and was first unnoticed. Then it was posed by Dr. Nauck in the popular "Illustrierte Zeitung" (June 1, 1850) and aroused great interest. Gauss read the problem in this paper and spent a lot of time on it. On 21 September 1850 the blind Dr. Nauck gave all solutions in the same paper. By that time Gauss had found only 72 solutions. By the way, Gauss used our algorithm. We know it from a letter to his friend Schumacher.

Exercises:

1. A knight is placed on any square of an $n \times n$ chessboard. It is to move so that it visits every square of the board exactly once. Write a backtrack algorithm for this "knight's tour" problem. Since it requires a lot of time you should work with $n = 5$. Use the variable `board:array[1..n,1..n]` `of integer` on which you store the number of the current move. Do not forget to set `board[x,y]:=0` when you have to backtrack.

2. Solve the knight's tour problem for the 8 × 8-board. Use divide-and-conquer, that is, solve the problem first for the lower left 4x5-board in Fig.46.4, extend it to the lower right 4 × 5-board, and finally, extend the solution again to the upper 8 × 3-board. One solution is shown in Fig. 46.4. The numbers give the sequence of visits of the squares of the board.

A. J. Schwenk constructed closed knight tours for all boards for which closed tours exist by generalizing the above method [Schwenk, 1991]. There the reader can also find proofs that no such closed tours exist for the remaining sizes.

60	63	50	53	56	43	48	45
51	54	59	62	49	46	57	42
64	61	52	55	58	41	44	47
3	8	15	20	27	22	36	40
16	11	4	9	34	39	28	23
7	2	19	14	21	26	31	36
12	17	10	5	38	33	24	29
1	6	13	18	26	30	37	32

Fig. 46.4 (Ex. 2)

3. *A Hamiltonian circuit* or *path* of a graph is a circuit or path which goes through every node exactly once. The exercises above ask us to find Hamiltonian paths in the graphs whose nodes are squares of the board, with two nodes connected if they are a knight-jump apart.

Finding a Hamiltonian circuit or path is one of the class of NP-complete problems. (See the end of Sect. 28, "Matched Pairs".) For these problems no method is known which will always work and is always much faster than trying out all possibilities. However, there are methods which find a Hamiltonian cycle quite quickly for "most" graphs that have one.

Write a program to find closed knight's tours using the following method used by Lajos Pósa to prove that most graphs of certain types have Hamiltonian cycles. Construct a path $H_0, H_1, \ldots, H_l$ by adding nodes until we can not extend it. Then H_l must be connected to some H_r with $r \leq l-2$. Then $H_0, \ldots, H_r, H_l, H_{l-1}, \ldots, H_{r+1}$ is also a path. We call the operation of replacing our previous path by this one a *rotation*. Try to extend the new path. If it cannot be done, rotate again. One usually has choices in making the extensions and rotations. When all nodes are in the path, perform further rotations to find a Hamiltonian cycle. A good way to avoid getting into a short loop is to insert a random element into the choices. Even so the process may get stuck.

4. Try to program Warnsdorff's rule (1823): The knight should always move to one of the cells from which it will command the fewest squares not already traversed. In case of several such moves choose any one of these. It also seems to work for $m \times n$- boards which have Hamiltonian paths. There is no proof that the rule always works. Sometimes we get into a dead end as a result of having made an unfortunate choice where choices were permitted. But we know of no example where the tour can not be completed without violating Warnsdorf's rule. The rule gives Hamiltonian paths, not cycles (re-entrant tours).

47. Permutations

a) *Permutations as Arrangements.* Take three distinct objects and call them 1,2,3. An arrangement of these objects in a row is called a *permutation* of the objects. There are six permutations of 1,2,3: 123, 132, 213, 231, 312, 321. The product rule tells us that there are $n!$ permutations of n distinct objects. There are excellent asymptotic formulas for $n!$, e. g.

$$n! \sim \sqrt{2\pi n}\left(\frac{n}{e}\right)^n \left(1 + \frac{1}{12n} + \frac{1}{288n^2}\right).$$

The formula is very good even without the last factor (Stirling's formula).

The following algorithm produces a random permutation of $1..n$.

```
for i:=1 to n do p[i]:=i;
for i:=n downto 2 do
begin
    r:=1+random(i);  copy:=p[i];
    p[i]:=p[r];  p[r]:=copy
end;
for i:=1 to n do write(p[i]:4);
```

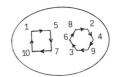

$f_1 = (1\ 5\ 7\ 10)(2\ 4\ 9\ 3\ 6\ 8)$

b) *Permutations as Rearrangements.* A permutation can also be regarded as a *rearrangement*, i.e. a function. For instance, the permutation 5 4 6 9 7 8 10 2 3 1 can be regarded as the function f_1 that maps $\{1,2,\ldots,10\}$ into the given rearrangement:

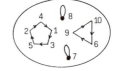

$f_2 = (1\ 3\ 5\ 2\ 4)(6\ 9\ 10)(7)(8)$

$$f_1 = \begin{pmatrix} 1 & 2 & 3 & 4 & 5 & 6 & 7 & 8 & 9 & 10 \\ 5 & 4 & 6 & 9 & 7 & 8 & 10 & 2 & 3 & 1 \end{pmatrix}.$$

Some other permutations of $1..10$ are

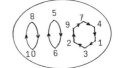

$f_3 = (1\ 3\ 2\ 9\ 7\ 4)(5\ 6)(8\ 10)$

$$f_2 = \begin{pmatrix} 1 & 2 & 3 & 4 & 5 & 6 & 7 & 8 & 9 & 10 \\ 3 & 4 & 5 & 1 & 2 & 9 & 7 & 8 & 10 & 6 \end{pmatrix}$$

$$f_3 = \begin{pmatrix} 1 & 2 & 3 & 4 & 5 & 6 & 7 & 8 & 9 & 10 \\ 3 & 9 & 2 & 1 & 6 & 5 & 4 & 10 & 7 & 8 \end{pmatrix}$$

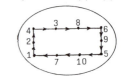

$f_4 = (1\ 2\ 4\ 3\ 8\ 6\ 9\ 5\ 10\ 7)$

$$f_4 = \begin{pmatrix} 1 & 2 & 3 & 4 & 5 & 6 & 7 & 8 & 9 & 10 \\ 2 & 4 & 8 & 3 & 10 & 9 & 1 & 6 & 5 & 7 \end{pmatrix}$$

Fig. 47.1

A permutation f can be represented by a graph in the plane. Choose n points and label them 1 to n. For each $i \in 1..n$ draw an arrow from i to $f(i)$. Fig. 47.1 shows the graphs of f_1 to f_4. They decompose into cycles. Below each graph we find the so called cycle notation of the permutation.

Fixed points, i.e. points with $f(i) = i$, are ignored but counted by the programs in this notation. The permutation f_4 is cyclic, since it has just one cycle.

c) *Permutation Programs.* Random permutations are excellent numerical material. Their processing leads to instructive programs.

Example 1. The inverse permutation. Let $X[1..n]$ be a permutation of $1..n$. We get the inverse permutation $Y[1..n]$ by means of the algorithm

```
for i:=1 to n do Y[[X[i]]:=i.
```

Example 2. The order of a permutation. Let f be a permutation of $1..n$, and let I be the identity permutation. The smallest positive integer p, so that $f^p = I$ is called the *order* or *period* of f. Fig. 47.2 shows that p is the least common multiple of all the cycle lengths.

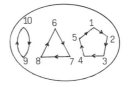

Fig. 47.2

```
program cycles;
var copy,first,i,n,r:integer;
    p:array[1..100] of integer;
begin randomize;
write('n='); readln(n);
for i:=1 to n do p[i]:=i;
for i:=n downto 2 do
begin
  r:=1+random(i); copy:=p[i];
  p[i]:=p[r]; p[r]:=copy
end;
i:=0;
repeat i:=succ(i); first:=i;
  if p[i]>0 then begin
    write('(');
    repeat
      r:=first; write(r:4);
      first:=p[first];
      p[r]:=-p[r]
    until first=i;
    writeln(')')
  end
until i=n
end.
```

Fig. 47.3

```
program jos_perm;
var k,n,s,x:real;
begin
write('n,k=');readln(n,k);
s:=0.0;
repeat
  s:=s+1; x:=k*s;
  while x>n do
    x:=int((k*(x-n)-1)/(k-1));
  write(x:0:0,' ')
until s=n
end.
```

Fig. 47.4

Example 3. Decomposition into cycles. Let $p[1..n]$ be a permutation of $1..n$. We want to decompose it into cycles. We start with i:=1 and set first:=i. Then we repeat first:=p[first] until i recurs. This closes the cycle. The whole sequence is printed. As soon as an element is printed, it must be marked, which can be done by reversing its sign. Before executing first:=p[first] we should make a copy r of first, because we need it for marking p[r]:=−p[r]. If a cycle is complete then we look for the next unmarked element and we start a new cycle. We are finished as soon as all elements are marked. The program cycles generates a random permutation of 1..n and decomposes it into cycles.

d) *The Josephus Permutation.* We return once more to the Josephus Problem: n persons are arranged in a circle. Then every k-th person is eliminated, beginning with number k. The problem is to find the number x of the s-th eliminated person. Fig. 5.6 shows a program which finds x. We want to print the sequence $J(n,k)$ of eliminations, which is called the *Josephus Permutation* of 1..n. By a slight modification of joseph1 we get the program jos_perm (Fig. 47.4).

Exercises for Section 47:

1. Modify the program cycles so that it counts the number of cycles.

2. Generate *rep* permutations of 1..n, count the number of cycles each time and estimate the expected number of cycles in a random n-permutation. It can be shown easily that this number is $H_n = 1 + 1/2 + 1/3 + \ldots + 1/n$.

3. We have n boxes, each one with its own key, which fits no other box. After mixing the keys at random we drop one key into each box. Now we break open k boxes. What is the probability $p(n,k)$ that we are now able to open the remaining boxes? This was a problem (with $k = 2$) in the Kürschák competition in Hungary in 1971. There are short ways of deriving the surprisingly simple formula for $p(n,k)$, but they are tricky. So we try to find the formula for it by simulation. But first we translate the problem into the language of permutations: Choose a permutation p of 1..n at random. What is the probability that the cycles of p containing 1..k cover all elements from 1..n? Write a simulation program for finding $p(n,k)$.

4. Consider the following variation of Exercise 3: Break one box and open all the boxes you can. Now break another box and open all the boxes you can, ..., until you have opened all boxes. How many boxes do you have to break on the average?

5. Show that the following Josephus permutations are cyclic:
   a) $J(n,2)$ for $n = 2, 5, 6, 9, 14, 18$;
   b) $J(n,3)$ for $n = 3, 5, 27, 89, 1139, 1219, 1921, 2155$;
   c) $J(n,4)$ for $n = 5, 10, 369, 609$;
   d) $J(n,7)$ for $n = 11, 21, 35, 85, 103, 161, 231$.

48. Games

We consider games for two players A and B, who move alternately. A always moves first but otherwise the rules are the same for A and B. The pieces on the boards will not be assigned to either player. A draw cannot occur.

We are given the starting state and the set M of legal moves. A player loses if he finds himself in a position from which no legal move can be made. We can think of each position as a vertex of a graph and each move as a directed edge. We consider only games with finitely many vertices and no directed circuit (a position can not repeat). This ensures that one of the players will lose.

It can be shown by induction on the number of moves that the set P of all positions can be partitioned into the set L of losing, and the set W of winning positions: $P = L \cup W$, $L \cap W = \emptyset$. A player finding himself in a position in L will lose provided his opponent plays correctly. A player finding herself in a position in W can force a win whatever her opponent does.

In order to win a player must always move so as to force his opponent into a position belonging to L. From each position in L every move must result in a position in W. From every position in W a move to a position in L must be possible. (See Fig. 48.1.) L must contain at least one final position f from which there is no move out. The player who leaves his opponent facing such a position has won the game. The problem is to identify the set L of losing positions.

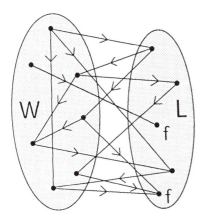

```
program nimiter8;
var i,n:integer;
    x:array[0..120] of byte;
begin write('n='); read(n);
for i:=0 to n+8 do x[i]:=0;
i:=0;
repeat
  if x[i]=0 then begin
    x[i+1]:=1; x[i+3]:=1;
    x[i+8]:=1 end;
  i:=i+1
until i>n;
for i:=0 to n do
  if x[i]=0 then write(i:4)
  else write('-':4)
end.
```

Fig. 48.1   Fig. 48.2

```
0 - 2 - 4 - 6 - - - - 11 - 13 - 15 - 17 - - - - 22 - 24
- 26 - 28 - - - - 33 - 35 - 37 - 39 - - - - 44 - 46 - 48 -
50 - - - - 55 - 57 - 59 - 61 - - - - 66 - 68 - 70 - 72 - -
- - 77 - 79 - 81 - 83 - - - - 88 - 90 - 92 - 94 - - - - 99 -
```

Output of `nimiter8` for $n = 100$.

1. Two players take turns to reduce a pile with initially n checkers. A move consists in removing m checkers, where m is a number from the set $\{1, 3, 8\}$ of legal moves. The winner is the one who sweeps the board clean. Find the set of losing positions.

2. This is the same as the preceding problem except the set of legal moves is the set of squares $\{1, 4, 9, 16, 25, 36, \ldots\}$. We present programs which print i or $|$ if $i \in L$ and "$-$" if $i \in W$. For the first problem we get a pattern with a period 11. L consists of all nonnegative integers of the form 11∗k, 11∗k+2, 11∗k+4, 11∗k+6. For the second problem there is no periodicity visible or to be expected.

```
program square_nim;
var i,k,n:integer; w:array[0..1000] of boolean;
begin write('n='); readln(n);
  for i:=0 to n do w[i]:=false; i:=0;
  for i:=0 to n do
    if not w[i] then
      for k:=1 to trunc(sqrt(n-i)) do w[i+k*k]:=true;
  for i:=0 to n do if w[i] then  write('-') else write('|');
readln end.
```

```
|-|--|-|--|-|--|-|--|-|------------|----|----|-------|----|----|--|-|----|--
----------|---------|-------------|---------|----|--|--|-|----|----|----|--
|-------------------|------|--|-|----|----|----|-------|-----|----|---------
-------------|----|-----|-----|-|--|-|-|----|----|--|------------------------
```

Fig. 48.3. Square_nim.
The pattern of losing (|) and winning (-) positions is shown up to $n = 300$.

3. *Wythoff's Game.* There is an old Chinese game with a beautiful mathematical theory. Here are the rules: On a table there are two piles of checkers. Two players move alternately. A move consists of taking any number of checkers from one pile or the same number of checkers from each pile. Winner is the one who takes the last checker.

We translate the game into the board game in Fig. 48.4. Initially a checker is placed in each of two rows. One checker may move forward any distance, or both checkers may move forward the same distance. The loser is the one who cannot move since both checkers are in column 0.

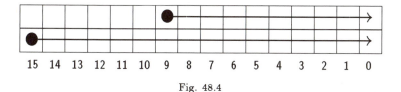

15 14 13 12 11 10 9 8 7 6 5 4 3 2 1 0

Fig. 48.4

Working with the board in Fig. 48.4 we get the following table of losing positions $(x(0), y(0))$, $(x(1), y(1)) \cdots$:

n	0	1	2	3	4	5	6	7	8	9	10	11	12	13	14	15	16	17	18
$x(n)$	0	1	3	4	6	8	9	11	12	14	16	17	19	21	22	24	25	27	29
$y(n)$	0	2	5	7	10	13	15	18	20	23	26	28	31	34	36	39	41	44	47

Fig. 48.5

This table suggests the following algorithm for constructing the set L of losing positions step-by-step: Suppose the losing positions $(x(i), y(i))$ for $i < n$ are known already. Then $x(n)$ is the smallest positive integer not yet part of a pair and $y(n) = x(n) + n$. Thus every positive integer occurs exactly once as a member of a pair and exactly once as a difference. We want to verify that L is indeed the set of losing positions.

Let W be the set of pairs not in L. It is easy to check that every move from L leads to a position in W. It remains to show that from any position in W we can move into L.

```
program wythoff;
var i,j,n:integer;
    a,x,y:array[0..300] of integer;
begin write('n='); readln(n);
for i:=0 to n do
begin
  a[i]:=i;x[i]:=0;y[i]:=0
end;
i:=0;  j:=0;
repeat j:=j+1;
  if a[j]=j then
    begin
      i:=i+1; x[i]:=j; y[i]:=j+i;
      a[i]:=0; a[j+i]:=0
    end
until j>=n;
for j:=0 to i do
  write(x[j]:7,y[j]:4)
end.
```

Fig. 48.6

Let $(p, q) \in W$, $p \leq q$. If $p = q$, we can move into $(0,0) \in L$ and win in one move. If $p \neq q$, let (p, p') or (p', p) be the element of L with one component p. If $p' < q$, we reduce q to p'. If $q < p'$, so that $p < q < p'$ and $0 < q - p < p' - p$, we reduce both piles by equal amounts, so as to obtain the element of L with the difference $q - p$.

This algorithm for the construction of the losing positions can be translated into the program in Fig. 48.6 which enables us to extend Fig. 48.5.

Exercises for Section 48:

1. Extend the array limits in Fig. 48.3 as far as your computer allows, run the program for large values of n and stop it from time to time to study the strange fluctuations of the elements of L.

2. Consider Wythoff's game.

   a) Write a program which transforms a position in W into a position in L.

   b) Write a program which plays against a human opponent.

   c) In 1907 the Dutch mathematician Wythoff showed that the losing positions are given by $x(n) = \text{trunc}(n*t)$, $y(n) = n+x(n)$, $t = (\sqrt{5}+1)/2$. Show this. *Hint:* Use Beatty's theorem (see Section 10.3) to show that the sequence $\{n + \lfloor tn \rfloor\}$ is the complementary sequence of $\{\lfloor tn \rfloor\}$.

   d) Write a program which plays against a human opponent, based on these formulas.

3. In a game with a pile of n checkers the set of legal moves is the set $M = \{1, 2, 3, 5, \ldots\}$ of Fibonacci numbers. Write a program which finds the losing positions.

   Try to solve the following four simple games without a computer.

4. Bachet's game. Initially there are n checkers on the table. The set of legal moves is $M = 1, 2, \ldots, k$. Winner is the one to take the last checker. Find the losing positions.

5. In problem #4 let $M = \{1, 2, 4, 8, \ldots\}$ (any power of 2). Find the set L.

6. In problem #4 let $M = \{1, 2, 3, 5, 7, 11, \ldots\}$ (1 and all primes). Find L.

7. In problem #4 let $M = \{1, 3, 5, 8, 13\}$. Find the set L.

8. Empirical exploration. Write the following program: Input is a finite sequence M of positive integers which are the legal moves. Experiment with the computer and try to make predictions about the set L of losing positions.

9. Write a recursive version of the program in Fig. 48.1 , which recognizes if an input n belongs to L.

10. Let $x(n) = \lfloor nt \rfloor$ with $t = (\sqrt{5} + 1)/2$. Consider the sequence $a(n) = x(n + 1) - x(n)$ for $n = 0, 1, 2, \ldots$. Its first terms are

$$1 \quad 2 \quad 1 \quad 2 \quad 2 \quad 1 \quad 2 \quad 1 \quad 2 \quad 2 \quad 1 \quad 2 \quad 2 \quad 1 \quad 2$$
$$12 \quad 122 \quad 12 \quad 122 \quad 122 \quad 12 \quad 122 \quad 12 \quad 122 \quad 122 \quad 12 \quad 122 \quad 122 \quad 12 \quad 122$$

The second line arises from the first by inflation: $1 \rightarrow 12$, $2 \rightarrow 122$. This seems to transform the sequence into itself, which would make the sequence self-similar.

a) Test this self-similarity on your computer as far out as possible.

b) What is the proportion of 2's among all digits? Are you surprised? [See the solution of Problem 3, Section 27.]

11. We start with 0 and perform repeatedly the substitutions $0 \rightarrow 1$, $1 \rightarrow 10$. We get 0, 1, 10, 101, 10110, 10110101, ... In this way we get an infinite binary sequence, which is more complicated than the Morse-Thue sequence. Investigate the sequence with the computer, especially self-similarity, proportion of 1's etc. Compare also with Mathematical Explorations no. 24 in Section 6, and problem 3 in Section 27 and its solution where the more difficult substitution $0 \rightarrow 001$, $1 \rightarrow 0$ was treated.

12. Initially there are 10000 checkers on the table. Two players move alternately. If it is your turn you may take any p^n checkers , where p is any prime and $n = 0, 1, 2, 3, \ldots$. Winner is the one to take the last checker. Who wins if he plays correctly?

13. A modification of Wythoff's game. You may remove an even number from a single pile, or equal numbers from both piles. By modifying the program Wythoff find the set L and closed formulas for its elements. Hint: There will be two formulas, depending on the parity of the initial number of checkers. In case of an odd initial number the final position will be $(0,1)$. It is enough to find the formulas empirically.

14. Start with $n = 2$. Two players A and B move alternately by adding to the current n a proper divisor of n. Goal is a number ≥ 1990. Who wins, A or B?

15. Yet another modification of Wythoff's game. You may remove any number of checkers from one pile, or m checkers from one pile and n checkers from the other pile if $|m-n| < a$. For $a = 3$ find with the PC the losing positions and guess closed formulas for these. You have met these formulas before in section 27, problem 3.

49. The Subset Sum Problem. The Limits of Computation.

Given are n positive reals and an additional real number G, the goal. Find a subset with sum as near to G as possible, without surpassing G. Of this famous and notorious problem we work on a special case:

Abby and Carl inherit n gold pieces with weights $\sqrt{1}, \sqrt{2}, \ldots, \sqrt{n}$ ounces. Make of them two heaps with weights differing as little as possible.

We get a good, but not necessarily the best solution by means of the FFD (First-Fit-Decreasing) heuristics: Arrange the weights in decreasing order and, starting at the beginning, assign the current weight to Abby, if her total weight stays below the goal G, otherwise to Carl. Here

$$G = \frac{\sqrt{1} + \sqrt{2} + \cdots + \sqrt{n}}{2}.$$

The program can be found in Fig. 49.1. It prints a binary word whose k-th digit is 1 if the coin of weight $\sqrt{k}$ is in Abby's set.

```
program FFD;
var i,n:integer; abby,carl,goal,sum:real;
    v:array[0..100] of real;
begin write('n='); readln(n); sum:=0; abby:=0; carl:=0;
write('binary word=');
for i:=1 to n do begin v[i]:=sqrt(i); sum:=sum+v[i] end;
goal:=sum/2;
for i:=n downto 1 do
  if abby+v[i]<goal then begin abby:=abby+v[i]; write(1) end
  else begin carl:=carl+v[i]; write(0) end;
writeln; writeln('abby=',abby:13:10);
writeln('carl=', carl:13:10); writeln('goal=',goal:13:10);
readln end.
```

Fig. 49.1

The FFD-algorithm yields such a good result that we get the suspicion that it may be the optimal solution. Let us estimate how close the optimal solution may be to our goal. We see that the sum of the seven largest pieces is so much

bin.wd.=11111110000000001000
abby = 30.811421944
carl = 30.854555867
goal = 30.832988906

Fig. 49.2 Printout for $n = 20$.

smaller than $G/2$ that we can add a small piece and still stay under $G/2$. Hence the share of each should consist of at least 8 pieces and, by symmetry, of at most 12. There are

$$\binom{20}{8} + \binom{20}{9} + \cdots + \binom{20}{12} = 772,616$$

subsets with 8 to 12 elements. Their weights lie in the range from 16 to 46 ounces. If the weights were uniformly distributed in this interval, they would split it into subintervals of length $\approx 30/770000 \approx 3.9 * 10^{-5}$. Thus it is very likely that the optimal solution is much better than the FFD-solution.

But how do you find the optimal solution? One could try all subsets, which is called *exhaustive search*. Step through all 2^{20} or 1048576 subsets and choose the one with weight as close to the goal as possible. In each step exactly one square root should be added or removed, so that the subset sum is easy to update. In other words, at each step just one bit of the 20 bit word encoding the set should be complemented. Can I step through all 2^{20} binary words by these elementary steps? In other words: does there exist a Hamilton path on the 20-dimensional cube?

The solution is provided by the Gray Code. It can be constructed recursively, up one dimension at each step, as shown in Figs. 49.3-4. The Gray Code

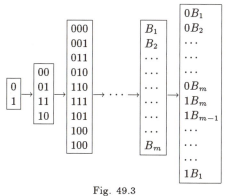

Fig. 49.3

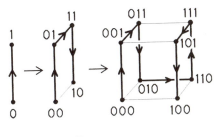

Fig. 49.4

can be conveniently represented by its transition sequence, that is the sequence of the positions of the bits which change as we go from one code word to the next. The positions are numbered from right to left. The transition sequences for dimensions 1, 2 and 3 are

$$T_1 = 1, \quad T_2 = 1, 2, 1, \quad T_3 = 1, 2, 1, 3, 1, 2, 1.$$

From the last two columns in Fig. 49.3 we get

$$T_{n+1} = T_n, \ n+1, \ T_n \text{ reversed}.$$

We can easily prove by induction that for all n

$$T_n = T_n \text{ reversed}$$

and so the recursive definition simplifies to

(1) $$T_1 = 1, \quad T_{n+1} = T_n, \ n+1, \ T_n.$$

Another way of obtaining the sequence T_n, easily verified using (1) and induction on n, is the following. At the m-th step, i.e. when we are changing the m-th binary word to obtain the $m+1$-st, we change the t-th bit where t is $1+$(the number of 0's at the end of the binary representation of m). We used this in the program gray.

We can generate the sequence T_n more efficiently by using a stack, which operates as follows: The stack initially contains $n, n-1, \ldots, 1$ (with 1 on top as in Fig. 49.5). The algorithm takes the top number i and puts it into the sequence. The numbers $i-1, i-2, \ldots, 1$ are then put on the stack.

By playing with a stack of four chips numbered 1,2,3,4 you can easily convince yourself that the method does indeed generate $T_4 = 1,2,1,3,1,2,1,4,1,2,1,3,1,2,1$. An easy induction exercise shows that the algorithm generates T_n for all n.

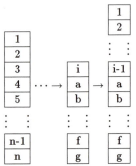

Fig. 49.5

The program gray_stack represents the stack described above by an array t[0..n+1]. It generates T_n more efficiently than gray, but is not so easy to comprehend. (Exercises 3 and 4.)

```
program gray;
var t,m,b:integer;
    max:real;
begin
write('max=');read(max);m:=1;
repeat b:=m; t:=1;
  while odd(b-1) do
  begin
    b:=b div 2; t:=t+1
  end;
  write(t,' '); m:=m+1
until m>max
end.
```

Fig. 49.6

```
program gray_stack;
var i,j,n:0..255;
    t:array[0..30] of 0..255;
begin
write('n='); readln(n);
for j:=0 to n+1 do t[j]:=j+1;
repeat
  i:=t[0]; t[0]:=1;
  t[i-1]:=t[i]; t[i]=i+1;
  if i<n+1 then write(i,' ')
until i=n+1
end.
```

Fig. 49.7

We are now equipped with an algorithm to step through all subsets of our 20-set efficiently. But 2^{20} subsets is a large number for an older microcomputer. We can cut the computation time in half by checking only the first 2^{19} subsets, which are the subsets not containing $\sqrt{20}$. The remaining sets are the complements of these, with the same distances from the goal. After some deliberation we hit on a trick which cuts the number of sets to be tried by an additional factor 16. We remove the subset $S = \{\sqrt{1}, \sqrt{4}, \sqrt{9}, \sqrt{16}\} = \{1, 2, 3, 4\}$. All integers from 0 to 10 can be represented as a sum of elements of S. Then we look for a subset with weight w such that $|G - w - k|$ is as small as possible, where k is the integer from $0..10$ which makes $|G - w - k|$ smallest. In Pascal the integer k which makes $|G - w - k|$ smallest has a name: $k = \mathbf{round}(G - w)$. So we are interested in finding a w such that $|G - w - \mathbf{round}(G - w)|$ is as small as possible, provided $0 \leq \mathbf{round}(G - w) \leq 10$. We get a solution by adding weights totaling $\mathbf{round}(G - w)$ from S to w.

If $G - w$ is exactly halfway between integers, then $k = \mathbf{round}(G - w)$ can be outside the range $0..10$ but an equally good value of k may be in it. Such a w would be excluded by the way $\mathbf{round}(x)$ is selected in borderline cases. However, in such a case w can be brought only to a distance 0.5 to G, and we already know that such a w is not a solution to our problem.

```
program subset_sum;
const n=16; {On Macintosh, use extended instead of real.}
var b,i,j:integer; s,w,G,min:real;
    d:array[1..30] of real; t:array[1..30] of integer;
begin s:=0; min:=1;
for j:=1 to n do
  begin d[j]:=sqrt(j+round(sqrt(j)));
  s:=s+d[j]; t[j]:=j; end;
G:=(10+s)/2; w:=0; i:=1;
while i<n do
begin
  i:=t[1]; w:=w+d[i]; d[i]:=-d[i];
  t[1]:=1; t[i]:=t[i+1]; t[i+1]:=i+1;
  if abs((G-w)-round(G-w))<=min then
    if (round(G-w)>=0) and (round(G-w)<=10) then
    begin
      for j:=1 to n do write(ord (d[j]<0));
      min:= abs((G-w)-round(G-w));
      writeln('   min= ',min:14:11);
    end;
end;
write(chr(7));readln;end.
```

Fig. 49.8

Program **subset_sum** needs some further explanation. We entered $n = 16$ for the 16 nonsquares $2, 3, \ldots, 20$. The variable s will hold $\sqrt{2} + \sqrt{3} + \cdots + \sqrt{20}$. Initially it must be set to zero. Initially min has a large value that does not occur but later on min will be the smallest value of $|G - w - \mathbf{round}(G - w)|$ found so far. We listed the noninteger square

```
1111011110000000   min= 0.22973562512
0111111110000000   min= 0.00180212356
1111000111110000   min= 0.00024840192
1010100101110000   min= 0.00013925636
0100001010010110   min= 0.00003584960
```

Output of subset_sum

roots using the formula $d[i] := sqrt(i+round(sqrt(i)))$. (Additional Exercises for Sections 1 to 18, Exercise 1.) We start with $00\ldots00$, i.e. the empty set. Now we incorporate the program gray_stack: $i:=t[1]$ is the current position of the Gray Code. The digit which is being complemented at this stage is the i-th. We add $d[i]$ to w and also change the sign of $d[i]$. As a result, when the i-th digit in the subset word is complemented the next time, we can update w by just adding $d[i]$ to w again.. Note the while $i<n$ instead of while $i<=n$. In this way we step through the sets not containing $\sqrt{20}$, which, as we noted, cuts the computation time in half.

There is a new feature of Pascal in this program. The statement write(ord(d[j]<0)) tells the computer to write the ordinal of true if $d[j]<0$ is true, else to write 0. In Pascal true is assigned the ordinal 1 and false has ordinal 0. Thus for each new minimum the corresponding binary word is printed. The remainder of the program should be clear.

In this way we get two optimal solutions for Abby:

$$a_1 = \sqrt{1} + \sqrt{2} + \sqrt{3} + \sqrt{4} + \sqrt{8} + \sqrt{9} + \sqrt{10} + \sqrt{12} + \sqrt{15} + \sqrt{16} + \sqrt{19}$$

and a_2, which is obtained from a_1 by replacing $\sqrt{2} + \sqrt{8}$ in a_1 by the equal quantity $\sqrt{18}$. Thus we have

$$a_1 = a_2 = 30.8329530597$$
$$goal = 30.8329889057.$$

The absolute and relative deviations from the goal are

$$|a_1 - goal| \approx 3.6 * 10^{-5}, \qquad a_1 - goal|/goal \approx 10^{-6}.$$

Our estimate $3.9 * 10^{-5}$ was quite good. The program subset_sum finds just one optimal solution a_2. Because of rounding errors Turbo Pascal does not recognize a_1 as optimal. See Ex. 7 for a discussion of roundoff errors.

Exercises:

1. Suppose k heirs inherit n gold pieces weighing $x_1, x_2, \ldots, x_n$. We want to separate the heritage into k heaps which are as nearly equal as possible. For this case the FFD-algorithm is defined as follows: First order the

items so that $x_1 \geq x_2 \geq \ldots \geq x_n$. Then proceed to pack the items in order, starting with x_1, which we place in the first heap H_1. In general, if there are heaps H_j which have room for x_i, i.e. the total size of items in H_j is not more than goal $-x_i$, then we place x_i into the heap with the smallest index j. Otherwise we place it into the heap with the least amount in it; if there are several such, we pick the one with the smallest index.

Write a program for this method and apply it to $k = 3$, $x_i = \sqrt{i}$, and $n = 20, 30$, and 40.

2. Rewrite the program subset_sum by using the program gray to determine the position i of the Gray Code digit to be changed in going from one subset to the next. The run time for Fig. 49.8 is 11 seconds. What time does the new program require?

3. Prove that the program gray works correctly.

4. Prove that the program gray_stack works correctly. (Not easy.) Execute this program for $n = 1, 2, 3, 4, \ldots$ until you understand how it works.

5. Here is an algorithm for the Gray Code.

```
1. Set c[i]:=0 and s[i]:=1 for i=1,..,n.
2. Write c[i]   for i=1,2,...,n.
3. Set i:=n.
4. Set c[i]:=c[i]+s[i].
5. if (c[i]=0) or (c[i]=1) goto step 2.
6. Set c[i]:=c[i]-s[i].
7. Set s[i]:=-s[i].
8. Set i:=i-1.
9. if i>0, goto step 4.
10. end.
```

Write a program, run it and prove its correctness by induction.

6. The golden permutation. We consider an interesting application of sorting. Let $t = (\sqrt{5} - 1)/2$. The n pairs of numbers $(i, \text{frac}(i * t))$, $i = 1 .. n$ are sorted so that their second coordinates are increasing. Then the first coordinates are printed. The results is the so-called "golden permutation" of $1 .. n$. Because of its amazing properties it is preferred by statisticians to a random permutation of $1 .. n$.

a) Write a program which prints the golden permutation of $1 .. n$ for $n = 100$. With suitable programs check the following surprising properties:

b) Form the difference between an element of the permutation and the previous one and reduce the result mod n. (The last element can be regarded as preceding the first one.) Only three different numbers will occur.

c) Two elements with difference 1 are separated by 37, 38, 60 or 61

elements are placed at the vertices of a regular 100-gon, only two shortest distances will be left, 38 and 39.

d) Suppose a random sample of 10 elements is to be chosen from a sequence of consecutive integers, for example $38 .. 77$. Then we start anywhere in the permutation, for example at 78, and choose successively those elements which lie in the interval $38 .. 77$. We get 57, 70, 49, 62, 41, 75, 54, 67, 46, 59. The sample is unusually uniformly distributed in $38 .. 77$. Note also that if we order the 10 numbers according to magnitude, only the differences 3 and 5 occur. In the list of numbers in the order of occurrence, only the differences 13, 21, 34 occur.

e) Consider on the number line the segment $38 .. 77$. If the points 57, 70, 49,... are marked successively on this segment the latest point falls in one of the currently largest free intervals or an endinterval.

f) Check the above properties for the inverse of the golden permutation. Explain the relations to the previous results.

g) Replace t in the above construction by $u = t^2$. Which properties remain valid?

h) Replace t in the above construction by another irrational number, such as $\sqrt{2}$, π, or e. Which properties remain valid?

i) Do a) to h) for values of n other than 100.

j) Sort the pairs of numbers $(i,\ 55i \bmod 89)$, $i = 0 .. 88$, so that the second coordinates are increasing. Check properties $b) - f)$ for this permutation. Note that 55, 89 are successive Fibonacci numbers. Try also two relatively prime integers which are not Fibonacci numbers.

7. Partition $\sqrt{1}, \ldots, \sqrt{25}$ into two heaps differing as little as possible. What changes are needed in the program subset_sum? My PC requires over 3 minutes, more than 16 times as much as for 20 square roots. My program gives for the least distance from the goal $min = 1.717E - 06$, attained for the binary word 01001110111101110100. Let us consider the roundoff error in this computation. We modify w by additions and subtractions a total of 2^{19} times. The sum is on the average about 40. Since variables of type **real** are represented with a mantissa of 39 binary digits in Turbo Pascal for IBM-compatibles, the average roundoff in each operation is roughly $40 \times 2^{-40} \approx 5\,E - 11$.

Each roundoff error in our computation is just added to the sum of the previous roundoff errors. We can give a rough estimate of the probable total from the theory of random walks. n steps of a symmetric random walk give an expected distance $\epsilon * \sqrt{n}$ from the correct value, where ϵ is the average error in one rounding. We get $2^{9.5} 10^{-10} \approx 0.0000004$

The word above and its complement yield $min = 1.7350750346E - 06$. So our estimate of the roundoff error is about 10 times too large, but still, only the first two significant digits of the computed value of min were correct. It is reasonable to assume that the different values of w in the neighborhood of the goal G do not get much closer to each other than they get to G. Thus they are likely to be separated by more than the roundoff error and we can be confident, although not absolutely certain, that we found the heaps which differ as little as possible.

In this problem one could avoid all further roundoff errors after the square roots have been rounded by using variables of type longint instead of real for the weights; this would also speed up the program. One can also make certain that we do not miss the optimal solution or solutions by printing out not just the sets which meet or improve the previous min but also those which come closer to it than the possible roundoff error.

8. Generate 20 random reals from the interval 0..5 and store them in x[1..20]. With the program subset_sum find exactly the subset which is nearest to the goal but below it. Use as the goal one half of the sum of the generated numbers. Do not use the randomize statement. In this way you can compare the results with those obtained by others who use the same random number generator. What simplifications are possible in the program subset_sum?

CHAPTER 6

Numerical Algorithms

Our aim is to compute the elementary transcendental functions. We will define them geometrically by means of the circle $x^2 + y^2 = 1$ and the hyperbola $xy = 1$. We will not use any calculus; an occasional use of the Pythagorean Theorem and similarity will suffice.

50. Powers with Integer and Real Exponents

If y is a nonnegative integer, we can find x^y recursively by using the formulas

$$x^y = \begin{cases} (x^2)^{y/2} & \text{if} \quad y \text{ is even} \\ x\, x^{y-1} & \text{if} \quad y \text{ is odd;} \end{cases}$$

the base of the recursion is when $y = 0$, since $x^0 = 1$.

If x, y are positive real numbers, we can compute x^y recursively by means of the relations

$$x^y = \begin{cases} (\sqrt{x})^{2y} & \text{if} \quad y < 1 \\ x\, x^{y-1} & \text{if} \quad y \geq 1. \end{cases}$$

The bases of this recursion are the relations $1^y = 1$ and $x^0 = 1$. The condition $x \approx 1$ (up to the roundoff error) is bound to be attained if we replace x by its square root enough times; if y happens to be an integer we get $y = 0$ before that. Our recursions are implemented in the programs natpow and **realpow**.

```
program natpow;
var x:real; y:integer;
function pow(x:real;y:integer):real;
begin
  if y=0 then pow:=1
  else if odd(y-1) then pow:=pow(x*x,y div 2)
      else pow:=x*pow(x,y-1)
end;
begin
  write('x,y='); readln(x,y); writeln(pow(x,y));
readln end.
```

Fig. 50.1

```
program realpow;
var x,y:real;
function pow(x,y:real):real;
begin
   if  (x=1) or (y=0) then pow:=1
   else  if y<1 then pow:= pow(sqrt(x),2*y)
         else pow:=x*pow(x,y-1)
end;
begin
   write('x,y=');  readln(x,y);  writeln(pow(x,y));  readln
end.
```

Fig. 50.2

I reformulate these computations as iterative programs. I want to find a^b for real a and integer b, $b > 0$. I introduce the variables x, y, z. Initially I set $z = 1$, $x = a$, $y = b$. Then

$$(1) \qquad\qquad 1 * a^b = z * x^y .$$

The idea is to drive y to 0 while keeping (1) invariant. This can be accomplished by means of the two transformations

$$(2) \qquad \begin{array}{lll} \texttt{x:=x*x,} & \texttt{y:=y/2,} & \texttt{z:=z} \quad \text{if y is even;} \\ \texttt{x:=x,} & \texttt{y:=y-1,} & \texttt{z:=z*x} \ \text{if y is odd.} \end{array}$$

The first reduces y quickly, but will result in an integer exponent only for even y. The second reduces y slowly, but can be used for odd y. At the end $y = 0$ and $z = a^b$. The iterative program **natpowit** is based on this idea.

```
program natpowit;
var a,x,z:real;  b,y:integer;
begin
   write('a,b=');  readln(a,b);  x:=a;  y:=b;  z:=1;
   while y>0 do begin
      if odd(y) then begin y:=y-1; z:=z*x; end
      else begin x:=x*x; y:=y div 2 end
   end;
writeln(z);  readln end.
```

Fig. 50.3

Suppose now that a, b are positive reals. We again keep (1) invariant but this time we drive x toward 1 by square root extractions. We use the transformations

$$(3) \qquad \begin{array}{lll} \texttt{x:=x,} & \texttt{y:=y-1,} & \texttt{z:=x*z} \quad \text{if } y \geq 1 ; \\ \texttt{x:=sqrt(x),} & \texttt{y:=2*y,} & \texttt{z:=z} \quad \text{if } y < 1. \end{array}$$

When $x \approx 1$ (within roundoff error), then $z \approx a^b$. Thus we get the iterative program **realpowit**.

```
program realpowit;
var a,b,x,y,z:real;
begin
  write('a,b='); readln(a,b); x:=a; y:=b; z:=1;
  while (x<>1) and (y>0) do begin
    if y>=1 then begin  y:=y-1 z:=z*x;  end
    else begin x:=sqrt(x); y:=2*y;  end
  end;
  writeln(z); readln end.
```

Fig. 50.4

Exercises:

1. Write a recursive function **superpower** which is defined by

$$m\wedge n = m^{m^{.^{.^{.^{m}}}}} = m^{m^{\wedge}(n-1)}; \quad m\wedge 0 = 1.$$

Evaluate **superpower(m,n)** for some small values of m and n.

2. a) What are the last three digits of 7^{999999}?
   b) What are the last five digits of 1987^{999999}?

   Hint: None of the preceding four powering programs is applicable here. So you should write your own powering program for this case. Write two programs, one recursive, the other iterative.

51. Design of a Square Root Procedure

For given $a > 1$, I want to find $z > 0$ such that

(1) $z * z = a$.

An important technique is to replace the unknown constant z by variables which are pushed toward z. So we replace z by the variables x, y such that

(2) $x > y$

and

(3) $x * y = a$.

These relations are easy to establish initially by setting $x = a$, $y = 1$. Now I move toward my goal (1) by keeping (2) and (3) invariant. That is, I must find a step which changes x, y so that the positive difference decreases, but (2) and (3) remain valid. This would be a step toward (1). We accomplish this by the step $(x, y) \leftarrow (x', y')$, where we set

$$x' = \frac{x + y}{2} \quad \text{and} \quad y' = \frac{2xy}{x + y} .$$

Then

$$a = x * y = \frac{x + y}{2} * \frac{2xy}{x + y} = x' * y'$$

so that (3) holds for (x', y'). Moreover, the arithmetic mean of two distinct positive numbers is larger than their harmonic mean; x' is the arithmetic mean of x, y and

$$y' = \frac{2}{\frac{1}{x} + \frac{1}{y}} = \frac{2xy}{x + y}$$

is their harmonic mean. Hence (2) remains valid:

$$x' - y' = \frac{x + y}{2} - \frac{2xy}{x + y} = \frac{(x - y)^2}{2(x + y)} > 0 .$$

From the last expression we also get

$$x' - y' = \frac{x - y}{2} \frac{x - y}{x + y} < \frac{x - y}{2}$$

So one step reduces the positive difference $x - y$ by at least the factor $\frac{1}{2}$. Do we gain just one bit per step? NO! This is a superfast procedure. Suppose

$$0 < x - y < 2^{-n};$$

then for the next difference we have

$$0 < x' - y' < \frac{2^{-2n}}{2(x + y)} < \frac{2^{-2n}}{4y} .$$

One step about doubles the number of correct digits. Thus we get the recursive procedure **root**.

```
program root; {initially x>y}
var x,y:real;

function r(x,y:real):real;
begin if x=y then r:=x  else r:=r((x+y)/2, 2*x*y/(x+y)) end;
begin
  write('x,y='); readln(x,y); writeln(r(x,y)); readln;
end.
```

Fig. 51.1

```
program rootit; {x>y}              program rootit1; {x>y}
var a,x,y,eps:real;                var a,eps,x,y:real;
begin                              begin write('a,eps=');
  write('a,eps=');                   read(a,eps); x:=(1+a)/2;
  readln(a,eps);                     repeat
  x:=a; y:=1;                          y:=x;x:=(x+a/x)/2;
    repeat x:=(x+y)/2; y:=a/x          writeln(x)
    until x<=y+eps;                  until abs(x-y)/x<=eps;
writeln(x); readln end.            readln end.
```

Fig. 51.2   Fig. 51.3

In the iterative version of the procedure we replace the simultaneous assignment $(x, y) \leftarrow (x', y')$ of the recursion by the two consecutive assignments $x \leftarrow (x + y)/2$; $y \leftarrow a/x$

Thus we arrive at program rootit. We could even eliminate the variable y by using $x \leftarrow \frac{1}{2}(x + a/x)$. But then we would not have a good stopping rule. So we have to reintroduce y as in program rootit1.

The program rootit works for all $a > 0$, not just for $a > 1$. The same is true for root and rootit1 (Exercise 12). Also try to understand the program in Fig. 51.4, which is a tricky way of listing approximations to square roots. The output shown is for $x = 2$, $y = 1$, eps=1E-07.

```
program root1; {x>y}
var x,y,eps:real;
function r(x,y:real):real;
begin
  writeln(x:20,y:20);
  if abs(x-y)/x<eps then r:=(x+y)/2
  else r:=r((x+y)/2, 2*x*y/(x+y))
end;
begin
  write('x,y,eps='); read(eps,x,y); x:=r(x,y); readln
end.
```

Fig. 51.4

```
        2.0000000000    1.0000000000
        1.5000000000    1.3333333333
        1.4166666667    1.4117647059
        1.4142156863    1.4142114385
        1.4142135624    1.4142135624
```

We just programmed the standard square-root algorithm used in high school, an instance of Newton's method for finding zeros of functions.

We next compute an exact formula for the relative error. Let $x = \sqrt{a}(1+\epsilon_0)$ where $\epsilon_0 > 0$ and $x' = (x + a/x)/2 = \sqrt{a}(1+\epsilon_1)$. We find that

$$\epsilon_1 = \frac{\epsilon_0^2}{2(1+\epsilon_0)} = \frac{\epsilon_0}{2}\frac{\epsilon_0}{1+\epsilon_0}.$$

We see again that when $\epsilon_0 \gg 1$, the error is halved in each step (linear convergence), and for small ϵ_0 the error is squared (quadratic convergence). For small ϵ_0 the number of correct digits approximately doubles at each step.

Does it pay to speed up quadratic convergence? We make two attempts. First we observe that for $x \gg 1$ we have

$$\sqrt{\frac{x+1}{x-1}} \approx 1 + \frac{1}{x}.$$

This can be verified by squaring.

We restrict our attention to the range $a > 1$. We utilize this approximate formula by writing a as $(x+1)/(x-1)$, where $x = (a+1)/(a-1)$ so that $\sqrt{a}$ can be written in the form

$$\sqrt{a} = \sqrt{\frac{x+1}{x-1}} = (1 + \frac{1}{x})\sqrt{\frac{y+1}{y-1}};$$

we get $y = 2x^2 - 1$, so that y is much larger than x. Hence we get the product representation

$$\sqrt{a} = \sqrt{\frac{x+1}{x-1}} = (1 + 1/q_1)(1 + 1/q_2)\cdots(1 + 1/q_n)\sqrt{\frac{q_{n+1}+1}{q_{n+1}-1}}$$

with $q_1 = x$, $q_{n+1} = 2q_n^2 - 1$, where the error factor

$$\sqrt{\frac{q_{n+1}+1}{q_{n+1}-1}} \approx 1 + \frac{1}{q_{n+1}}$$

converges superfast to 1. With $x = 3$ we get $a = 2$ and

$$\sqrt{2} = (1 + 1/3)(1 + 1/17)(1 + 1/577)(1 + 1/665857)\cdots.$$

It is easy to show that $\epsilon_{n+1} \approx \epsilon_n^2/2$, which is no improvement over Newton's method.

A seemingly faster approximation is based on writing a as $(x+3)/(x-1)$ and using the identity

$$\sqrt{\frac{x+3}{x-1}} = (1 + \frac{2}{x})\sqrt{\frac{y+3}{y-1}},$$

where $y = x^3 + 3x^2 - 3$. With $x = 5$ we get

$$\sqrt{2} = (1 + \frac{2}{5})(1 + \frac{2}{197})(1 + \frac{2}{7761797}) \cdots$$

$$x_1 = 1.4$$
$$x_2 = 1.414213\underline{198}$$
$$x_3 = 1.41421356237309504880\ldots.$$

The incorrect digits are underlined. Thus we have 2, 7, 21 correct digits. Each step triples the number of correct digits. We have cubic convergence. It is easy to show that $\epsilon_{n+1} \approx \epsilon_n^3/4$. The curtain that hides the unknown digits is rolled back 50% faster, but unfortunately this is negated by more work at each step. So it does not seem to pay to speed up quadratic convergence.

Remark. There is a simple exact formula for the error in the n-th approximant in Newton's method. One sets

$$x_0 = \sqrt{a}\frac{1 + \omega}{1 - \omega}; \quad \text{then} \quad x_n = \sqrt{a}\frac{1 + \omega^{2^n}}{1 - \omega^{2^n}}.$$

In spite of the fact that the computation here is somewhat shorter we prefer to set $x_0 = \sqrt{a}(1 + \epsilon_0)$, because the interpretation of ω is not as natural as that of ϵ_0. Indeed

$$\omega = \frac{x_0 - \sqrt{a}}{x_0 + \sqrt{a}}.$$

Exercises:

The next 10 exercises investigate the quadratic equation by putting it into the form $x = f(x)$ and iterating. This method of solving an equation numerically is discussed in some calculus textbooks, e.g. Gilbert Strang's and Al Shenk's.

1. We want to solve $x^2 - 4x - 1 = 0$ by iteration. So we "solve" for x and get $x^2 = 4x + 1$ or $x = 4 + 1/x$ and use the recurrence

$$x_1 = 4, \qquad x_{n+1} = 4 + \frac{1}{x_n}.$$

Show that x_n converges to a root of $x^2 - 4x - 1 = 0$ and that at each iteration the absolute error becomes about 18 times smaller.

2. We want to solve $x^2 = 10$ iteratively. Now $x = 10/x$ leads nowhere since the sequence of iterates of $f(x) = 10/x$ has period 2. So we set $x = z - 3$ and get $z = 6 + 1/z$. Show that the sequence $z_1 = 6$, $z_{n+1} = 6 + 1/z_n$ converges to a root of $z^2 - 6z - 1 = 0$ with linear rate $q \approx 1/38$. That is, we get on the average about 1.6 additional decimals per iteration.

3. Take the quadratic $x^2 - 2x + 1 = 0$ (with just one root), which we transform into $x = 2 - 1/x$ and $x_1 = 2$, $x_{n+1} = 2 - 1/x_n$. Show by induction that $x_n = 1 + 1/n$ with very slow convergence to the fixed point $s = 1$.

4. In the context of studying iterates of rational functions we adopt the definitions

$$\frac{\text{nonzero}}{0} = \infty, \qquad \frac{\text{nonzero}}{\infty} = 0;$$

this eliminates the need for restrictions on the domains of definition of the functions.

Take a quadratic without real solutions: $x^2 - 2x + 2 = 0$, or $x = 2 - 2/x$. The sequence $x_1 = 2$, $x_{n+1} = 2 - 2/x_n$ has period 4 (Fig.51.5).

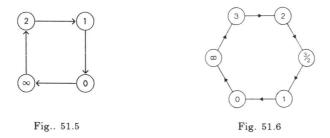

Fig.. 51.5   Fig. 51.6

The quadratic $x^2 - 3x + 3 = 0$, or $x = 3 - 3/x$ leads to the sequence $x_1 = 3$, $x_{n+1} = 3 - 3/x_n$ with period 6 (Fig.51.6). The equation $x^2 - x + 2 = 0$ or $x = 1 - 2/x$ results in the sequence $x_1 = 1$, $x_{n+1} = 1 - 2/x_n$. Computer output seems to indicate that the sequence does not converge at all. It seems that the points of the sequence are everywhere dense on the line, i. e. the sequence visits every interval infinitely often. Since computer arithmetic is finite we will ultimately have overflow or cycling behavior.

5. We formulate some questions about the quadratic $x^2 - px + q = 0$ and the associated sequence

$$(1) \qquad\qquad x_1 = p, \qquad x_{n+1} = p - \frac{q}{x_n}$$

   a) When does the sequence (1) converge?
   b) To which root of $x^2 - px + q = 0$ does it converge if it converges at all?
   c) How fast is the convergence rate?
   d) How can we speed up convergence?

6. Let r, s be the roots of $x^2 - px + q = 0$ and let x_1, x_2 be the the first two terms of (1). Set $x_1 = r(1 + \epsilon_1)$, $x_2 = r(1 + \epsilon_2)$, where ϵ_1 and ϵ_2 are the relative deviations of x_1 and x_2 from r.

a) Show that

$$\epsilon_1 = \frac{s}{r}, \quad \frac{1}{\epsilon_2} = \frac{r}{s} + \frac{r}{s}\frac{1}{\epsilon_1}, \dots, \quad \frac{1}{\epsilon_n} = \frac{r}{s} + \frac{r^2}{s^2} + \dots + \frac{r^n}{s^n}.$$

b) What do you get for ϵ_n if $r = s$?

c) What do you get for ϵ_n for $r \neq s$?

d) Show that if $|r| > |s|$ we have, for large n, $\epsilon_n \approx (1 - s/r)(s/r)^n$. Thus x_n converges to the root with larger absolute value with constant rate s/r. This will be so even for complex values of p and q.

e) What happens if $|r| = |s|$, but $r \neq s$?

f) What happens if r/s is an nth root of unity, i.e. $(r/s)^n = 1$ and $(r/s)^m \neq 1$ for $m < n$?

7. We are not satisfied with linear convergence. So we transform $x^2 - px + q = 0$ into $(x - p/2)^2 = p^2/4 - q$. Show that Newton's method $x_{n+1} = \frac{1}{2}(x_n + a/x_n)$ leads to

$$x_{n+1} = \frac{x_n^2 - q}{2x_n - p}$$

with quadratic convergence for $p^2 \neq 4q$ and linear convergence for $p^2 = 4q$.

8. Show that, for $r \neq s$ and any starting approximant x_0,

(2)
$$x_0 = \frac{s(x_0 - r) - r(x_0 - s)}{(x_0 - r) - (x_0 - s)}, \quad x_1 = \frac{s(x_0 - r)^2 - r(x_0 - s)^2}{(x_0 - r)^2 - (x_0 - s)^2},$$

$$x_n = \frac{s(x_0 - r)^{2^n} - r(x_0 - s)^{2^n}}{(x_0 - r)^{2^n} - (x_0 - s)^{2^n}}.$$

9. Deduce from (2) the following

Theorem. Let $x^2 - px + q = 0$, where p and q may be complex numbers, have roots r, s. Let

$$g(x) = \frac{x^2 - q}{2x - p}.$$

We generate the sequence x_0, $x_1 = g(x_0)$, $x_2 = g(x_1)$, ...

If $p^2 = 4q$ i.e., $r = s$, then x_n converges linearly to $x = p/2$.

If $p^2 \neq 4q$, i.e., $r \neq s$ and x_0 is closer to r than to s then x_n converges quadratically to r. If the location of x_0 in the complex number plane is on the perpendicular bisector of r and s, then there is no convergence. The sequence lies on the perpendicular bisector. It is periodic only for some initial values x_0.

10. *Quadratic equations and matrices.* Let $x^2 - px + q = 0$, $x_1 = p$, $x_{n+1} = p - q/x_n$ with linear convergence to the root with largest absolute value. Then

$$x_1 = \frac{u_1}{v_1} = \frac{p}{1}, \dots, \quad x_{n+1} = \frac{u_{n+1}}{v_{n+1}} = \frac{pu_n - qv_n}{u_n}.$$

Consider the matrix

$$Q = \begin{bmatrix} p & -q \\ 1 & 0 \end{bmatrix}$$

and set

$$\begin{bmatrix} u_0 \\ v_0 \end{bmatrix} = \begin{bmatrix} 1 \\ 0 \end{bmatrix}, \quad \begin{bmatrix} u_1 \\ v_1 \end{bmatrix} = \begin{bmatrix} p \\ 1 \end{bmatrix}, \quad \cdots, \quad \begin{bmatrix} u_{n+1} \\ v_{n+1} \end{bmatrix} = \begin{bmatrix} pu_n - qv_n \\ u_n \end{bmatrix}.$$

Then $\begin{bmatrix} u_n \\ v_n \end{bmatrix} = \begin{bmatrix} p & -q \\ 1 & 0 \end{bmatrix}^n \begin{bmatrix} 1 \\ 0 \end{bmatrix}$. So the first column of Q^n is $\begin{bmatrix} u_n \\ v_n \end{bmatrix}$ and the second is $\begin{bmatrix} -qu_{n-1} \\ -qv_{n-1} \end{bmatrix}$.

To get large powers of Q rapidly, we square Q repeatedly. Show that we arrive at Newton's method, i.e. $x_{2n} = (x_n^2 - q)/(2x_n - p)$. Apply the result to $x^2 - 4x - 1 = 0$ with matrix $Q = \begin{bmatrix} 4 & 1 \\ 1 & 0 \end{bmatrix}$.

11. In the cycle-finding algorithm of Fig. 27.2, replace the function f by

```
function f(u:real):real;
begin f:=1-2/u end;
```

In addition throw away the global variables a, m and redeclare the variables x, y, z, c, t to be reals. If you start with $x = 1$, then you will trace the nonperiodic sequence $x_1 = 1$, $x_{n+1} = 1-2/x_n$. Run the program until you detect a cycle, which will occur if there is no overflow or underflow, and your computer is very fast.

12. Why do the programs in Fig. 51.1 to 51.3 work for all $a > 0$, not just $a > 1$?

13. (Lenstra). Consider the sequence

$$x_0 = 1, \quad x_n = (1 + x_0^2 + x_1^2 + \cdots + x_{n-1}^2)/n \quad \text{for} \quad n = 1, 2, 3, \ldots.$$

This sequence is also generated by the formulas

$$x_0 = 1, \quad x_1 = 2, \quad x_n = \frac{x_{n-1}(x_{n-1} + n - 1)}{n} \quad \text{for} \quad n > 1.$$

We would like to know if all the terms x_n are integers.

a) Find $x_0, x_1, \ldots, x_9$. They are integers!

b) How do you find out if the 12-digit number x_9 is indeed an integer?

c) Using muMATH83, an early algebraic manipulation program, I computed $x_0, \ldots, x_{19}$ one by one, and they were integers. x_{19} had several thousand digits. The number of digits roughly doubles at each step. In trying to compute x_{20} the computer signalled: ALL SPACES EXHAUSTED. With newer programs, such as the programmable version of *Derive*, the members of the sequence can be generated automatically but memory is still exhausted soon after x_{20}.

In fact the sequence does not consist of integers only. Can you think of a way to find a term which is not an integer?

d) It turns out that x_n is an integer for $n \leq 42$, but not for $n = 43$. How could you prove this?

14. (Boyd and van der Poorten). Consider the sequence

$$x_0 = 1, \quad x_n = \frac{1}{n}(1 + x_0^3 + x_1^3 + \ldots + x_{n-1}^3), \quad n = 1, 2, 3, \ldots.$$

It seems to be even better than the sequence in Ex. 13; $x_2, x_3, \ldots, x_{89}$ seem to be integers. There are some doubts about powers of 2, 3, 5 and 7 in the denominator. Investigate as in 13.

15. *Superfast reciprocation.* Given $a > 0$; find z such that

(1) $$az = 1.$$

Here we replace the unknown constant z by a variable x. We try to drive the error y in the equation

(2) $$ax = 1 - y$$

to 0. We assume that we can find an initial value x such that the y in (2) satisfies $|y| < 1$. If we multiply x by $1 + y$ the error y in (2) is replaced by y^2. Thus we get quadratic convergence. Write a program fastrec based on this idea. (This trick can also be used for fast reciprocation of power series. In each step the number of correct coefficients doubles. It is Newton's method for power series.

52. The Natural Logarithm

The natural logarithm $\ln x$ is defined for $x > 0$ as the area under the hyperbola $xy = 1$ from 1 to x (Fig. 52.1).

If we stretch the hyperbola $y = 1/x$ horizontally by the factor h and compress it vertically by the

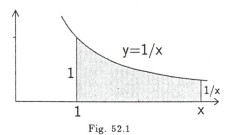

Fig. 52.1

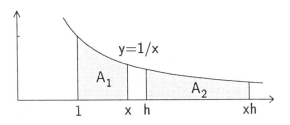

Fig. 52.1a

same factor, we get back to the curve $xy = 1$ since the product of the two coordinates of any point is unchanged. In Fig. 52.1a we see a region A_1 and the region A_2 resulting from it by stretching horizontally and compressing vertically by the factor h. The two regions have the same area because the compression cancels the effect of the stretching. In terms of our notation this means

$$(1) \qquad \ln x = \ln xh - \ln h \quad \text{or} \quad \ln xh = \ln x + \ln h.$$

This property of the hyperbola's area function, which we now regard as the basic property of a logarithm, was actually noticed decades before logarithms were invented.

We want to construct an algorithm which efficiently computes $\ln x$. The trapezoids in Fig. 52.2 have the same area $s(x) = (x - 1/x)/2$. If we apply this to the trapezoids in Fig. 52.3 we find that each has the area $s(\sqrt{x})$ and their sum is $2s(\sqrt{x})$.

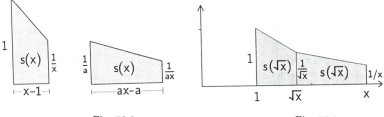

Fig. 52.2   Fig. 52.3

Starting with the first approximation $s(x)$ to $\ln x$ we repeatedly replace x by $\sqrt{x}$ and multiply by 2. After n iterations we approximate $\ln x$ by 2^n trapezoids of equal area. Thus we get the program cancel.

```
program cancel;
var i,n:integer; x,p,log:real;
function s(x:real):real;
  begin s:=(x-1/x) end;
begin
  write('x,n='); readln(x,n); p:=0.5;
  for i:=1 to n do begin
    x:=sqrt(x); p:=2*p; log:=p*s(x); writeln(log:20:12) end;
readln end.
```

0.70710678118	0.69314719178	0.69314193726	0.69140625000
0.69662139950	0.69314717874	0.69313812256	0.68750000000
0.69401475784	0.69314717501	0.69313049316	0.68570000000
0.69336401383	0.69314716756	0.69311523438	0.68750000000
0.69320138506	0.69314715266	0.69311523438	0.68750000000
0.69316073135	0.69314712286	0.69311523438	0.62500000000
0.69315056817	0.69314706326	0.69311523438	0.50000000000
0.69314802717	0.69314694405	0.69287109375	0.50000000000
0.69714739155	0.69314670563	0.69238281250	0.00000000000
0.69314723183	0.69314575195	0.69140625000	

Fig. 52.4. (Read the output vertically.)

But alas this naive approach leads to catastrophic cancellation. We have run the program for $x = 2$ and $n = 39$ and find that after 39 steps we arrive at zero instead of the correct value $\ln 2 = 0.69314718056$. In $x - 1/x$ both terms approach 1 if we repeatedly replace x by $\sqrt{x}$. The roundoff error in the individual terms becomes a larger and larger fraction of the difference, so the number of correct digits begins to decrease after a while. In the last computation apparently both terms were rounded to 1.

In the 11th output the error is less than 2 units in the 9th digit. At this stage we are approximating the area with 2^{11} trapezoids. We have taken square roots 11 times to compute the area of one trapezoid and then we multiplied by 2^{10}. This last step magnifies the roundoff error by 2^{10} but since Turbo Pascal uses 39 bit mantissas for reals, the effect of the roundoff error is still quite small at this stage. So the method can give reasonable results if used judiciously. However, we can do much better by rewriting our formulas to avoid the subtraction of nearly equal quantities. We introduce the function $c(x) = (x + 1/x)/2$. Then

$$c(x) = 2\big(c(\sqrt{x})\big)^2 - 1 , \qquad s(x) = 2s(\sqrt{x})\,c(\sqrt{x}) .$$

We used the letters c and s to emphasize the similarity to familiar trigonometric formulas. Solving these formulas for the functions of $\sqrt{x}$ gives

(2)
$$c(\sqrt{x}) = \sqrt{\frac{1+c(x)}{2}}\,, \quad 2s(\sqrt{x}) = \frac{s(x)}{c(\sqrt{x})}\,.$$

If we use this formula, the transition from $s(x)$ to $2s(\sqrt{x})$ is accomplished by dividing by the factor $c(x)$ which approaches 1. We stop the computation when $c(x)$ is within ϵ of 1. Program lniter puts this algorithm into effect. For eps=1E-10 and x = 2 it gives the output ln(x) = 0.69314718057. This is correct except for the last digit, which should be 6.

```
program lniter;
var eps,s,c,x:real;
begin
   write('eps,x='); readln(eps,x);
   s:=(x-1/x)/2; c:=(x+1/x)/2;
   repeat c:=sqrt((1+c)/2); s:=s/c
   until abs(c-1)<=eps;
   writeln('ln(x)=',s:20:11);
readln end.
```

Fig. 52.5

53. The Inverse Trigonometric Functions

Given the coordinates (c, s) of the point C on the unit circle in Fig. 53.1, we want to compute the angle $\alpha = \sphericalangle BOC$ without using trigonometric tables.

An angle can be measured in degrees by assigning to the half turn the measure $180°$. This is somewhat arbitrary and comes from Old Babylonian astronomy. Many computations in mathematics and science are simpler if one uses the *radian* measure, which is the length of the arc intercepted by the angle on the unit circle. Thus $180°$ is equal to π radians.

We shall find the radian measure of the angle

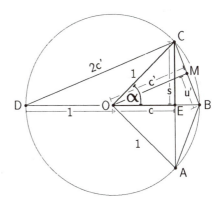

Fig. 53.1

$$\alpha = \arcsin s = \arccos c = \arctan \frac{s}{c} = \operatorname{arccot} \frac{c}{s}$$

by computing the length of the arc BC. Into the arc AC we inscribe 1, 2, 4, 8,...equal chords. Half of the sum of their lengths will converge to the length of the arc BC. We compute the formulas for one doubling and apply them as many times as necessary.

We start with the chord $AC = 2s$ which is our first approximation to the arc. On duplication we get two chords, each with distance c' from O and length $BC = 2u'$. We need to express c', u' in terms of c and s. The reason we denote MB by u' and not s' is that we wish to use s' to denote the length of the next polygonal approximation to the length of the arc BC, so that s' will denote $2u'$.

Since O and M are the midpoints of BD and BC, DC is parallel to OM and twice as long. Thus $CD = 2c'$. $\triangle CDE \sim \triangle BOM$, and so $2c'/(1+c) = 1/c'$, or

$$(1) \qquad\qquad c' = \sqrt{\frac{1+c}{2}}\,.$$

To find a relation between s and u' we express twice the area of the right triangle BCD in two ways and we get the equation $2c'2u' = 2s$ and hence

$$(2) \qquad\qquad 2u' = s' = s/c'\,.$$

Our formula (1) tells us that when we double the number of sides in our polygon, the distance of the sides from the center changes from c to $c' = \sqrt{(1+c)/2}$ and the length of the polygon is multiplied by $1/c'$. So to compute $\arccos c$, we start with c and $s = \sqrt{1-c^2}$. Then we compute c' from (1) and $s' = s/c'$ from (2) repeatedly. The quantity c' approaches 1 and we stop when it is within ϵ of 1. Program arcit implements the algorithm; it stops when c differs from 1 by at most ϵ.

```
program arcit; var eps,c,s:real;
begin write('eps,c='); readln(eps,c); s:=sqrt(1-c*c);
  repeat c:=sqrt((1+c)/2); s:=s/c until abs(1-c)<=eps;
writeln('arc=',s:12:10,' 2s=',2*s:12:10); readln end.
```

Fig. 53.2

Once the values of $1-c$ are small, they are multiplied by a factor $\approx 1/4$ as can easily be seen by computing $(1-c')/(1-c)$. For input $c = 0$ we get $s = \pi/2$. As a check we also printed $2s$ which is π for this input.

The idea of this section is due to Archimedes, the greatest mathematician of antiquity. His aim was to compute π, i.e. the circumference of the circle. The inscribed polygons give lower bounds for π. He also gave a method for computing circumscribed regular polygons, to furnish upper bounds for π. We adapted his approach to arcs and left out the circumscribed polygons. See the next Exercise.

We hope the reader is pleased and surprised by this simple and rapid way of computing the inverse trigonometric functions. What is even more surprising is that the formulas (2) and (3) for computing $\arccos c$ are the same as the formulas (2) in Section 52 which we used to compute $\ln x$. The only difference is that the starting values of c and s in that computation are calculated in a different way; in particular, there they satisfy the relation $c^2 - s^2 = 1$ whereas in the trigonometric computation we have $c^2 + s^2 = 1$. If one examines further the implications of the fact that the same algorithm yields both the natural logarithm and the inverse cosine, one arrives at Euler's formula $e^{i\alpha} = \cos\alpha + i\sin\alpha$ which is usually obtained by looking at the Taylor series of these functions.

Exercise:

1. We can introduce circumscribed polygons almost without additional computation. In Fig. 53.1 we draw the tangent at B. Let F be its intersection with OC. Let $t = BF$. Then $t : 1 = s : c$, or $t = s/c$. Write a program which computes both lower and upper bounds for α, using inscribed and circumscribed polygons.

54. The Function exp

Let exp denote the inverse of the function $\ln$, i.e. $x = \exp y \iff y = \ln x$. We rewrite the functional equation $\ln xh = \ln x + \ln h$ of Section 52 in terms of exp. Let $u = \ln x$, $v = \ln h$. If we apply the function exp to both sides of our equation we get

$$(1) \qquad \exp(u + v) = \exp u \, \exp v \,.$$

In particular, we get the duplication formula

$$(2) \qquad \exp 2x = (\exp x)^2 \,.$$

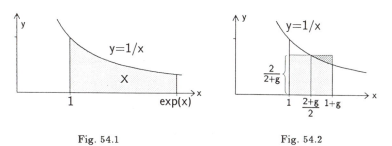

Fig. 54.1   Fig. 54.2

We can compute $\exp(x)$ directly from its definition in Fig. 54.1. We first obtain an approximation to $\exp x$ when x is close to 0. Then $\exp x$ is close to 1. We shall be able to represent numbers of interest more accurately if we work with $g(x) = \exp x - 1$.

By definition, the area under the hyperbola between 1 and $g(x)$ is x. We see from Fig. 54.2 that when x and hence $g(x)$ are small, the area under the curve is well approximated by the "midpoint rule":

$$(3) \qquad \frac{2g(x)}{2 + g(x)} \approx x \quad \Rightarrow \quad g(x) \approx \frac{2x}{2 - x}.$$

The relative error in $g(x)$ is as small as we like if x is sufficiently small. We will say more about this error in section 57.

We go from small values of x to larger ones by means of the duplication formula for $g(x)$ which we get from (2):

$$(4) \qquad g(2x) = g(x)(2 + g(x)).$$

Note that if we know $g(x)$ with a certain accuracy and $|g(x)|$ is small, the relative error in $2 + g(x)$ is much smaller than the relative error in $g(x)$. So if we compute $g(2x)$ using (4), the relative error is only a little greater than the relative error in $g(x)$. So we can divide x by a power of 2 to make the quotient so small that the relative error in (3) is very small. Then we can get back to x by means of the duplication formula (4) without increasing the relative error too much. The program **expo** implements this idea; n is the number of times we want to divide x by 2 before applying (3). With $n = 16$, $x = 1$ we get $1 + g = 2.7182818285$. All 11 digits are correct.

```
program expo;
var i,n:integer; x,g,p:real;
begin write('n,x'); readln(n,x);
    p:=1; for i:=1 to n do p:=2*p;
    x:=x/p; g:=2*x/(2-x);
    for i:=1 to n do g:=g*(2+g);
writeln(1+g:20:10); readln;end.
```

Fig. 54.3

55. The Cosine

Here the situation is similar to that of the function exp. If we try to compute $\cos x$ by means of an obvious algorithm then we are punished by severe cancellation errors. We first try a naive approach and use two obvious ideas.

a) If I know $c(x) = \cos x$ for the arc x, then I can easily find $c'(x) = c(2x)$ for the arc $2x$. Indeed, in Fig. 55.1, $(1+c')/2c = 2c/2$, so $4c^2 = 2(1+c') \Rightarrow c' = 2c^2 - 1$; thus

$$(1) \qquad c(2x) = 2c^2(x) - 1 \qquad \text{(duplication formula for } c(x) = \cos x\text{)}.$$

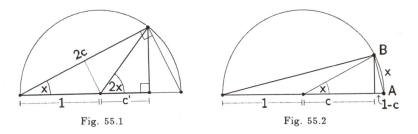

Fig. 55.1   Fig. 55.2

b) For sufficiently small x we know $c(x)$ with any precision we want. Indeed, in Fig. 55.2, $(1-c)/AB = AB/2$ or $1-c = \frac{1}{2}AB^2$. For small x, the percentage difference between the arc x and the line segment AB is very small, hence

$$(2) \qquad (1-c) \sim \frac{1}{2}x^2 \quad \Rightarrow \quad c(x) \approx 1 - \frac{1}{2}x^2.$$

To compute $c(x)$ we find the cosine of $x/2^n$ by means of (2) and apply the duplication formula (1) n times.

It turns out that errors grow rapidly in this calculation and it is not difficult to find the reason. Suppose that at some stage the quantity we have computed is $c + \epsilon$ where c is the correct value. One more application of the duplication formula gives

$$c' + \epsilon' = 2(c + \epsilon)^2 - 1 + \text{ new roundoff error}$$
$$= 2c^2 - 1 + 4c\epsilon + 2\epsilon^2 + \text{new roundoff error.}$$

Since c is close to 1 through most of the computation, the error is multiplied by about 4 at each step. Nevertheless, the method is usable, because when x is small, the error in (2) is only about $-x^4/24$. (This error estimate follows from Taylor's formula, which can be found in calculus books.) We need to calculate cosines only for acute angles. So $(s-1)/4$ divisions of our original x by 2 will produce a value for which (2) gives the cosine with an accuracy of about s binary digits. We then apply the duplication formula $(s-1)/4$ times and lose about $(s-1)/2$ bits. The final result will have about $s/2$ correct bits.

Since $c(x)$ is close to 1 through most of the computation, the difference $f(x) = c(x) - 1$ is much smaller and can therefore be represented with a much smaller error in floating point arithmetic. We get from (1)

$$(3) \qquad f(2x) = 2f(x)(2 + f(x)).$$

We can see in the same way as before that the error increases by a factor of about 4 each time we apply (3), but this time the *relative* error does not change much as long as $f(x)$ is small, so that there is no great loss of significant digits.

How many times should we divide x by 2 before we apply (2) and start the duplications? In terms of $f(x)$ the Taylor remainder formula we mentioned above is

$$(4) \qquad \frac{f(x)}{\frac{1}{2}x^2} \approx 1 - \frac{x^2}{12} \quad \text{for small } |x|.$$

If we are using a mantissa of s bits, the relative error due to just one rounding can be about 2^{-s}. By (4) the relative error we introduce when we approximate $f(x)$ by $-\frac{1}{2}x^2$ is about $-\frac{1}{12}x^2$. This will be no more than a single roundoff error if $\frac{1}{12}x^2 < 2^{-s}$. So no further gain of accuracy can be expected if we continue halving after we have $|x| < \sqrt{12/2^s}$. With this method we would not lose much accuracy if we performed more halvings than necessary.

Program cos implements the algorithm. For more convenient use it also performs a conversion from degrees to radians.

```
program cos;
var x,f,eps,e:real; n,p:integer;
begin write('x(degrees), eps = '); read(x,eps);
    x:=x*pi/180; e:=sqrt(12*eps);
    p:=0; while abs(x)>e do begin x:=x/2; p:=p+1 end;
    f:=-x*x/2; for n:= 1 to p do f:=2*f*(2+f);
writeln(' cos x=',1+f:15:12); readln  end.
```

Fig. 55.3

56. Archimedes' Integration of the Parabola

A great variety of problems in the exact sciences can be reformulated as the determination of an area or *integral* as it is called in calculus. *The integral of the function f from a to b, denoted $\int_a^b f(x)\,dx$, is the area bounded by the x-axis, the curve and the verticals $x = a$ and $x = b$.* Areas under the x-axis count as negative in this work. This makes the formulas in what follows valid irrespective of the sign of $f(x)$. $\int_a^b f(x)\,dx$ is defined even if $b < a$, as $-\int_b^a f(x)\,dx$.

In the following we shall say "parabola" when we should really say "graph of a polynomial of degree ≤ 2". (Such a graph could be a straight line; also, a parabola whose axis is not parallel to the y-axis is not the graph of a quadratic polynomial.) Archimedes determined the area under an arc of a parabola exactly. We need his result and reproduce part of his argument.

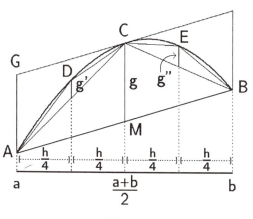

Fig. 56.1

In Fig. 56.1, let P be the area of the parabolic segment cut off by the chord AB, and let Δ_1 be the area of the triangle ABC. A and B are at equal distances from the vertical through C. We want to show that

$$(1) \qquad P = \frac{4}{3}\Delta_1.$$

We see from Fig. 56.1 that Δ_1 is the sum of the areas of the triangles CMA and CMB. These have a common base of length g. The sum of their heights is $h = b - a$. Hence $\Delta_1 = gh/2$. Let the equation of the parabola be $y = c_0 + c_1 x + c_2 x^2$. A little computation shows that the distance g from M to C is

$$(2) \qquad g = |c_2|\left(\frac{h}{2}\right)^2.$$

Next we construct two smaller triangles ADC and CEB whose bases are chords of the parabola and whose vertices lie on the parabola vertically above the midpoints of the chords. It follows as above that the lengths g' and g'' satisfy

$$g' = g'' = |c_2|\left(\frac{h}{4}\right)^2 = \frac{g}{4} .$$

The two triangles ADC and CEB can be decomposed into four triangles with bases g', g'' of length $g/4$ and the sum of the altitudes is again h. Thus the sum Δ_2 of their areas is $\Delta_1/4$. We continue halving the subintervals. At the nth step we add 2^n new triangles whose areas add up to $\frac{1}{4}\Delta_{n-1}$. The resulting polygons fill out more and more of the parabolic segment. Thus

$$(3) \qquad P = \Delta_1 + \frac{\Delta_1}{4} + \frac{\Delta_1}{16} + \cdots.$$

Summing this infinite geometric series gives (1).

Archimedes also proved that the tangent at C is parallel to the chord AB. This has many consequences for the geometry of the parabola but we shall not need to use it.

57. Numerical Integration

Next we look at some approximate formulas for the area A under a curve $y = f(x)$ between $x = a$ and $x = b$.

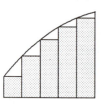

a) *Rectangular Approximation.* We can approximate A by rectangles whose height is the function value at the left endpoint of the base, see Fig. 57.1. This is clearly not the best choice for the heights but it will give a particularly transparent illustration of the extrapolation to the limit which we will discuss in the next section.

Fig. 57.1

Let R_n be the sum of the areas of the rectangles when we use n rectangles; for definiteness assume their widths are equal. Write

$$(1) \qquad R_n = A + e_n^R$$

where e_n^R is the error in the approximation by n rectangles. Cut each subdivision in half. Then

$$(2) \qquad R_{2n} = A + e_{2n}^R.$$

For large n there is a simple relation between e_n^R and e_{2n}^R. If we take a sufficiently small piece of any smooth curve and enlarge it, then the piece will look more and more like a straight line segment because the direction does not change much from one end of a short piece to the other, and direction remains the same under enlargement.

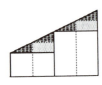

Fig. 57.2 shows that, for straight lines, $e_{2n}^R = e_n^R/2$ exactly. For any smooth function we have

$$(3) \qquad e_{2n}^R \approx e_n^R/2.$$

Fig. 57.2

If $f(x)$ is increasing over part of the interval and decreasing over the rest of it, then the errors in the increasing part will be negative and the ones in the decreasing part will be positive. It is possible that these will largely cancel out. After the subdivision the individual errors will be replaced by two errors whose sum is about half of the previous error but it could be that when we add all these, the cancellation is less exact than before, and (3) may not be valid.

Our aim in this section and the next will be to derive formulas in which the main error term is eliminated. The validity of that computation is not affected by the fact that for some special instances of f, a and b the main error term has the value 0.

We shall compute more and more accurate approximations to A by bisecting the intervals we already have. We shall refer to doubling the number of intervals as one step in the approximation process. Note that one of these steps involves evaluating and summing n new function values

so that the amount of work doubles with each "step". (3) tells us that
the error is multiplied by a factor of about $1/2$ in each step. We say the
rectangular approximation converges with linear rate $1/2$.[†]

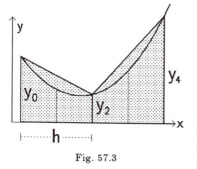

Fig. 57.3

b) *Trapezoidal approximation.* We
approximate the area by a sum of trapezoids as shown in Fig. 57.3. The formula expressing this is

$$T_n = h\left(\frac{1}{2}y_0 + y_2 + y_4 + \cdots + \frac{1}{2}y_{2n}\right),$$

where $y_0, y_1, y_2, \cdots, y_{2n}$ are the y-values
at intervals of $\frac{1}{2}h$, starting from the left
endpoint. (The notation is designed to
make it easy to compare and combine
the trapezoidal approximation with the midpoint approximation, discussed below.)

We observed already that if we enlarge a small piece of a smooth curve
we get something close to a straight line segment, because the direction
of a curve does not change much over a short distance. However, it does
change a little. For most parts of most curves given by simple formulas,
the *curvature* changes continuously, i.e. it changes little over a short
stretch. That is why a parabola through 3 closely spaced points of a curve
is extremely close to the curve along that arc. More exact information
about the difference can be obtained from Taylor's theorem. (It might
seem more natural to approximate a segment of a curve by an arc of a
circle. That would be just as close but a parabola has a simpler equation.)

We are ready to consider how the error in the trapezoidal rule changes
when we double the number of trapezoids. Let T_n be the area under the
polygon when we have n trapezoids. Write

$$(4) \qquad\qquad T_n = A + e_n^T.$$

Our convention is that the area between a chord and the curve is negative
if the curve is below the chord. Hence $e_n^T = -$(sum of the areas between
the sides of the polygon and the curve). Similarly, we write

$$(5) \qquad\qquad T_{2n} = A + e_{2n}^T.$$

In case the curve is a parabola, Fig. 56.1 and Archimedes' Theorem enable
us to say exactly how e_n^T changes when we subdivide the interval. The

---

[†] Our meaning of "linear" in this context is different from what is meant by an
algorithm of *linear time complexity* in computer science. The time complexity
of our algorithm is exponential since each bit of the result requires twice as
much work as the previous one.

difference between the area P bounded by the whole arc and its chord, and its two halves and their chords is Δ_1. From Sect. 56, equ. (3) we see that this means the sum of the two small areas is $\frac{1}{4}$ of the large one. So we conclude that for the trapezoidal rule

$$(6) \qquad\qquad\qquad e_{2n}^T \approx \frac{1}{4} e_n^T.$$

We say that the trapezoidal rule converges linearly with rate $\frac{1}{4}$.

If the curve is convex down over part of the interval (a, b) and concave over the rest, (6) may not hold for the reason explained in small print in a), but again we need not be concerned about this.

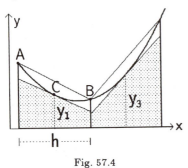

Fig. 57.4

c) *Midpoint Approximation* We get an approximation by rectangles which is clearly better than the one in a) if we take the height of each rectangle to be the function value for the midpoint of the base. The following formula expresses this:

$$M_n = h(y_1 + y_3 + y_5 + \cdots),$$

where we used the same notation as in the Trapezoidal rule. The difference between the midpoint approximation and the area under the curve can be seen in Fig. 57.4. We have not drawn the rectangles with the y-value at the midpoint as height. Instead , we have drawn trapezoids bounded on top by the tangents at the midpoints. These have the same area as the corresponding rectangles, and the difference between the area under the curve and the area of a trapezoid is very easy to see. With our sign convention, the error is $-$ the area between the tangent and the curve, since the error is negative when the curve is above the tangent and the area is positive then.

If the curve is a parabola then Archimedes' Theorem again enables us to determine the error exactly. For a parabola the area between the chord AB and the curve is $\frac{4}{3}\Delta_1$ where Δ_1 is the area of triangle ABC. The area bounded by the chord AB, the tangent at C and the verticals through A and B is $2\Delta_1$. Thus we get:

Two-thirds of the area of the quadrilateral bounded by the chord AB, the verticals at A and B and the tangent at C is between the parabola and the chord and one-third is between the parabola and the tangent.

We conclude that for a smooth curve the error in the midpoint rule is approximately half the error in the trapezoidal rule and it has the opposite sign. In particular, the midpoint rule also converges linearly, with convergence factor $\frac{1}{4}$.

d) *Simpson's rule.* What we have learned about the errors in the midpoint and the trapezoidal rule tells us that if we form the weighted mean of the midpoint and the trapezoidal approximations with weights $\frac{2}{3}$ and $\frac{1}{3}$, we get an approximation which is exact for arcs of parabolas. This approximation is *Simpson's rule.* The Simpson approximation for the area over an interval of length h is $\frac{h}{6}(y_0 + 4y_1 + y_2)$. For n equal intervals the formula becomes

$$(7) \quad S_n = \frac{h}{6}(y_0 + 4y_1 + 2y_2 + 4y_3 + 2y_4 + \cdots + 2y_{2n-2} + 4y_{2n-1} + y_{2n}).$$

Note that here h denotes the width of the intervals corresponding to the trapezoidal and midpoint approximations from which we derived Simpson's formula; if it denoted the difference in the abscissas of consecutive function values then the denominator in front would be 3.

If a curve is smooth it is approximated much better by arcs of parabolas through triplets of equally spaced points than it is by straight line segments on which the trapezoidal and midpoint methods are based. Simpson's rule assigns the weights $\frac{2}{3}$ and $\frac{4}{3}$ to consecutive y-values. From a naive point of view it seems unreasonable to assign to, say, the 7th function value twice the weight of the 6th. Why would one value in the middle of the range be twice as important as the next one? Indeed, if the curve is not smooth and the above argument does not apply, e.g. if the function represents daily temperature data, Simpson's rule will probably be less accurate than the trapezoidal rule. However, we shall see that for smooth curves it is very good.

Program `tramisim1` computes π by computing 4 times the area under the positive quadrant of the unit circle. The input `eps` is used in the stopping criterion. When the change in the value `simp` given by Simpson's

```
program tramisim1;
var a,b,eps,h,trap,mid,simp,simpold,x:real;
function f(x:real):real; begin f:=4*sqrt(1-x*x) end;
begin write('a,b,eps= ');readln(a,b,eps);
   writeln('trapezoid':13,'midpoint':15,'simpson':15);writeln;
   h:=b-a; trap:=(f(a)+f(b))*h; mid:=0;
   repeat trap:=(trap+mid)/2;
      mid:=0; x:=a+h/2;
      while x<=b do begin mid:=mid+f(x); x:=x+h end;
      mid:=mid*h; simpold:=simp; simp:=(trap+2*mid)/3;
      writeln(trap:15:10, mid:15:10, simp:15:10);
      h:=h/2
   until abs((simp-simpold)/simp)<eps;
readln; end.
```

Fig. 57.5

rule is less than eps times simp we stop. The accuracy of single-precision computation with Turbo Pascal is good enough to come to a conclusion with a much smaller value of eps than the one we input in our sample computation, but that would have produced a longer output than we need to see what happens. (It would also have taken a long time to compute.) With input $a = 0$, $b = 1$, $eps = 0.00001$ we get the output

trapezoid	midpoint	simpson
2.0000000000	3.4641016151	2.9760677434
2.7320508076	3.2593673286	3.0835951549
2.9957090681	3.1839292206	3.1211891698
3.0898191444	3.1566869313	3.1343976690
3.1232530378	3.1469518079	3.1390522179
3.1351024229	3.1434914027	3.1406950761
3.1392969128	3.1422646720	3.1412754189
3.1407807924	3.1418303735	3.1414805131
3.1413055830	3.1416767225	3.1415530093
3.1414911527	3.1416223804	3.1415786378

We are disappointed; Simpson's rule is not that much more accurate than the other two. The reason is this: We argued that Simpson's rule is particularly good because it integrates arcs of parabolas exactly, and most curves can be approximated very closely with arcs of parabolas. As $x \to 1$, the tangent of the circular arc approaches the vertical. The graph of a quadratic function is nowhere vertical and therefore a quadratic function can not approximate the circle near $x = 1$ very closely.

We can calculate the area of the circle without approaching the vertical part by calculating 12 times the area of a 30° segment of the unit circle. One such segment is the area bounded by the y-axis, the arc of the unit circle between $x = 0$ and $x = 0.5$ and the line $y = \sqrt{3}x$. The following output was obtained using program tramisim1, but with the function definition replaced by

```
begin f:=12*(sqrt(1-x*x)-sqrt(3)*x) end;
```

and input $a = 0$, $b = 0.5$ and eps = 1E-9:

trapezoid	midpoint	simpson
3.0000000000	3.2113988080	3.1409325386
3.1056994040	3.1594684463	3.1415454322
3.1325839251	3.1460923941	3.1415895711
3.1393381596	3.1427196081	3.1415924586
3.1410288839	3.1418745201	3.1415926414
3.1414517020	3.1416631282	3.1415926528
3.1415574151	3.1416102728	3.1415926535

This time Simpson's rule is much more accurate than the other two. The error decreases by a factor 1/16 at each step. In textbooks on numerical analysis it is proved that Simpson's rule converges with that rate for sufficiently smooth curves which do not have vertical tangents.

58. Extrapolation to the Limit and Romberg Integration

The integration methods of the previous section, when applied to sufficiently smooth functions, all had the following property: after a few steps, halving the intervals approximately multiplied the error by a factor which did not depend on the function being integrated. Now if the error were multiplied by, say, exactly $1/2$ in each step, then one could calculate the error and hence the exact value of the integral from the result of two consecutive steps. (If the factor corresponding to each step is unknown but exactly the same in each step then the limit can be computed from three steps.) So instead of doing more and more work (remember each step involves twice as much computation as the previous one) to obtain more and more predictable results, we can attempt to predict the limit of the sequence. We call this procedure *extrapolation to the limit*. It was first proposed by Alexander Craig Aitken for iterations in general and applied to quadrature problems by Lewis Fry Richardson. Let us apply it to some of the methods we gave before.

a) *The rectangular rule.* From the formulas (1), (2) and (3) of the previous section we find that extrapolating to the limit gives

$$A \approx 2R_{2n} - R_n = h(y_1 + y_3 + y_5 + \cdots + y_{2n-1}).$$

Here $h = (b - a)/n$. This is the midpoint rule approximation M_n.

b) *The trapezoidal rule.* For this rule the error is about 4 times smaller when we halve the intervals. Extrapolation to the limit gives

$$A \approx \frac{4T_{2n} - T_n}{3} = \frac{h}{6}(y_0 + 4y_1 + 2y_2 + 4y_3 + \cdots + 4y_{2n-1} + y_{2n}),$$

which is just the Simpson's rule formula S_n.

The formulas we got so far by extrapolation to the limit are familiar and very good. If we extrapolate the midpoint rule to the limit, we get $A \approx (4M_{2n} - M_n)/3$. The latter expression can be written as $2S_{2n} - S_n$. This formula is less accurate than Simpson's rule with the same number of meshpoints. It would be too much to expect that extrapolation to the limit will always deliver the best possible formula of its kind.

We next present a systematic process which starts with the trapezoidal method and repeatedly extrapolates to the limit. We want to find $A = \int_a^b f(x)\,dx$, i. e. the area under $y = f(x)$ between $x = a$ and $x = b$. Let $T_{00} = \frac{h}{2}(f(a) + f(b))$. Repeated bisection of the interval $[a, b]$ gives our first sequence

(1) $T_{00}, T_{10}, T_{20}, \cdots$

with limit A. Note that now the first index counts the number of times we have subdivided, while in our previous notation it denoted the number of subintervals. In the present notation, the number of subintervals is 2^n.

We are now going to sketch the background of the Romberg method, for readers who know calculus. We begin by stating the Euler-MacLaurin summation formula (see, e.g., Handbook of Mathematical Functions, edited by Milton Abramowitz and Irene Stegun, p. 806). This interesting formula can be regarded as an expression for the error in the trapezoidal rule for a function $f(x)$ which has continuous derivatives up to order $2k+2$ in the interval (a, b). It says that

$$(2) \qquad A = T_{n0} + a_1 4^{-n} + a_2 \left(4^{-2}\right)^n + \cdots + a_k \left(4^{-k}\right)^n + e_{k+1}(n)$$

where

$$(3) \qquad |e_{k+1}(n)| \le c_{k+1}(4^{-(k+1)})^n,$$

and the a_i and c_{k+1} do not depend on n. We mention that if $f(x)$ is a polynomial of degree $\le 2k$, then $e_{k+1}(n) = 0$. This means that for such polynomials we can find the exact area using a single trapezoid and the coefficients a_i.

We write down the formula for the a_i, although for our purposes it suffices to know that they do not depend on n:

$$a_i = \alpha_i (b - a)^{2i} (f^{(2i-1)}(b) - f^{(2i-1)}(a)).$$

Here $\alpha_1 = -1/12$, $\alpha_2 = 1/720$, $\alpha_3 = -1/30240, \ldots$ are, of all things, the coefficients of the odd powers in the Taylor series of $\frac{1}{2}\cot\frac{x}{2} - \frac{1}{x}$, with every second sign reversed.

We needed to refer to derivatives to explain the origin of (2), but from now on we just assume that for each value of k formula (2) holds for the function and the interval (a, b) we have chosen. No familiarity with calculus is needed to follow the rest of our procedure.

We first compute terms of the sequence (1). They differ from A by the error given by (2). We form a new sequence in which the largest term in the error, the term with 4^{-n}, is eliminated:

$$T_{n,1} = T_{n+1,0} + \frac{T_{n+1,0} - T_{n0}}{3} = A + a * 4^{-2n} + \cdots.$$

We form additional sequences, each time eliminating the leading term in the error of the preceding sequence:

$$T_{n2} = T_{n+1,1} + \frac{T_{n+1,1} - T_{n1}}{15} = A + b * 4^{-3n} + \cdots,$$

$$T_{n3} = T_{n+1,2} + \frac{T_{n+1,2} - T_{n2}}{63} = A + c * 4^{-4n} + \cdots,$$

etc.

T_{00}	T_{10}	T_{20}	T_{30}	T_{40}	T_{50}	$\cdots$
T_{01}	T_{11}	T_{21}	T_{31}	T_{41}	$\vdots$	$\vdots$
T_{02}	T_{12}	T_{22}	T_{31}	$\vdots$	$\vdots$	$\vdots$
T_{03}	T_{13}	T_{23}	$\vdots$	$\vdots$	$\vdots$	$\vdots$
T_{04}	T_{14}	$\vdots$	$\vdots$	$\vdots$	$\vdots$	$\vdots$
T_{05}	$\vdots$	$\vdots$	$\vdots$	$\vdots$	$\vdots$	$\vdots$

We represent the quantities which can be computed from, say, the first 6 entries of (1) as a triangular array. The first row converges with factor $1/4$. Each succeeding row converges four times faster than the preceding one. This is the Romberg method, one of the best numerical integration methods. Fig. 58.2 implements it. The number of elements of the first row it computes is the input max. Instead of using such an input, we could have programmed it to compute larger and larger triangles until the results no longer change significantly. The function we put in the program is the one we used in the last section to compute π by taking 12 times the area of a $30°$ sector of a unit circle.

```
program romberg;
var a,b,h,m,x:real;
    t:array[-1..10,-1..10] of real; i,j,c,max:integer;
function f(x:real):real;
begin f:=12*(sqrt(1-x*x)-sqrt(3)*x) end;
begin write( 'a,b,max=' ); readln(a,b,max);
    for i:=-1 to 10 do
    for j:=-1 to 10 do t[i,j]:=0;
    i:=0; j:=0; h:=b-a; m:=0;
    t[i-1,0]:=(f(a)+f(b))*h;
    repeat
        t[i,0]:=(t[i-1,0]+m)/2;
        m:=0; x:=a+h/2;
        while x<=b do begin m:=m+f(x); x:=x+h end;
        m:=m*h; i:=i+1; h:=h/2
    until i>max; c:=0;
    for j:=1 to  max do
    begin c:=4*c+3;
        for i:=0 to max-j do
        begin
            t[i,j]:=t[i+1,j-1]+(t[i+1,j-1]-t[i,j-1])/c
        end
    end;
    for j:=0 to max do begin
        for i:=0 to max-j do write(t[i,j]:16:10);
        writeln end;
    readln end.
```

Fig. 58.2

With $a = 0$, $b = 0.5$, max $= 5$ we get

3.0000000000 3.1056994040 3.1325839251 3.1393381596 3.1410288839 3.1414517020
3.1409325386 3.1415454322 3.1415895711 3.1415924586 3.1415926414
3.1415862917 3.1415925137 3.1415926511 3.1415926535
3.1415926125 3.1415926533 3.1415926536
3.1415926535 3.1415926536
3.1415926536

The result is excellent indeed; the next to the last entry in the first column is based on only 17 function values and is correct to practically 11 significant digits.

For comparison, we try to compute π by computing 4 times the area under a quarter circle by replacing the function definition in program **romberg** by `f:=4*sqrt(1-x*x)`. With input $a = 0$, $b = 1$, max $= 5$ we get

2.0000000000 2.7320508076 2.9957090681 3.0898191444 3.1232530378 3.1351024229
2.9760677434 3.0835951549 3.1211891698 3.1343976690 3.1390522179
3.0907636490 3.1236954374 3.1352782356 3.1393625212
3.1242181642 3.1354620895 3.1394273511
3.1355061834 3.1394429011
3.1394467493

Our sophisticated processing of the data produces little improvement here because the vertical tangent at the end violates the assumptions which justify the method.

Finally, we apply Romberg's method to the computation of $\ln x$ discussed in Section 52. There we divided the interval $[1, x]$ into 2^n subintervals by points which formed a geometric progression. The sum of the areas of the trapezoids with these bases was $T_{n0} = 2^{n-1}(x^{1/2^n} - x^{-1/2^n})$. This is not quite the same situation for which we derived Romberg's method, because the bases of the trapezoids in this procedure are not equal. Nevertheless, let us see what Romberg's method gives.

We modify program **romberg** by adding p to the list of variables of type **real**. We replace the part of the program which computes the first line of the triangular array (the lines from `i:=0;`...to **until**..) by the following lines:

```
i:=0;  j:=0;  p:=0.5;  x:=b;
repeat t[i,0]:=p*(x-1/x);  i:=i+1;  p:=2*p;  x:=sqrt(x);
until i>max;  c:=0;
```

To see what Romberg's method can do, we use **extended** instead of **real** variables in the computation, so that roundoff errors should not spoil our results, and we print out more digits. With $a = 1$, $b = 2$, max $= 4$ we get

0.750000000000000   0.707106781186548   0.696621399498013   0.694014757842346
                                                                          0.693364013830721
0.692809041582063   0.693126272268501   0.693145877290457   0.693147099160180
0.693147420980931   0.693147184291920   0.693147180618162
0.693147180534952   0.693147180559848
0.693147180559946

Fig. 58.3 Romberg's scheme applied to the method of Section 52.

Note that if one goes only up to $n = 3$ (8 intervals) the trapezoidal approximation (first row) is wrong in the third digit already, but Romberg's method gives 10 correct digits! The number in the bottom row is completely correct; the last digit is rounded up. Using only 17 function values, Romberg's method manages to give here an accuracy which corresponds to determining an area the size of a city to within a square millimeter. This seems to be exceptional (see the exercise).

Exercise:

Compute $\ln 2$ by means of the regular Romberg method, i. e. using trapezoids of equal widths, and compare the results with Fig. 58.3.

59. One Thousand Decimals of e

If we truncate the series for e after the n-th term, we get

$$e_n = 1 + \frac{1}{1!} + \frac{1}{2!} + \cdots + \frac{1}{n!}$$

with an error

$$
\begin{aligned}
f_n = e - e_n &= \frac{1}{(n+1)!} + \frac{1}{(n+2)!} + \cdots \\
&= \frac{1}{(n+1)!}\Big(1 + \frac{1}{n+2} + \frac{1}{(n+2)(n+3)} + \cdots\Big) \\
&< \frac{1}{(n+1)!}\Big(1 + \frac{1}{n+2} + \frac{1}{(n+2)^2} + \cdots\Big) = \frac{1}{(n+1)!}\,\frac{1}{1 - \frac{1}{n+2}} \\
&= \frac{1}{(n+1)!}\,\frac{n+2}{n+1}\,.
\end{aligned}
$$

Since $f_{450} < 452/(451!\,451) \approx 1.3\,10^{-1003}$, 450 terms are sufficient for an accuracy of 1000 decimals. Using $n = 450$ the following algorithm will produce 1000 decimals of e:

```
e:=1; while n>0 DO begin e:=1+e/n; n:=n-1 end;
```

So we set $d[0]:=1$. Next we divide by 450 with the grade school method getting $1/450=0.00222\ldots2$. These digits are stored in $d[0..1000]$. Then we set d[0]:=d[0]+1. Now we divide by 449. That is, at the i-th step we divide d[i] by 449, the integer part of the quotient q goes to d[i], and the remainder r is multiplied by 10 and added to d[i+1]. This is repeated until we reach $n = 0$. Now we do the carries, starting at the end. That is u:= d[i] div 10 is added to d[i-1] and d[i] is reduced mod 10. Here is the complete program:

```
program exp1;
const a=450; b=1000; bplus1=1001;
var i,n,q,r,u:integer; d:array[0..bplus1] of integer;
begin d[0]:=1;
  for i:=1 to b do d[i]:=0;
  for n:=a downto 1 do begin
    for i:=0 to b do begin
      q:=d[i] div n; r:=d[i] mod n;
      d[i]:=q; d[i+1]:=d[i+1]+10*r
    end;
    d[0]:=d[0]+1
  end;
  for i:=b downto 1 do begin
    u:=d[i] div 10; d[i]:=d[i] mod 10;
    d[i-1]:=d[i-1]+u
  end;
  for i:=0 to b do begin
    write(d[i]);
    if i mod 6=0 then write(' '); if i mod 60=0 then writeln
  end
end.
```

2.71828	18284	59045	23536	02874	71352	66249	77572	47093	69995	95749	66967
62772	40766	30353	54759	45713	82178	52516	64274	27466	39193	20030	59921
81741	35966	29043	57290	03342	95260	59563	07381	32328	62794	34907	63233
82988	07531	95251	01901	15738	34187	93070	21540	89149	93488	41675	09244
76146	06680	82264	80016	84774	11853	74234	54424	37107	53907	77449	92069
55170	27618	38606	26133	13845	83000	75204	49338	26560	29760	67371	13200
70932	87091	27443	74704	72306	96977	20931	01416	92836	81902	55151	08657
46377	21112	52389	78442	50569	53696	77078	54499	69967	94686	44549	05987
93163	68892	30098	79312	77361	78215	42499	92295	76351	48220	82698	95193
66803	31825	28869	39849	64651	05820	93923	98294	88793	32036	25094	43117
30123	81970	68416	14039	70198	37679	32068	32823	76464	80429	53118	02328
78250	98194	55815	30175	67173	61332	06981	12509	96181	88159	30416	90351
59888	85193	45807	27386	67385	89422	87922	84998	92086	80582	57492	79610
48419	84443	63463	24496	84875	60233	62482	70419	78623	20900	21609	90235
30436	99418	49146	31409	34317	38143	64054	62531	52096	18369	08887	07016
76839	64243	78140	59271	45635	49061	30310	72085	10383	75051	01157	47704
17189	86106	87396	96552	12671	54688	95703	50354				

Fig. 59.1. One Thousand Decimals of e.

We next discuss ways to compute π to many decimals. This is harder to program than the computation of e. Since 1976 we know a quadratically convergent algorithm for π (Salamin-Brent algorithm). Its derivation is difficult but the algorithm is simple:

```
a:=1; x:=1; b:=1/sqrt(2); c:=1/4;
for i:=1 to n do {n is the number of iterations}
begin
   y:=a; a:=(a+b)/2; b:=sqrt(b*y); c:=c-x*sqr(a-y); x:=2*x
end;
writeln(sqr(a+y)/4/c:30:20);
```

Salamin-Brent Algorithm.

In the meantime several additional highly efficient algorithms for π have been discovered, mostly by the brothers J. M. and P. B. Borwein. Here is one of the most efficient:

$$y_0 = \sqrt{2} - 1; \quad a_0 = 6 - 4\sqrt{2};$$

$$y_{n+1} = \frac{1 - \sqrt[4]{1 - y_n^4}}{1 + \sqrt[4]{1 - y_n^4}};$$

$$a_{n+1} = (1 + y_{n+1})^4 a_n - 2^{2n+3} y_{n+1}(1 + y_{n+1} + y_{n+1}^2).$$

Borwein's quartic algorithm.

Here $1/a_n$ converges to π. $1/a_1$, $1/a_2$, $1/a_3$, $1/a_4$ give 8, 41, 171, 694 correct digits. The algorithm converges quartically, i.e. at each iteration the number of correct digits quadruples. It was used in 1988 by Kanada (Japan) to compute 201,326,000 places of π. $1/a_{15}$ would give more than two billion decimal digits of π. Nine iterations of the Salamin-Brent algorithm will give π well beyond 1000 decimals. But we must do all our computations with the precision we want, since these algorithms, unlike Newton's Method for square roots, are not self correcting. This is quite inconvenient and pays off only if we want ultra high precision.

A microcomputer suffices for computing many thousands of digits of π, but it is easier to do it with a language which has arithmetic with arbitrarily large numbers built in, such as ISETL, Derive, Maple or Mathematica, instead of Pascal. Moreover, in the Turbo Pascal implementation of Pascal all the variables combined may not occupy more than 64K bytes (32K on the Macintosh), irrespective of how much memory the computer has.

Still, it is possible to compute a few thousand digits of π even with Turbo Pascal and without programming full multi-digit multiplication or division. One can use the Taylor series for arctan x:

$$(1) \qquad \arctan x = x - \frac{x^3}{3} + \frac{x^5}{5} - \frac{x^7}{7} + \cdots.$$

The simplest formula to use would be $\pi = 4 \arctan 1$, but the series this gives is Leibniz's series, the slowly convergent series discussed in Exercises to Sections 7-9, Ex. 10. The following identities enable one to use (1) with much smaller values of x.

$$\pi = 16 \arctan \frac{1}{5} - 4 \arctan \frac{1}{239} \qquad \text{(Machin 1706)}$$

$$\pi = 24 \arctan \frac{1}{8} + 8 \arctan \frac{1}{57} + 4 \arctan \frac{1}{239} \quad \text{(Störmer 1896)}$$

$$\pi = 48 \arctan \frac{1}{18} + 32 \arctan \frac{1}{57} - 20 \arctan \frac{1}{239} \quad \text{(Gauss).}$$

We can compute $r \arctan(1/a)$ by means of the recurrences

$$u_{-1} = ra, \qquad u_{2n+1} = -u_{2n-1} \frac{2n-1}{(2n+1)a^2},$$

$$s_{-1} = 0, \qquad s_{2n+1} = s_{2n-1} + u_{2n+1}.$$

Then

$$r \arctan \frac{1}{a} = \lim_{n \to \infty} s_{2n+1}.$$

Ramanujan discovered a new family of series for π, one of which is

$$1/\pi = 2\sqrt{2} \sum_{n=0}^{\infty} \frac{(4n)!}{4^{4n}(n!)^4} \frac{1103 + 26390n}{99^{4n+2}}.$$

It converges with amazing speed. We have

$$\pi_0 = 3.14159\underline{273}, \quad \pi_1 = 3.14159\,26535\,8979\underline{387},$$

$$\pi_2 = 3.14159\,26535\,897932\,38462\,6\underline{490}$$

$$\pi_3 = 3.14159\,26535\,89793\,23846\,26433\,83279\,5\underline{55}$$

with 7, 16, 24, 32 correct digits. The incorrect digits are underlined. This series gives about 8 decimal places accuracy per term. A similar series was used in 1989 by G. V. and D. V. Chudnovsky of Columbia University to find π to 1,011,196,691 places. Their method utilizes sophisticated theoretical considerations. As of 1991, G. and D. Chudnowsky have computed 2.16 billion digits of π.

We wish to point out that one needs to use sophisticated methods just to multiply two huge numbers. Using the grade school method we need $O(n^2)$ operations. With this method one can not carry out even a single multiplication of two 10^9 digit numbers. In Knuth's volume 2 you can find out how to multiply with far fewer operations.

For more on computing π, see FOCUS, The Newsletter of the MAA, October 1989 and the article by Jonathan and Peter Borwein in the Notices of the Am. Math. Soc., October 1992, 825-829.

Finally, a remark about the randomness of the digits of π. There exists a very strong test for randomness based on random walks. It is the probabilistic law of the iterated logarithm. None of the usual RNG's satisfies this test. Yet the random walk based on the digits of π seems to meet it well. So the digits of π seem to come from an extremely good RNG. This is strange. According to a definition of Kolmogorov and Chaitin, a finite sequence is random if its shortest description (program) is about as long as the sequence itself. But the digits of π are produced by a very short program. So according to this definition the digits of π are extremely non-random. G. Chudnovsky believes that there should be hidden regularities in π, not yet discovered, at least in some base. For instance, take the big giants $9\wedge(9\wedge(9\wedge 9))$ and $2\wedge(2\wedge(2\wedge(2\wedge(2\wedge 2))))$. These huge numbers have very simple descriptions. Yet both the decimal and binary representations of the first number may look completely random and satisfy all the usual tests for randomness, as smaller similar numbers like 9^{9999} seem to indicate. In base 9 it is a 1 followed by 0's. The decimal expansion of the second number could look random by analogy with smaller powers of 2 while its binary representation is 1 followed by a huge number of 0's.

Exercise:

(Roy North) Compute π by adding the first 500 terms of Leibniz' series accurately. Find the first and the second wrong digit. Do the same with 5000 terms. Can you explain your observations? (Hint: Look at Section 6.4 for ideas on transforming the tail of Leibniz' series.)

CHAPTER 7

Miscellaneous Problems

60. A Problem from Geometry

Start with a circle of radius $r(2) = 1$ and do the following:

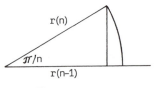

for n:=3 to max **do begin**
draw a regular n-gon around
the circle with radius $r(n - 1)$
and around the n-gon a
a circle with radius $r(n)$ **end**.

Figure 60.1

The sequence $r(n)$ of radii is monotonically increasing. Is it bounded? In case the answer is "yes", what is the limit of $r(n)$ for $n \to \infty$?

Fig. 60.1 shows that between $r(n)$ and $r(n-1)$ there is the relation

$$\frac{r(n-1)}{r(n)} = \cos(\frac{\pi}{n}) \text{ or } r(n) = \frac{r(n-1)}{\cos(\pi/n)}, \quad r(3) = \frac{r(2)}{\cos(\pi/3)} = 2 .$$

```
program radius;
var i,n:integer;  r:real;
begin
  write('n=');readln(n);r:=2;
  for i:=4 to n
  do r:=r/cos(pi/i);
  writeln('r=',r)
end.
```

Fig. 60.2

```
program radius1;
var i,n,r:real;
begin
  write('n=');readln(n);r:=2;i:=3;
  while i<n do
    begin i:=i+1;r:=r/cos(pi/i)
  end;
  writeln('r=',r)
end.
```

Fig. 60.3

The program `radius` is valid up to n=maxint, `radius1` is valid beyond that limit. From the table in Fig. 60.4 we cannot decide with certainty if the $r(n)$ have a limit. Only theory can decide this. In fact,

$$r(n) = \frac{2}{\cos(\pi/3)} * \frac{1}{\cos(\pi/4)} * \frac{1}{\cos(\pi/5)} * \cdots * \frac{1}{\cos(\pi/n)} .$$

n	$r(n)$
10	5.426745
100	8.283143
1000	8.657231
10000	8.695744
20000	8.697890
30000	8.698605
50000	8.699177
60000	8.699320

Fig. 60.4

Each factor of this product is greater than 1. So the sequence $r(n)$ increases monotonically. But the individual factors approach 1 from above. It is not clear if the approach is fast enough so the product stays finite. The PC cannot decide this. Slow convergence or divergence cannot be detected with a PC. We should develop some theory to decide this which we will not do here.

Exercise

It can be shown that $r(n)$ indeed converges to some fixed value. Try to find as many correct digits of this limit 8.7000366252... as you can, for instance, by extrapolation, or by other means.

61. Coupled Difference Equations. The Forward Declaration

How many n-words from the alphabet $A = \{0, 1, 2, 3, 4\}$ can be formed with the restriction that neighboring digits differ by exactly 1?

Let x_n be the number of these words, and let y_n, z_n, u_n be the number of n-words starting with 0 (or 4), 1 (or 3), and 2. Then we see from Fig. 61.1 that

(1) $x_n = 2y_n + 2z_n + u_n$ (3) $z_n = y_{n-1} + u_{n-1}$

(2) $y_n = z_{n-1}$ (4) $u_n = 2z_{n-1}$

with the boundary conditions $y_1 = 1$, $z_1 = u_1 = 1$, $x_1 = 5$. Suppose I define the functions x_n, y_n, z_n, u_n in this order. Then in the definition of x_n also y_n, z_n, u_n will occur, which have not yet been defined. So they must be announced by a `forward` declaration, as shown in Fig. 61.2. Note that the formal parameter lists are in the forward declaration and are not repeated in the headings of the function definitions.

```
program coupled_recurrences;
var i,n:integer;
function y(n:integer):integer;
forward;
function z(n:integer):integer;
forward;
function u(n:integer):integer;
forward;
function x(n:integer):integer;
begin
  x:=2*y(n)+2*z(n)+u(n)
end;
function y;
begin
  if n=1 then y:=1
  else y:=z(n-1)
end;
function z;
begin if n=1 then z:=n
else z:=y(n-1)+u(n-1) end;
function u;
begin if n=1 then u:=1
else u:=2*z(n-1)
end;
begin write('n='); readln(n);
  for i:=1 to n do
    writeln(i:5, x(i):9, y(i):8,
            z(i):8, u(i):8);
  readln;
end.
```

Fig. 61.2

Fig. 61.1

n	x_n	y_n	z_n	u_n
1	5	1	1	1
2	8	1	2	2
3	14	2	3	4
4	24	3	6	6
5	42	6	9	12
6	72	9	18	18
7	126	18	27	36
8	216	27	54	54
9	378	54	81	108
10	648	81	162	162
11	1134	162	243	324
12	1944	243	486	486
13	3402	486	729	972
14	5832	729	1458	1458
15	10206	1458	2187	2916
16	17496	2187	4374	4374
17	30618	4374	6561	8748

Fig. 61.3 $(n = 17)$.

Exercises:

1. Study the table in Fig. 61.3 carefully and find closed formulas for x_n, y_n, z_n, u_n. Hint: Consider even and odd n separately.

2. How many n-words from the alphabet $\{0, 1, 2, 3\}$ do not contain the two-words 01 or 10? Hint: Draw a graph with four states. If there is a two-way connection between two states, just draw one connecting line without an arrow (two-way street). Write a program for the coupled difference equations.

3. How many n-words from the alphabet $\{0, 1, 2\}$ are such that neighbors differ by at most 1?

62. Continued Fractions

Let x be a real number. If x is not an integer then we can write

$$x = a_0 + \frac{1}{x_1} \,, \quad a_0 = \lfloor x \rfloor \,, \quad x_1 = \frac{1}{x - a_0} > 1 \,.$$

If x_1 is not integral, we apply the same transformation to x_1

$$x_1 = a_1 + \frac{1}{x_2} \,, \quad a_1 = \lfloor x_1 \rfloor \,, \quad x_2 = \frac{1}{(x_1 - a_1)} > 1 \,.$$

For rational numbers we get

$$x = a_0 + \cfrac{1}{a_1 + \cfrac{1}{a_2 + \cfrac{1}{a_3 + \cddots + \cfrac{1}{a_n}}}} = a_0 + 1/(a_1 + 1/(a_2 + 1/(a_3 + \cdots + 1/a_n)) \cdots)$$

This expression is called a *simple continued fraction*. It is denoted by $[a_0; a_1, a_2 \cdots, a_n]$. The a_i are the *partial quotients*. If we stop with a_i we get the *i-th convergent* $r_i = p_i/q_i$. We can get the p_i, q_i recursively:

$$p_0 = a_0 \,, \qquad\qquad q_0 = 1$$
$$p_1 = a_0 a_1 + 1 \,, \qquad\quad q_1 = a_1$$
$$\cdots \qquad\qquad\qquad \cdots$$
$$p_k = p_{k-1} a_k + p_{k-2} \qquad q_k = q_{k-1} a_k + q_{k-2} \,, \quad k = 2, \ldots, n.$$

Indeed, $p_0/q_0 = r_0$, $p_1/q_1 = a_0 + 1/a_1 = r_1$, $p_2/q_2 = r_2$ and induction shows that $p_k/q_k = r_k$ for $k = 0, 1, 2, \cdots, n$.

Let

$$\Delta_k = p_{i-1} q_k - q_{k-1} p_k, \qquad k = 1, 2, \cdots, n.$$

Then $\Delta_1 = p_0 q_1 - q_0 p_1 = -1$ and induction shows that

$$\Delta_k = p_{k-1} q_k - q_{k-1} p_k = (-1)^k.$$

For irrational x, the continued fraction does not terminate.

Continued fractions are easily evaluated (beginning at the tail end), even with a pocket calculator. The following algorithm prints the sequence a_i for input x.

```
readln(x);
while x<>int(x) do
begin
  write(trunc(x)); x:=1/(x-trunc(x))
end;
```

This is not a complete program and it contains some flaws. But still it pays to run it after obvious completions. By trying some inputs like $x = 2.71828$ or $x = 3.14159$ we observe a steady stream of many thousands of partial quotients a_i, until `trunc(x)` exceeds `maxint`. The first impression is that many a_i are equal to one. We will investigate this observation in the exercises.

It is easiest to use a counter to stop the stream after 20 to 30 steps. Then we get

$$x = \pi \approx 3.1415926536 = [3; 7, 15, 1, 292, 1, 1, 1, \underline{4, 2, 10, 4, 1, 1, 1, 1, 1, 1, 4}, \cdots]$$

$$x = \sqrt{2} \approx 1.4142135624 = [1; 2, 2, 2, 2, 2, 2, 2, 2, 2, 2, 2, 2, 2, 2, \underline{5, 19, 3}, \cdots]$$

$$x = \frac{\sqrt{5} + 1}{2} \approx 1.61803398875$$

$$= [1; 1, 1, 1, 1, 1, 1, 1, 1, 1, 1, 1, 1, 1, 1, 1, 1, 1, 1, 1, 1, 1, 1, 1, 1, 1, 1, 1, \underline{4}, \cdots]$$

Because of rounding errors the underlined quotients are not correct. We know that $\sqrt{2} = [1; \overline{2}]$ and $(\sqrt{5} + 1)/2 = [1; \overline{1}]$. Here a bar over a digit means that it is to be repeated indefinitely. We also know the simple continued fraction expansion of e:

$$e = [2; 1, 2, 1, 1, 4, 1, 1, 6, 1, 1, 8, 1, 1, 10, 1, 1, 12, 1, 1, 14, 1, 1, 16, \cdots]$$

But the simple continued fraction of π does not seem to have any regularity:

$$\pi = [3; 7, 15, 1, 292, 1, 1, 1, 2, 1, 3, 1, 14, 2, 1, 1, 2, 2, 2, 2, 1, 84, 2, 1, 1, 15,$$
$$3, 13, 1, 4, 2, 6, 6, 99, 1, 2, 2, 6, 3, 5, 1, 1, 6, \cdots]$$

Continued Fractions and Euclid's Algorithm. We apply Euclid's Algorithm to two positive integers $n_0, n_1, \ n_0 \geq n_1 > 0$:

$$n_0 = q_1 n_1 + n_2$$
$$n_1 = q_2 n_2 + n_3$$
$$\cdots$$
$$n_{k-2} = q_{k-1} n_{k-1} + n_k$$
$$n_{k-1} = q_k n_k \ .$$

From these equations we easily get

$$\frac{n_0}{n_1} = [q_1; q_2, \cdots, q_k] \ .$$

Example:

$$2.71828 = 2 * 100000 + 71828$$
$$100000 = 1 * 71828 + 28172$$
$$71828 = 2 * 28172 + 15484$$
$$28172 = 1 * 15484 + 12688$$
$$15484 = 1 * 12688 + 2796$$
$$12688 = 4 * 2796 + 1504$$
$$2796 = 1 * 1504 + 1292$$

$$1504 = 1 * 1292 + 202$$
$$1292 = 6 * 202 + 80$$
$$202 = 2 * 80 + 42$$
$$80 = 1 * 42 + 38$$
$$42 = 1 * 38 + 4$$
$$38 = 9 * 4 + 2$$
$$4 = 2 * 2$$

$$\frac{271828}{100000} = [2; 1, 2, 1, 1, 4, 1, 1, 6, 2, 1, 1, 9, 2].$$

Simple Continued Fraction for $\sqrt{d}$. Suppose d is not a square. There is an algorithm for representing $\sqrt{d}$ as a simple continued fraction:
Set $a_0 = \lfloor\sqrt{d}\rfloor$, $b_1 = a_0$, $c_1 = d - a_0^2$ and find the integers a_{n-1}, b_n, c_n successively using the formulae

$$a_{n-1} = (a_0 + b_{n-1}) \text{ div } c_{n-1}, \qquad b_n = a_{n-1}c_{n-1} - b_{n-1},$$
$$c_n = (d - b_n^2) \text{ div } c_{n-1}.$$

Now we look at the sequence (b_2, c_2), (b_3, c_3), (b_4, c_4), and find the smallest index s for which $b_{s+1} = b_1$, $c_{s+1} = c_1$; the representation of $\sqrt{d}$ as a simple continued fraction then is

$$\sqrt{d} = [a_0; \overline{a_1, a_2, \cdots a_s}].$$

We have quoted almost literally from *W. Sierpinski, Elementary Theory of Numbers.* This outstanding book is not easily available, and it is very expensive. Luckily we can refer to Knuth's volume 2, 2nd edition. In 4.5.3, Exercise 12 he states an algorithm for the general quadratic irrationality $(\sqrt{d} - b)/c$; and in the answers to the exercises he gives a proof that the algorithm works. There you will also find the proof that $\sqrt{d}$ has a periodic expansion.

```
program contfrac;
var a,a0,b,b1,c,c1,d:integer;
begin
   write('d=');readln(d);a0:=trunc(sqrt(d));
   a:=a0; b:=a; b1:=a; c:=d-a*a; c1:=c; write(a0,' ');
   repeat
      a:=(a0+b) div c; write(a,' ');
      b:=a*c-b; c:=(d-b*b) div c
   until (b=b1) and (c=c1)
end.
```

Fig. 62.1

$$\sqrt{7} = [2; \overline{1, 1, 1, 4}]$$

$$\sqrt{43} = [6; \overline{1, 1, 3, 1, 5, 1, 3, 1, 1, 12}]$$

$$\sqrt{54} = [7; \overline{2, 1, 6, 1, 2, 14}]$$

$$\sqrt{76} = [8; \overline{1, 2, 1, 1, 5, 4, 5, 1, 1, 2, 1, 16}]$$

$$\sqrt{94} = [9; \overline{1, 2, 3, 1, 1, 5, 1, 8, 1, 5, 1, 1, 3, 2, 1, 18}]$$

$$\sqrt{1000} = [31; \overline{1, 1, 1, 1, 1, 6, 2, 2, 15, 2, 2, 6, 1, 1, 1, 1, 1, 62}]$$

$$\sqrt{919} = [30; \overline{3,5,1,2,1,2,1,1,1,2,3,1,19,2,3,1,1,4,9,1,7,1,3,6,2,11,1,1,1,29,\cdots,60}]$$

$$\sqrt{991} = [31; \overline{2,12,10,2,2,2,1,1,2,6,1,1,1,1,3,1,8,4,1,2,1,2,3,1,4,1,20,6,4,31,\cdots,62}]$$

The periods of $\sqrt{919}$ and $\sqrt{991}$ have 62 and 60 terms, respectively. Each period always ends with $2\lfloor\sqrt{d}\rfloor$. The period without the last term is a palindrome; it reads the same forward and backwards. This observation allows us to restore all digits in the periods of $\sqrt{919}$ and $\sqrt{991}$. There are no rounding errors in this algorithm.

Best Approximation. The following theorem is also known under the name *law of best approximation*:

If the rational number r/s is closer to an irrational number x than the n-th convergent $r_n = p_n/q_n$ of x, then $s > q_n$.

We could easily prove this theorem, but we do not care to do so. Instead we refer to Section 19, where we already dealt with the subject from a more general point of view. We state without proof some more theorems about best approximation:

a) For any convergent p/q of x we have

$$\left| x - \frac{p}{q} \right| < \frac{1}{q^2}.$$

b) Of any two consecutive convergents at least one, call it p/q, satisfies the inequality

$$\left| x - \frac{p}{q} \right| < \frac{1}{2q^2}.$$

c) If p/q is in lowest terms and

$$\left| x - \frac{p}{q} \right| < \frac{1}{2q^2},$$

then p/q is a convergent of x.

Exercises:

1. Run the program contf in Fig. 62.2 for $x = 0.314159$, $x = 0.271828$ and some other values with at least 6 significant digits. It will print the frequency z of the occurrence of the partial quotient $a_i = 1$ for $i = 1$ to n, and it will print the proportion z/n. Because of rounding errors we will compute the average of $a_i = 1$ for many different rationals from $(0,1)$. There is a deep result which says that for rationals (and irrationals) the proportion of 1's is on the average close to a constant p_1; z/n will be an estimate of p_1. The same deep result tells us that the proportions of $a_2, a_3, a_4, \ldots$ will be close to some constants $p_2, p_3, p_4, \ldots$. Estimate these constants and compare the results with the solutions. ($x = \text{int}(x)$) practically never occurs. It is just a precaution. Find a value of x for which $x = \text{int}(x)$. To speed up the program you should rewrite the program by using longint variables. See *Continued Fractions and Euclid's Algorithm* earlier in this Section.

```
program contf;
var n,z,x:real;
begin write('x=');readln(x);n:=0;z:=0;
  while (x<>int(x)) and (x<32767.0) do
  begin n:=n+1;
    if trunc(x)=1 then   z:=z+1;
    x:=1/(x-trunc(x))
  end;
  writeln('z=',z:0:0,'n=':20,n:0:0,'z/n=':20,z/n)
end.
```

Fig. 62.2

2. We want to test the results in Exercise 1 empirically. Repeat the following random experiment $n = 10000$ times: For $m = 10^8$ choose at random two integers a, b from $1..m$. Find by means of the Euclidean algorithm the relative frequencies of the partial denominators $q = 1, 2, 3, 4$ averaged over all n rationals a/b. Compare with the theoretical results in the solutions. To speed up the program it is advisable to use Longint variables throughout.

3. We can apply $\sqrt{d} = [a_0; a_1, a_2, \cdots, a_n]$ to find all integer solutions of Pell's equations $x^2 - dy^2 = 1$ and $x^2 - dy^2 = -1$. They are to be found among the pairs (p, q). Check all pairs (p, q) until you hit on the smallest solution. Which pair seems to be the smallest solution? Consider the cases of even and odd s separately.

63. Shrinking Squares: An Empirical Study

We label the four successive vertices of a square by four nonnegative real numbers a, b, c, d. Each midpoint is labeled by the absolute value of the difference between the labels of its neighbors. This process is repeated with the midpoints until we get the vector $(0,0,0,0)$, as in Fig. 63.1. That is, we apply repeatedly the transformation

$$T : (a, b, c, d) \mapsto (|a - b|, |b - c|, |c - d|, |d - a|).$$

Fig. 63.2 shows a program that prints the succesive labels and the number of steps until it stops.

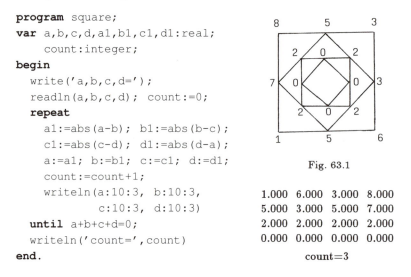

```
program square;
var a,b,c,d,a1,b1,c1,d1:real;
    count:integer;
begin
    write('a,b,c,d=');
    readln(a,b,c,d); count:=0;
    repeat
        a1:=abs(a-b); b1:=abs(b-c);
        c1:=abs(c-d); d1:=abs(d-a);
        a:=a1; b:=b1; c:=c1; d:=d1;
        count:=count+1;
        writeln(a:10:3, b:10:3,
                c:10:3, d:10:3)
    until a+b+c+d=0;
    writeln('count=',count)
end.
```

Fig. 63.1

```
1.000  6.000  3.000  8.000
5.000  3.000  5.000  7.000
2.000  2.000  2.000  2.000
0.000  0.000  0.000  0.000
        count=3
```

Fig. 63.2

After playing with the program **square** for some time we get the impression that we always reach $(0, 0, 0, 0)$ after not too many steps. But there is an exception. We are guided by the fact that if we have a geometric sequence, the differences of the consecutive terms form a geometric sequence with the same common ratio. Label the vertices of the square $1, t, t^2, t^3$ where t is some number > 1. The three differences $t - 1$, $t^2 - t$ and $t^3 - t^2$ also advance by the same factor. To bring the fourth difference into this geometric progression, we need

$$t^3 - 1 = t(t^3 - t^2) \quad \text{or} \quad t^3 = t^2 + t + 1.$$

Now forming the absolute differences reproduces the original configuration, multiplied by $t - 1$. Hence the algorithm does not stop. It turns out that the only initial values for which the algorithm does not stop are those obtained from the above by an affine mapping $y = ax + b$ applied to each component of the quadruple, or by reversing the order in which the numbers are written. [This is shown in M. Gardner, *Riddles of the Sphinx*, NML vol. 32, MAA (1987), pp. 160–163. See also R. Honsberger, *Ingenuity in Mathematics*, NML vol. 23, MAA (1970), pp. 82-83.] The program **bis** (see Fig. 3.4) tells us that $t = 1.8392867552\ldots$. Since the computer works with a decimal approximation of this irrational number, the program will stop anyway.

Let us now investigate the more general mapping

$$T: (x_0, x_1, \cdots, x_{n-1}) \mapsto (|x_0 - x_1|, |x_1 - x_2|, \cdots, |x_{n-1} - x_0|).$$

After the first step the numbers will be nonnegative. Also for any $c > 0$, $(x_0, x_1, \cdots, x_{n-1})$ and $(cx_0, cx_1, \cdots, cx_{n-1})$ have the same life expectancy or cycling behavior. So instead of rational numbers it suffices to take nonnegative integers as initial values and we assume their gcd is 1.

The program **polygon** starts with n random integers from $\{0, 1, \ldots, 99\}$ and applies repeatedly the mapping T to this n-tuple, until the all-zero n-tuple is reached. By playing with this program we make the following discoveries:

```
program polygon;
var i,n,s,count:integer; x:array[0..1024] of integer;
begin
  write('n='); readln(n); count:=0;
  for i:=0 to n do
    x[i]:=random(100);
  repeat count:=count+1;s:=0;
    for i:=0 to n-1 do
      x[i]:=abs(x[i]-x[i+1]);
    x[n]:=x[0];
    for i:=0 to n-1 do
      write(x[i]:5);
    writeln;
    for i:=0 to n-1 do
      s:=s+x[i]
  until s=0;
  writeln('count=',count)
end.
```

29	12	17	14	30	28	5	22
17	5	3	16	2	23	17	7
12	2	13	14	21	6	10	10
10	11	1	7	15	4	0	2
1	10	6	8	11	4	2	8
9	4	2	3	7	2	6	7
5	2	1	4	5	4	1	2
3	1	3	1	1	3	1	3
2	2	2	0	2	2	2	0
0	0	2	2	0	0	2	2
0	2	0	2	0	2	0	2
2	2	2	2	2	2	2	2
0	0	0	0	0	0	0	0

$n = 8$ **count=13**

Fig. 63.3

a) The algorithm always stops if n is a power of 2, see Fig. 63.3.

b) For $n \neq 2^r$ the algorithm almost never stops. So it must run into a cycle.

c) On the cycle just two numbers survive: 0 and some positive integer.

d) By our earlier discovery we may just as well use 0 and 1. But then

$$|a - b| = (a + b) \bmod 2 = a \textbf{ xor } b.$$

The right side can be used to speed up Turbo Pascal computations. The expression $(a + b) \bmod 2$ is useful in algebraic work, since addition mod 2 has the algebraic properties of ordinary addition, whereas expressions with absolute value signs can seldom be simplified. In particular it will be convenient to use the additivity of $T \bmod 2$, i.e.

$$T(x_0 + y_0, x_1 + y_1, \ldots, x_{n-1} + y_{n-1}) =$$
$$T(x_0, x_1, \ldots, x_{n-1}) + T(y_0, y_1, \ldots, y_{n-1})$$

because this implies that, if c is the cycle length of both n-tuples on the right, it is also the cycle length or a multiple of the cycle length of the n-tuple on the left.

e) If n is odd, then $(1,1,0,...,0)$ always lies on a cycle.

We use the observations d) and e) to find the maximal cycle length $c(n)$ for odd n efficiently. For any n-tuple x, the elements of Tx have sum 0 mod 2. Hence the n-tuples of a cycle must all have sum 0 mod 2. Every such n-tuple is the element-wise sum modulo 2 of n-tuples with two neighboring 1's and 0's in all other places. This and the remark we made after d) above imply that the cycle length of $(1,1,0,\ldots,0)$ is $c(n)$. Program cycle finds $c(n)$ for odd n by applying T until this n-tuple reappears. Check the following table:

n	3	5	7	9	11	13	15	17	19	21	23	25	27	29	31
$c(n)$	3	15	7	63	341	819	15	255	9709	63	2047	25575	13797	475107	31

n	33	35	37	39	41	43	45	47	49	51
$c(n)$	1023	4095	3233097	4095	41943	5461	4095	8388607	2097151	255

f) Our data suggest that $n \mid c(n)$.

The program cycle has two gotos. The program cycle1 has no goto and has almost the same efficiency. Is it more comprehensible than cycle?

We adapt our "hare-and-turtle" algorithm (cycle finding algorithm, see Section 27) for study of cycle length for even n (program cycle2). This is one of the longest programs in our book, but it is easy to understand once you have understood the program findcycle.

```
program cycle;
label 0;
var c:longint; i,n:integer;
    x:array[0..1024] of byte;
begin
  write('n='); readln(n);
  x[0]:=1; x[1]:=1;
  x[n]:=1; c:=0;
  for i:=2 to n-1 do x[i]:=0;
0:c:=c+1;
  for i:=0 to n-1 do
    x[i]:=x[i] xor x[i+1];
  x[n]:=x[0];
  if x[0]+x[1]<2 then goto 0;
  for i:=2 to n-1 do
    if x[i]>0 then goto 0;
  write('cycle length= ',c)
end.
```

Fig. 63.4

```
program cycle1;
var c:longint; i,n,s:integer;
    x:array[0..1024] of byte;
begin
  write('n='); readln(n);
  x[0]:=1; x[1]:=1;
  x[n]:=1; c:=0;
  for i:=2 to n-1 do x[i]:=0;
  repeat
    s:=0;
    repeat c:=c+1;
      for i:=0 to n-1 do
        x[i]:=x[i] xor x[i+1];
      x[n]:=x[0]
    until x[0]+x[1]=2;
    for i:=2 to n-1 do s:=s+x[i]
  until s=0;
  write('cycle length c= ',c);
end.
```

Fig. 63.5

By means of the program cycle2 we make the following discoveries:

g) $c(2n) = 2c(n)$,

h) The number of catchup steps is always equal to the cycle length.

i) The tail length can vary from 0 (pure cycle) to 2^r, where r is the maximum exponent, such that $2^r \mid n$.

j) $c(n) = n$ for $n = 2^r - 1$.

Thus for odd n we have $c(2^r n) = 2^r c(n) = 2^r n q(n)$. We need more information about $q(n)$. Very often we have

(1) $$q(n) = 2^m - 1 \quad \text{and} \quad c(n) = n(2^m - 1).$$

The first two exceptions are $n = 23$ and $n = 35$. But here we have $c(23) = 2^{11} - 1$ and $c(35) = 2^{12} - 1$. Thus $q(23) \mid (2^{11} - 1)$, $q(35) \mid (2^{12} - 1)$. What about m in (1)? It seems to be the order of 2 mod $c(n)$, i.e. the smallest m such that $2^m \equiv 1 \bmod c(n)$. That is

$$c(n) \mid (2^m - 1).$$

Now try to find proofs for some of our discoveries; most are difficult to prove.

```
program cycle2;
label 0,1,2,3;
var c:longint; i,j,n:integer; x,y,z:array[0..1024] of byte;
begin write('n='); readln(n); randomize;
  for i:=0 to n-1 do begin
    x[i]:=random(2); y[i]:=x[i]; z[i]:=x[i]
  end;
  x[n]:=x[0]; y[n]:=y[0]; z[n]:=z[0]; c:=0;
0:for i:=0 TO n-1 do   y[i]:=y[i] xor y[i+1];
  y[n]:=y[0];
  for j:=1 TO 2 DO
  begin
    for i:=0 to n-1 do z[i]:=z[i] xor z[i+1]; z[n]:=z[0];
  end;
  c:=c+1;
  for i:=0 to n-1 do if y[i]<>z[i] then goto 0;
  writeln('catchup steps = ',c);
  c:=0;
for i:=0 to n-1 do if y[i]<>x[i] then goto 1;
  writeln('pure cycle '); goto 3;
1:c:=c+1;
  for i:=0 to n-1 do y[i]:= y[i] xor y[i+1];
  y[n]:=y[0];
  for i:=0 to n-1 do if y[i]<>z[i] then goto 1;
  writeln('cycle length c= ',c);
  for i:=0 to n do z[i]:=x[i];
  c:=0;
2:c:=c+1;
  for i:=0 to n-1 do
    begin y[i]:=y[i] xor y[i+1]; z[i]:=z[i] xor z[i+1] end;
  y[n]:=y[0]; z[n]:=z[0];
  for i:=0 to n-1 do if y[i]<>z[i] then goto 2;
  writeln('tail length= ',c);
3:end.
```

Fig. 63.6

Exercises:

1. Show that we always arrive at the all-zero n-tuple for $n = 4$ and $n = 2^r$.

2. Show that on a cycle just two numbers survive: 0 and some other number $a > 0$.

3. Show that for odd n the n-tuple $(1,1,0,\ldots,0)$ always lies on a cycle and $(1,0,\ldots,0)$ never lies on a cycle.

4. Show that $c(2n) = 2c(n)$.

5. Show that $c(n) = n$ for $n = 2^r - 1$.

6. Show that $n \mid c(n)$ for all n.

7. Suppose $t = 1.8392867552$. How many steps do you need to reduce $(1, t, t^2, t^3)$ to $(0, \ldots, 0)$?

8. *Bulgarian Solitaire.* a) Start with a triangular number $t = n(n + 1)/2$ of cards. Initially they are distributed into any number k of stacks. A move consists in taking one card from each stack and forming from these cards a new stack with k cards. One-card stacks will disappear during this move. You win the game as soon as there are on the table n stacks with 1,2, 3,..., n cards respectively. Write a program which plays the game and stops as soon as you have won. Count the number of moves and try to find empirically the maximum number of moves until you win.

b) Investigate the case that t is not a triangular number. Then a win is no longer possible. Still you can find some theorems about cycle structure.

64. Self Avoiding Random Walks

Fig. 64.1 shows a 11×11 square lattice. How many lattice paths are there from $(0,0)$ to $(10,10)$, which do not pass through the same point twice? Nobody knows this, and it may well be that nobody will ever know this number exactly. But physicists and chemists would very much like to know the number of self avoiding paths from $(0,0)$ to (m, n). It would give them valuable insight into the structure of long molecules.

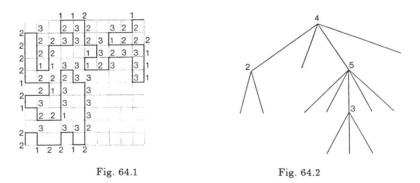

Fig. 64.1   Fig. 64.2

Although we do not know the number of self avoiding paths from $(0,0)$ to $(10,10)$ or (m, n) exactly, we can estimate this number by a nontrivial simulation. We start with a lemma due to Knuth. Fig. 64.2 shows a tree which is supposed to be oriented downward. Each node i is marked by the number X_j of branches leaving that node. We want to estimate by random sampling the number of paths through the tree. We start at the root and pick a path by going from node to node, picking a branch at random at each node, until we reach an endnode. If the number of branches at the i-th node of this path is X_i, then the probability of obtaining the path we have traversed is $p = 1/X$, where $X = X_1 X_2 \cdots X_n$. Thus, if we write down the definition of the expectation of the random variable X, we get

$$E(X) = \sum_{\text{All paths}} 1 = \text{\# of paths through the tree.}$$

Thus counting the paths is reduced to finding the expected or average value N of the random variable X.

I generated a random walk from $(0,0)$ to $(10,10)$ by means of a die, writing at each node the number of choices that were available. The result is shown in Fig. 64.1. The value of X for this path is $2^{32} 3^{35} \approx 3.6 * 10^{21}$.

Twenty students have generated 20 self avoiding random walks and the results are shown in Fig. 64.3. A comparison of these numbers is quite instructive. Practically all path probabilities have different orders of magnitude. On the one hand there is a path with probability $> 10^{-9}$, but on the other hand there are two paths with probability less than 10^{-24}. The reason for getting such paths must be that the overwhelming majority of paths are very low probability paths.

The huge fluctuations prevent us from getting a good estimate $\widehat{N}$ of

Walk no.	Length l	Probability of path	$\widehat{N}$
1	44	$2^{-15}\ 3^{-20}$	$1.1\ 10^{14}$
2	32	$2^{-9}\ 3^{-20}$	$1.6\ 10^{12}$
3	64	$2^{-23}\ 3^{-32}$	$1.6\ 10^{22}$
4	28	$2^{-15}\ 3^{-9}$	$6.4\ 10^{8}$
5	52	$2^{-18}\ 3^{-26}$	$6.7\ 10^{17}$
6	54	$2^{-17}\ 3^{-24}$	$3.7\ 10^{16}$
7	58	$2^{-25}\ 3^{-24}$	$9.5\ 10^{18}$
8	46	$2^{-22}\ 3^{-18}$	$1.6\ 10^{15}$
9	78	$2^{-31}\ 3^{-29}$	$1.5\ 10^{23}$
10	30	$2^{-8}\ 3^{-18}$	$9.9\ 10^{10}$
11	62	$2^{-25}\ 3^{-27}$	$2.6\ 10^{20}$
12	48	$2^{-12}\ 3^{-26}$	$1.0\ 10^{16}$
13	54	$2^{-19}\ 3^{-21}$	$5.5\ 10^{15}$
14	32	$2^{-8}\ 3^{-20}$	$8.9\ 10^{11}$
15	32	$2^{-10}\ 3^{-16}$	$4.0\ 10^{11}$
16	84	$2^{-33}\ 3^{-30}$	$1.7\ 10^{24}$
17	38	$2^{-11}\ 3^{-19}$	$2.4\ 10^{12}$
18	24	$2^{-6}\ 3^{-16}$	$2.8\ 10^{9}$
19	84	$2^{-46}\ 3^{-22}$	$2.2\ 10^{24}$
20	40	$2^{-12}\ 3^{-17}$	$5.3\ 10^{11}$

Fig. 64.3

N. But suppose these are the only data available. Then we should take their arithmetic mean which gives the estimate

$$\widehat{N} \approx 0.2 * 10^{24}.$$

Knuth has generated several thousand paths, getting

$$\widehat{N} \approx (1.6 \pm 0.3) * 10^{24}.$$

To get this average when, as our table shows, most trials yield much smaller X-values, a small proportion of the trials must give values much larger than the ones in our table. To obtain a confidence interval such as Knuth's one needs a sample large enough so that its mean is not too much influenced by a few large X-values, and one also needs some information about the largest X-values that can occur.

Exercises:

1. The number of shortest paths from $(0,0)$ to $(10,10)$ is $\binom{20}{10} = 184756$. Estimate this number by means of Knuth's lemma.

2. The number of shortest lattice paths from $(0,0)$ to $(10,10)$ which may touch the line $y = x$, but not cross it, is $\frac{2}{11}\binom{20}{10} = 33592$. They lie below $y = x$ or touch it. Estimate this number by means of Knuth's lemma.

3. The 8×8 chessboard can be covered by 32 2×1 dominoes in $2^4 901^2 = 12988816$ different ways. Try to estimate this number by means of Knuth's lemma.

4. *A Project.* Write a program which draws self avoiding random walks on an $n \times n$ lattice and computes an estimate $\widehat{N}$ of N.

65. A Very Difficult Problem

Show that if a, b, q are positive integers with $a^2 + b^2 = q(ab + 1)$, then q is a perfect square.

This problem was proposed by the German Federal Republic at the XXIX. IMO in Canberra, 1988. The Australian Problem Comittee liked it very much, but nobody on the Comittee could solve it. Among the members of the Comittee were George Szekeres and his wife, both famous problem solvers and problem creators. So it was proposed to the four most eminent number theorists of Australia. Each one worked on the problem for six hours, but no one was able to solve it. To make the story short the problem was chosen for the Olympiad and eleven high school students produced complete solutions. The future of mathematics looks bright!

Suppose we are mathematicians of average ability, but we have a computer at our disposal. We will show that then this problem becomes comparatively simple.

As a first step we collect material. Then we study it to find some clues. Finally we see how to generate all solutions. This generation process suggests an elegant proof.

Because of symmetry in a and b we may assume $a \leq b$.

a) It is easy to see that $a = b$ for $a = b = q = 1$ only. So we may even assume $a < b$.

b) Write a program which generates all solutions with $a \leq 150$, $b \leq 1000$.
We get the following result:

a	1	2	3	4	5	6	7	8	8	9	10	27	30	112
b	1	8	27	64	125	216	343	30	512	279	1000	240	112	418
q	1	4	9	16	25	36	49	4	64	81	100	9	4	4

c) A look at the table suggests the solution $(a, b, q) = (c, c^3, c^2)$. Indeed:

$$c^2 + c^6 = c^2(c^4 + 1).$$

We have found one solution for each square.

d) Let us look at the triples giving the same q, for instance $q = 4$:

$$
\begin{array}{cccccc}
2 & 8 & 30 & 112 & a & b \\
8 & 30 & 112 & 418 & b & a_1 \\
4 & 4 & 4 & 4 & q & q
\end{array}
$$

The second component for each triple (a, b, q) is the first component of the succeeding triple. This suggests the transformation

$$(a, b, q) \mapsto (b, a_1, q).$$

The diophantine equation $a^2 + b^2 = q(ab + 1)$, or

(0) $a^2 - qba + b^2 - q = 0$

is a quadratic in a and has two solutions a, a_1, which satisfy

(1) $a + a_1 = qb$

(2) $a * a_1 = b^2 - q$

Equation (1) shows that with a also a_1 is an integer, and we have

(3) $a_1 = qb - a$.

Since $q \geq 2$ and $b > a$, we have $a_1 > b$. So for any b, the two solutions a, a_1 of (0) straddle b. By symmetry the two solutions b, b_1 of (0) for a given a straddle a. Thus one can get ever larger pairs a, b of integers which satisfy (0) for a fixed q.

We shall now prove that q is a square by going downwards in this family of solutions. The original equation $a^2 + b^2 = q(ab + 1)$ shows that ab can not be negative. Hence a and b must have the same sign (possibly one is 0). Using the straddling property we can alternately replace the larger of a, b by a smaller nonnegative integer. In the end one of the numbers must become 0 and q is the square of the other.

Exercises:

1. Write a program, which prints the table of triples a, b, q in 65, b).

2. Let a and b be positive integers such that $ab + 1$ divides $a^2 + ab + b^2$. Show that $(a^2 + ab + b^2)/(ab + 1)$ is the square of an integer. Explore the situation as in the IMO-Problem.

3. Consider the more general term $q = (a^2 + rab + b^2)/(sab + t)$ with $r = -1$, 0 or 1, and positive integers s, t. What condition between s and t must be satisfied in order that
   a) q be a square ,   b) qt be a square .

4. *Empirical Exploration.* Start with a stack of n coins, each having "head" up. Now turn over the top i-substack and place it again on top for $i = 1, 2, \cdots, n, 1, 2, \cdots, n, 1, 2, \cdots$. For each i check if the initial state (all "heads" on their top) has occured. If yes, then stop. Let $f(n)$ be the number of reversals until stop. Write a program which finds $f(n)$. Can you discover a formula for $f(n)$?

Additional Exercises for Sections 1-65:

1. In a city of n persons someone thinks of a joke. Anybody who hears the joke continues telling it until he/she tells it to someone who heard it already. Then he/she stops telling it. Find by simulation the percentage of the population who never hear the joke. It is practically independent of n. Assume a totally mixing population: Any person meets any other person with the same probability.

2. Start with 1 and 2 in the computer memory and generate additional numbers as follows: Take any two stored numbers a, b, compute the new number $c = ab + a + b$ and store it. What numbers do you get in this way?

3. Let d_n be the number of fixed-point-free permutations of $\{1, 2, \ldots, n\}$.
   a) Prove that $d_n = (n-1)(d_{n-1} + d_{n-2})$ and write an inefficient recursive program based on this recursion. b) From the formula in a), derive the recursion $d_n = nd_{n-1} + (-1)^n$. Write an efficient recursive program based on this recursion. (A permutation p is fixed-point-free if $p(i) \neq i$ for all i from $1..n$.)

4. Studying the Goldbach Count one observes that, except for very small numbers, a number $6n$ has always greater Goldbach Count than its even left and right neighbors. Show that this is not correct. Find a "contact" below 2000 and a "crossing" between 80000 and 80100. We have a "contact" if $6n$ and $6n - 2$ or $6n + 2$ has the same count as $6n$. We have a "crossing" if $6n - 2$ or $6n + 2$ has larger count than $6n$.

5. From the Lady's and Gentleman's Diary 1861, Problem 1987: *Three points being taken at random in space as corners of a plane triangle, determine the probability that it shall be acute.* In 1862 Stephen Watson gave the

answer 33/70. Check by simulation if this answer is correct. Choose the points at random   a) in the unit cube;   b) in the unit sphere.

6. S. W. Golomb defined the "self-describing sequence" $f(1), f(2), \ldots$ as the only nondecreasing sequence of positive integers with the property that it contains exactly $f(k)$ occurrences of k for each k. Check by hand the following values:

n	1	2	3	4	5	6	7	8	9	10	11	12	13	14	15	16	17	18	19	20	21	22	23
$f(n)$	1	2	2	3	3	4	4	4	5	5	5	6	6	6	6	7	7	7	7	8	8	8	8

a) Write a program which computes this sequence for $n = 1$ to 10000.

b) Plot $y = \ln f(n)$ versus $x = \ln n$. You will find that a straight line $y = a + bx$ is approached asymptotically.

c) From this derive the asymptotic formula $f(n) \sim cn^d$.

d) Find approximate values for c and d by substituting $n = 5000$ and $n = 10000$.

7. *Knuth Numbers.* In [1988] Knuth defined the sequence

$$K_0 = 1; \quad K_{n+1} = 1 + \min(2K_{\lfloor n/2 \rfloor}, \ 3K_{\lfloor n/3 \rfloor}).$$

a) Write a recursive program which computes the Knuth Numbers up to 10000.   b) Check if $K_n > n$ in this interval.

8. *Number of comparisons in merge-sort.* To sort n numbers $(n > 1)$ we divide them into (almost) equal parts $\lfloor n/2 \rfloor$ and $\lceil n/2 \rceil$. After each part has been sorted (by the same method applied recursively), we can merge the numbers into one sorted set by doing at most $n - 1$ comparisons. Let $f(n)$ be the total number of comparisons. Then

$$f(1) = 0, \ f(n) = f(\lfloor n/2 \rfloor) + f(\lceil n/2 \rceil) + n - 1, \quad n > 1 \ .$$

a) Write a recursive program, which for input n computes $f(n)$.

b) Try to guess a closed formula for $f(n)$.

9. *Sequences which omit powers.* Check that the sequence

$$b_n = \lfloor n + \left((n - 0.5)^{1/m} + n - 0.5 \right)^{1/m} \rfloor$$

omits m-th powers. Can you prove this?

10. Two rectangles R1 and R2 with sides parallel to the coordinate axes are given by means of the coordinates of their lower left and upper right vertices. Write a program which finds the area of R1 not contained in R2. The program should work for any two rectangles in the plane.

11. Three persons 1, 2, 3 possessing initially a, b, c chips respectively play the following game: In each round each person stakes one chip. Then a 3-sided symmetric die is rolled. On outcome i person #i gets all three chips. The game is over as soon as any of the players runs out of chips.

Denote by $f(a, b, c)$ the expected duration of the game (number of rolls). Try to find $f(a, b, c)$ by intelligent simulation. I will not tell you the result, although I know it. It is not to be found in any book I know of.

12. On the nonnegative integers we define a pair of recursively intertwined functions $f(n) = n - g(f(n-1))$ and $g(n) = n - f(g(n-1))$, $n > 0$, $f(0) = 1$, $g(0) = 0$. Compute these functions in the interval 0..max by means of recursive and iterative programs. For the recursive program use the **forward** declaration. Try to find closed expressions for $f(n)$ and $g(n)$.

13. Pick a finite sequence of integers $a_1 < a_2 < \cdots < a_k$ to be the first k elements of a sequence. Then if the sequence is defined up to n where $n \geq a_k$, $n + 1$ is in the sequence if it is not a sum of two (not necessarily distinct) elements in the sequence. Find the first max elements of the sequence starting with (1,2).

14. *Pile Games.* (See Additional Exercises for Sections 1 to 41.) Start with three piles of a, b, c chips. In one step an ordered pair of piles is chosen at random and a chip is moved from the first pile to the second pile. If some pile becomes empty, continue with two piles. Stop when only one pile is left.
   a) Find by simulation the expected number $g(a, b, c)$ of steps until stop.
   b) Generalize to four piles with a, b, c, d chips. Try to guess and then check by simulation the expected duration of the game.
   c) Guess the corresponding answer for n piles and try to prove your guess as at the end of the Drash Course in Probability. See also *Arthur Engel, The Computer Solves the Three Tower Problem.* To appear early 1993 in the *American Mathematical Monthly.*

15. Generate 10000 triples (a, b, c) of variables each one uniformly distributed in (0,1) and estimate the probability of getting 3 numbers which are the side lengths of a triangle. Solve the corresponding problem for quadrilaterals and pentagons. Try to guess the general law.

16. *Records and antirecords in a random permutation.* Generate and count the records and antirecords in a random permutation $x_1 x_2 \ldots x_n$ of 1..n. An element x_k is a record if $x_i < x_k$ for $i < k$. The index k is the position of the record. The integer k is the position of an antirecord if $x_k < x_j$ for $j > k$. Try to find asymptotic formulas for these numbers in a random n-permutation.

17. Write a program, which splits the sequence $1, 2, \ldots, 2n$ at random into two subsequences $a_1 < a_2 < \ldots < a_n$ and $b_1 > b_2 > \cdots > b_n$ and compute $|a_1 - b_1| + |a_2 - b_2| + \cdots + |a_n - b_n|$. Comment on the result. Can you prove the result?

18. Find all representations of the numbers of the form a, aa, aaa, $aaaa$, where a is a digit, as sums of squares $x_1^2 + x_2^2 + \ldots$ with the x_i in arithmetic progression: m, $m + d$, $m + 2d$, $\ldots$.

19. *Empirical Exploration.* Fill the array $x[1..n]$ with $n = 2^k$ by random integers from $\{-1, 1\}$, and set $x[n + 1] := x[1]$. Apply repeatedly the transformation

```
for i:=1 to n do y[i]:=x[i]*x[i+1];
for i:=1 to n do x[i]:=y[i]; x[n+1]:=x[1];
```

a) Try to prove what you see eventually.
b) What happens if $n \neq 2$? Study cycle length $c(n)$ in this case.
c) When do you get a pure cycle?
d) Generalize to $y[i] := x[i] * x[i + 1] * x[i + 2]$ and generally to $y[i] := x[i] * x[i + 1] * \cdots * x[i + p]$.

20. There is a checker at point $(1,1)$ of the lattice (x, y) with x, y positive integers. One may make the following moves: double one coordinate, or subtract the smaller from the larger coordinate and replace the larger by this difference. Which lattice points can the checker reach?

21. Construct a sequence of positive integers starting with $a_1 = 1$, $a_2 = 2$, so that any positive integer can be uniquely represented as the difference of two numbers of the sequence.
Hint: Write the sequence in pairs. Suppose we can construct k pairs. Let us construct the $(k + 1)$-th pair. We consider all differences, that are realized by the k pairs, and let d be the smallest difference not yet realized. Then we set $a_{2k+1} = 2a_{2k} + 1$, $a_{2k+2} = a_{2k+1} + d$.

22. What does the algorithm on the right do? Here eps is a small number, say $\mathrm{eps} = 10^{-5}$, so that smaller numbers may be neglected; a, b are real inputs. Find the output $p = f(a, b)$. First play with different inputs and try to guess the answer. For a proof we give you a strong hint: Initially $Q = p^2 + q^2 = a^2 + b^2$. What happens

```
input a,b, eps;
p:=max(a,b); q:=min(a,b);
while q>=eps do
begin
    r:=sqr(q/p); s:=r/(r+4);
    p:=(2*s+1)*p; q:=s*q
end;
writeln(p);
```

to Q after one traversal of the loop? This cubically convergent algorithm is due to Cleve Mohler and Donald Morrison, IBM J. Res. Dev.,1983.

23. Which numbers a) from 1 to 100 b) from 1 to 10000 have a palindromic square? Find some palindromic squares, whose squares are also palindromes.

24. An n-digit natural number is an *Armstrong* number if it is equal to the sum of the n-th powers of its digits. Find all Armstrong numbers with 2, 3, 4, 5 digits.

25. Find all natural numbers $< 10^6$ which are palindromes in both the decimal and binary notation.

26. a) Generate a random permutation $x_1 \cdots x_{2n}$ of $1..2n$ and check for the

occurrence of the event $|x_i - x_{i+1}| = n$, $i \in 1..2n - 1$. Repeat this experiment 10000 times and estimate the probability that this event occurs at least once. Guess the asymptotic value.

b) Let $p(n)$ be the probability that $|x_i - x_{i+1}| = n$ never occurs. Then one can show that $p(n) = p(n-1) + p(n-2)/(2n-1)/(2n-3)$. Check that $p(1) = 0$, $p(2) = 1/3$. Use this recurrence to find $p(n)$ for $2n = 10000$ and check your guess. (XXX. IMO 1989.)

27. All odd primes are of the form $4n + 1$ or $4n + 3$. Let $\pi_1(x)$ and $\pi_3(x)$ be the numbers of primes $< x$ of type $4n + 1$ and $4n + 3$, respectively. It seems that always $\pi_1(x) < \pi_3(x)$. But this is not true. Find the first x such that $\pi_1(x) > \pi_3(x)$.

28. How many of all 6-words from 000000 to 999999 have the property that the digital sums of the first and second halfs are equal? (See Ex.16 after Section 5.) Let this number be denoted by N_6^{10}, where 10 stands for the number base b and 6 for the length $2s$. One can show that

$$N_6^{10} = \frac{1}{\pi} \int_0^\pi (\sin(10x)/\sin x)^6 \, dx, \quad \text{and} \quad N_{2s}^b = \frac{1}{\pi} \int_0^\pi (\sin(bx)/\sin x)^{2s} \, dx$$

a) Find N_6^{10} by means of the program tramisim in Fig. 58.5.

b) We consider for N_6^{10} the approximate formula

$$
(1) \qquad N_6^{10} = \frac{1}{n} \sum_{i=1}^n f(x_i) \, , \quad x_i = x_0 + i\pi/n,
$$
$$
f(x) = (\sin(10x)/\sin x)^6, \quad x_0 \neq k\pi/2, \quad k \in \mathbb{Z} \, .
$$

The restriction avoids $0/0$. Show empirically that for $n \geq 28$ the formula (1) is exact.

c) From the general formula $N_{2s}^b = \sum_{i=0}^m (-1)^i \binom{2s}{i} \binom{(s-i)b+s-1}{2s-1}$, $m = \lfloor s(1-1/b \rfloor$ find some values of N_{2s}^b for different b and s, and try to discover empirically for which n in terms of s and b the analogue of formula (1) is exact.

29. Let $N(2n)$ be the number of $2n$-digit decimal words with the equisum property, and let $S(i, k)$ be the number of i-digit words with digit sum k. Show that

$$S(n+1, k) = \sum_{i=0}^{\min(k,9)} S(n, k-i), \quad \text{and} \quad N(2n) = \sum_{k=0}^{9n} S(n, k)^2.$$

Write a program, which computes a table of the numbers $S(n, k)$, and find $N(2n)$.

30. In the money changing problem of section 13 we have assumed $d[1] = 1$. Dropping this assumption we get the Frobenius Problem. Modify the

program **change** into **change4**, which uses any set of coins to pay the amount n by these coins. Check: For $n = 30000$, $k = 4$ and coins of denominations 131, 247, 353, 661 you will get $a(n, k) = 638$.

31. Write a program **change5**, which uses **change4** of the preceding problem. It should find G, the largest not representable integer. *Hint*: $i = G$ if i has 0 representations and $d[1]$ numbers after i have more than 0 representations. Experiment with 2, 3, 4 types of coins. For 2 types of coins you will get quickly a formula for G. From 3 upward the problem seems hopeless. If your program is correct you will get $G = 4464$ for the set of coins $\{131, 247, 353, 661\}$.

32. To each vertex of a pentagon we assign an integer x_i with sum $s = \sum x_i > 0$. If any of the numbers is negative, we pick a negative x_i, add it to both of its neighbors and replace x_i by $|x_i|$. Decide if the process always stops.

This was the most difficult problem of the IMO 1986. You are not expected to solve it but you can find empirically the largest possible number $f(n)$ of steps until stop starting with $(n, n, 1 - 4n, n, n)$, n a positive integer. As a check of your program we give $f(3) = 50$.

A Crash Course in Probability

Fig. A1 shows a spinner with perimeter 1:

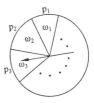

$$(1) \qquad p_i \geq 0, \quad \sum_{i \geq 1} p_i = 1.$$

Spinning it once is a *one stage random experiment*. ω_i denotes the *outcome* that the spinner comes to rest in sector i. The set

Fig. A1

$$\Omega = \{\omega_1, \omega_2, \omega_3, \ldots\}$$

of all possible outcomes is the *sample space*.

Spinning the spinner n times is an *n-stage experiment*, the outcomes being n-words from the alphabet Ω. Suppose that in n spins, ω_i is obtained F_i times. Then F_i is the *absolute frequency* and $f_i = F_i/n$ is the *relative frequency* of ω_i.

The symmetry of the circle implies that if the relative frequency of ω_i always approaches the same value, that value has to be the proportion of the perimeter assigned to ω_i, i.e. p_i. Indeed, Las Vegas records of various spinning devices confirm that for large n, $f_i \approx p_i$, although the approach of the relative frequency to p_i with increasing n is slow and erratic. We call p_i the *probability* of the outcome ω_i.

We should mention that it is a profound problem to come up with a *logical* argument that averages over many repetitions of a random experiment *must* approach limits; they could conceivably fluctuate aimlessly forever.

We call every subset A of Ω an *event*. Suppose a spin results in ω. If $\omega \in A$ then we say that *the event A has occurred*.

For every event the frequency of occurrences of A is the sum of the frequencies of the outcomes in A. Hence the probability of the event A is

$$(2) \qquad P(A) = \text{sum of the probabilities of all outcomes in } A.$$

If $|\Omega| = n$ and $p_i = 1/n$ for all i, i.e. the outcomes are equally likely, we have

$$(3) \qquad P(A) = \frac{|A|}{|\Omega|} = \frac{\#\text{of elements in } A}{\#\text{of elements in } \Omega} = \frac{\#\text{of favorable cases}}{\# \text{ of possible cases}}.$$

One would think that a more natural mathematical description of the spinner would be to take the set of all angles as the set of outcomes. However, that model requires a more complicated theory; not only are there infinitely many outcomes, but each has probability 0. We study only *discrete probability,* which is the theory of random experiments in which every possible outcome has positive probability. Usually we consider a *random process* in which we perform a *series* of random experiments. A random process can be pictured by a *directed graph,* with vertices called *states.* At each state there is a spinner, which determines the *transition probabilities* to the neighboring states. Fig. A2 shows such a graph. A "particle" starts in some state, say state 0. Then it performs a walk from state to state, according to the transition probabilities. For example, in state 3 there is the spinner in Fig. A3 which decides where to go from there. Such a walk is called a *random walk.* One of the infinitely many possible evolutions could be the path (word, outcome) 01201232123K. With this terminology we can say:

Discrete probability studies random walks on graphs.

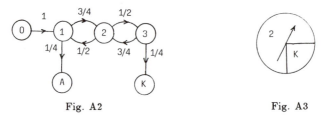

Fig. A2   Fig. A3

To see how path probabilites are obtained from transition probabilities, consider the following example. From each urn in Fig. A4 (from left to right) we select at random a letter and we write the letters in the order in which they were drawn. What is the probability of the outcome (path, word) SOS in Fig. A5?

Suppose we repeat the experiment a large number, N, times. Fig. A6 shows the computation of the approximate numbers of paths which reach the two intermediate states and the final state. The example tells us that the probability of SOS is

$$P(SOS) = \frac{1}{4} \frac{1}{3} \frac{2}{5} = \frac{1}{30}.$$

A general formulation of what we have just observed is:

P1. The probability of a path is equal to the product of the probabilities along the path.

This is the *first path rule.*

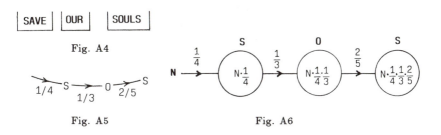

Fig. A4

Fig. A5   Fig. A6

A second example illustrates another rule. From the urn in Fig. A7 we draw two letters at random without replacement. What is the probability that the second letter will be "a"? Tracing the three paths which give "a" in the second draw (see Fig. A8) we get

| b a n a n a |

Fig. A7

$$P(a \text{ at second draw}) = \frac{1}{2}\frac{2}{5} + \frac{1}{3}\frac{3}{5} + \frac{1}{6}\frac{3}{5} = \frac{1}{2}.$$

In general, we have the *second path rule:*

P2. The probability of getting from S to F is the sum of the probabilities of all paths leading from S to F.

This is just a reformulation of (2) into the language of paths. Here we think of the possible outcomes as all ordered pairs of letters that can be drawn and $A = \{aa, ba, na\}$.

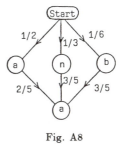

Fig. A8

The object of the mathematical theory of probability is to compute the probabilities of more complicated events associated with random experiments from known probabilities of the possible outcomes of a single experiment. For this purpose a simplified concept of probability suffices.

The only thing we need to introduce into the mathematical theory is that the probabilities p_i of the outcomes ω_i of a single experiment are nonnegative real numbers satisfying (1), and that the probabilities of compound events are given by the rules $P1$ and $P2$.

That individual outcomes are unpredictable, while long-term averages approach limits, is a subtle and mysterious fact, but to compute probabilities we need only the very simple consequences of this fact which we have stated above! Mathematicians were very pleased when it became clear that the mathematical theory could be derived from axioms as simple as these. We ought to add however that the relative-frequency meaning of probability enables us to grasp easily many facts of probability theory which are much less easy to formulate or prove by means of the axioms.

Next we look at examples which bring up additional concepts.

Bold Gamble. You have 1 dollar but you desperately need 5 dollars. So you go into a casino where you can win as much as you bet with probability p and you lose your bet with probability $q = 1 - p$. You decide to use the *bold strategy*: at each stage you stake so much money as to minimize the distance to your goal. Find the probability of going broke.

Fig.   A9 shows the graph of the game.   You start at state 1. You stake all your money.   With probability p you move to state 2 and with probabilty q you are ruined.   In state 2 you stake all your money, in state 4 you stake one dollar, in state 3 two dollars.

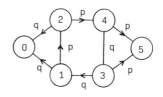

Fig. A9

The probability B of going broke is the sum of the probabilities of all paths leading to state 0. You can lose directly, or via one, two, ... cycles. Thus $B = q + pq + p^2q^2(q + pq) + p^4q^4(q + pq) + \dots$ This is a geometric series with sum

$$B = \frac{q + pq}{1 - p^2q^2} \ .$$

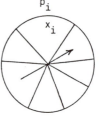

Fig. A10

Often the random experiment produces a *random variable* X.   This means that the possible outcomes are real numbers x_1, x_2, ... whose probabilities are p_1, p_2, ... You can always interpret the x_i as gains in a roulette game (Fig. A10).

Suppose you spin the spinner in Fig.   A10 N times, where N is large.   We can compute the average gain as follows.   We shall win x_i approximately Np_i times, so the total gain will be approximately

$$Np_1x_1 + Np_2x_2 + Np_3x_3 + \dots .$$

If we play long enough, the average gain per game is going to be close to $1/N$-th of this:

(4) $E(X) = x_1p_1 + x_2p_2 + x_3p_3 + \dots = \sum_{i \geq 1} x_ip_i \ .$

The quantity $E(X)$ defined by (4) is called the *expectation* or *expected value* of X. Note that long term averages and the complicated concept of playing "long enough" enter only into explaining the meaning of $E(X)$; as far as mathematical work is concerned, $E(X)$ is just the quantity given by (4).

Suppose that besides X, another random variable X' with values x'_1, x'_2, ... and corresponding probabilities p'_1, p'_2, ... is defined. If a is any real constant then the definition of expectation as a long-term average tells us that the expectations of the random variables aX and $X + X'$ are:

(5a) $\quad E(aX) = aE(X)$, $\qquad$ (5b) $\quad E(X + X') = E(X) + E(X')$.

We wish to show how these formulas can be derived from (4), which serves as the definition of expectation in the mathematical theory.

Formula (5a) is an immediate consequence of (4). To derive (5b) we need to set up a notation for an experiment in which one value of each of two random variables X, X' is obtained. (The same approach is needed if the experiment involves obtaining two values of the same random variable.)

An outcome of the two-variable experiment is that we get a value x_i for X and and a value x'_j for X'. Let p_{ij} denote the probability of this outcome. In general these quantities can not be computed from p_1, p_2, ... and p'_1, p'_2, ...

For example, suppose we have an urn with 4 balls, two with 0 written on them and two with 1. X and X' are obtained by drawing 2 balls without replacement. Assuming that each ball is equally likely to be the first and each of the remaining ones is equally likely to be the second we get

$$p_1 = p_2 = p'_1 = p'_2 = \frac{1}{2} \quad \text{and} \quad p_{11} = p_{22} = 1/6, \qquad p_{12} = p_{21} = 1/3.$$

The only relations between the individual probabilities and the joint probabilities are the following. The event $X = x_i$ occurs if we have any one of the outcomes $X = x_i$ and $X' = x'_1$ or x'_2 or x'_3, ... Hence by (2)

(6a) $\quad p_{i1} + p_{i2} + \ldots = p_i$ $\quad$ and similarly $\quad$ (6b) $\quad p'_{1j} + p'_{2j} + \ldots = p'_j$.

Applying formula (4) to $X + X'$ we get

$$E(X + X') = \sum_{i,j} (x_i + x'_j) p_{ij}.$$

If we separate out the terms involving the x_i and apply (6a) we get $E(X)$ and the remaining terms combine to give $E(X')$.

We return to random walks on directed graphs. Usually there is a set A of *absorbing* states, i.e. states you cannot leave. The remaining states are the set I of *interior* states. We are interested in two problems.

If we start in some state i, what is the probability p_i of reaching some subset W of A?

Starting in a state i, what is the expected time (# of transitions), E_i, to reach A?

For $i \in W$ we have $p_i = 1$, for $i \in L = A \setminus W$ (the complement of W in A) we have $p_i = 0$. If we are in state $i \in I$, let p_{ik} denote the probability that the next move will be to state k. We have

$$(7) \qquad p_i = \sum p_{ik}p_k,$$

where the sum is over all neighbors k of i. Indeed: you go from i to k with probability p_{ik}, and from there you reach W with probability p_k.

For all $i \in A$ we have $E_i = 0$ and for $i \in I$ we get

$$(8) \qquad E_i = 1 + \sum p_{ik}E_k \quad \text{(The sum is over all neighbors } k \text{ of } i\text{.)}$$

Indeed, you make one step. The step is to k with probability p_{ik}, and from there the expected time to absorption is E_k.

Expectations can be infinite. For instance, in a symmetric random walk on a line, starting at O, the expected number of steps to return to O is infinite. The system of equations $E_i = 1 + \frac{1}{2}E_{i-1} + \frac{1}{2}E_{i+1}$ has no solution, unless we consider $E_i = +\infty$ a solution. (What happpens is that as time goes on, the particle wanders over ever longer stretches of the line. One can show that it is almost certain to come back eventually to the origin or, equivalently, to any vertex it has been to, but the average time it takes to return to any one of them must increase without limit as there are more and more vertices waiting to be revisited.)

Our easy to apply rules solve quite complicated problems. But we need practice in representing a problem by a graph and in applying formulas (7) and (8).

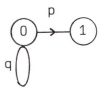

Fig. A11

1) Fig. A11 shows a graph. What is the expected time m to get from 0 to 1 ? The expected time to get from 1 to 1 is 0. So we get

$$(9) \qquad m = 1 + qm, \quad q = 1 - p, \text{ and } m = 1/p.$$

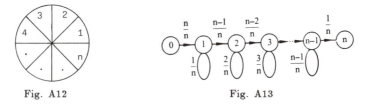

Fig. A12   Fig. A13

2) Spin the spinner in Fig. A12 until all n outcomes have occurred. What is the expected number of spins ?

Look at the graph in Fig. A13. We are in state i if we have collected i

outcomes. By (9) the expected time to get from i to $i+1$ is $n/(n-i)$. The expected time to get from 0 to n is, because of (5),

(10) $$E(n) = \sum_{i=1}^{n} n/(n-i) = n(1 + \frac{1}{2} + \frac{1}{3} + \ldots + \frac{1}{n}).$$

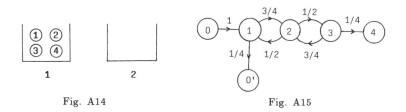

| ① ② |
| ③ ④ |

1   **2**

Fig. A14   Fig. A15

3) Fig. A14 shows two urns labeled 1 and 2. At the start urn 1 contains four balls. We repeatedly select at random one of the balls and move it to the other urn. The system is in state i if the second urn contains i balls. We stop if either state 0 recurs or state 4 is reached for the first time. The random process can be represented by the graph in Fig. A15. Let p_0 be the probability of being absorbed in 4 if we start in 0. Our set W consists of the state 4. The equations (7) are

$$p_0 = p_1, \quad p_1 = \frac{3}{4}p_2, \quad p_2 = \frac{1}{2}p_1 + \frac{1}{2}p_3, \quad p_3 = \frac{3}{4}p_2 + \frac{1}{4}$$

Solving this we get $p_0 = \frac{3}{8}$.

Let E_i be the expected time until stop, starting in state i. Equations (8) become

$$E_0 = 1 + E_1; \quad E_1 = 1 + \frac{3}{4}E_2; \quad E_2 = 1 + \frac{1}{2}E_1 + \frac{1}{2}E_3; \quad E_3 = 1 + \frac{3}{4}E_2.$$

For E_0 this system of equations gives the value 8.

4) Consider the following procedure. Start with $x = 3$ coins. While $x > 0$ toss all coins and eliminate those which show heads. What is the expected number of tosses until $x = 0$?

Let a, b, c be the expected number of tosses for $x = 3, 2, 1$. From the graph in Fig. A16 we get

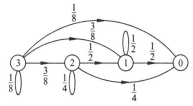

Fig. A16

$a = 1 + a/8 + 3b/8 + 3c/8$, $\quad b = 1 + b/4 + c/2$, $\quad c = 1 + c/2$, with the result $c = 2$, $b = 8/3$, $a = 22/7$.

5) Another example of the use of our theory is the solution to the Three Pile Game (Additional Exercises for Sections 1 to 41, ex. 2): We have three piles of a, b, and c chips, respectively. Each second two different piles X, Y are chosen at random and a chip is moved from X to Y. Let T be the waiting time until one pile becomes empty. Try to find by simulation a formula for the expected time $E(T) = f(a, b, c)$.

By simulation we can arrive at the guess

$$f(a, b, c) = \frac{3abc}{a + b + c}.$$

Our theory enables us to verify this formula. The six neighbors of the state (a, b, c) are $(a, b + 1, c - 1)$, $(a, b - 1, c + 1)$, $(a + 1, b, c - 1)$, $(a - 1, b, c + 1)$, $(a + 1, b - 1, c)$, $(a - 1, b + 1, c)$. Equation (8) becomes

$(12a)$ $f(a, b, c) = 1 + \dfrac{1}{6} \sum f(x, y, z)$ (sum over all neighbors of (a, b, c))

$(12b)$ $f(a, b, 0) = f(a, 0, c) = f(0, b, c) = 0$ (boundary conditions).

We can verify that the function (11) is a solution of (12) by substitution. But is it the only solution? Suppose there is another solution $g(a, b, c)$. We consider $h(a, b, c) = f(a, b, c) - g(a, b, c)$. At an interior point (i.e. a point where a, b, c are all positive) this difference satisfies

$$h(a, b, c) = \frac{1}{6} \sum h(x, y, z).$$

In words: the value of h at an interior point is the average of the values at the six neighboring points.

Let h_{max} be the maximum of h. If $h = h_{max}$ at an interior point then we must have $h = h_{max}$ at all neighbors of that point also. We can go from point to point and obtain $h = h_{max}$ at every point. Since $h = 0$ on the boundary, $h = 0$ everywhere.

If $h = h_{max}$ only on the boundary we still have $h_{max} = 0$ but now we can say only that $h \leq 0$ everywhere. However, we can apply the same argument to the minimum of h and get $h \geq 0$ everywhere so that we again get $h = 0$. Hence h is 0 everywhere, and hence every solution g of (12) is equal to f.

A superb presentation of discrete probability theory, for those who wish to immerse themselves in the subject, is William Feller's Introduction to the Theory of Probability, vol. 1.

Solutions

Sections 1-5

1. a) f(20)/f(19)=4.7326261527   b) Same answer as in a). Use Fig. 3.4 after changing 1E−09 into 1E−08.

2.
```
program bisiter;
var a,b:real;
function f(x:real):real;
begin f:=x*x*x-x-1 end;
function bis(a,b:real):real;
const eps=1E-09;
var m,y:real; faneg:boolean;
begin faneg:=f(a)<0.0;
   repeat
     m:=(a+b)/2.0; y:=f(m);
     if (y<0.0)=faneg then a:=m else b:=m
   until abs(a-b)<eps;
   bis:=m
end;
begin
   write('a,b='); readln(a,b); writeln(bis(a,b))
end.
```
The iterative program `bisiter` finds a root of the continuous function f in the interval $[a, b]$, if $f(a) * f(b) < 0$. The boolean variable **faneg** is true if $f(a)$ is negative. The name should remind you of this. Study it in detail.

3. a) 0.73908513226 b) 1.7632228322 c) 0.56714328937 d) 2.9999997616

4. $x_1 = -1.8793852418$, $x_2 = 3.4729635529$, $x_3 = 1.5320888860$.

5. b) We conjecture $g^2(n) - 5 * f^2(n) = 4 * (-1)^n$. Proof by induction.

6. a) True. b) True. Proof by induction.

7. a) $f(1/3) = pf(2/3) = p(p + qf(1/3))$, $f(1/3) = p^2/(1 - pq)$.
   b) $f(2/5) = pf(4/5) = p^2 + pqf(3/5) = p^2 + p^2q + pq^2f(1/5)$
      $= p^2 + p^2q + p^2q^2 f(2/5)$, $f(2/5) = p^2(1 + q)/(1 - p^2q^2)$.
   c) $f(1/1984) = p^{11}/(1 - p^4q)$. For $p = q = 1/2$ we get $f(x) = x$.
   d) The longer the period of the binary expansion of x the harder it is to find $f(x)$.

8.
```
program joseph2;
var a,b,x,n,k,s,q:real;
function ceil(x:real):real;
begin if x=int(x) then ceil:=x else ceil:=int(x)+1 end;
begin writeln(' n,k,s,' ); readln(n,k,s);
   q:=k/(k-1); x:=1+k*(n-s); a:=ceil(x);b:=a;
   while b<=n*k do begin a:=b; b:=ceil(q*a) end;
   x:=1+k*n-a;
   writeln(' the ' s:0:0,'-th elim person has # ',x:0:0)
end.
```

11.   We get an unbranching recursion if we compute u values in blocks of two. This reduces the number of operations to $O(\log n)$. (The time complexity is greater because just the number of digits in $u(n)$ is $O(n)$.)

In Turbo Pascal a function value can not be a pair of numbers. We use a recursive *procedure* and get the function values from the lower level of recursion by means of *variable parameters* (see p. 122).

```
program  fastfib;                    if odd(n) then
var  n,u,v:longint;                    begin  t:=2*u*v+v*v;
procedure u2(n: longint;                 v:=v*v+sqr(u+v); u:=t end
             var u,v:longint);       else begin t:=u*u+v*v;
var t: longint;                         v:=2*u*v+v*v; u:=t end;
begin                                end;
if n=1 then                          end;
  begin u:=0; v:=1; end              begin
else begin                           writeln('n='); readln(n);
  u2(n div 2, u, v);                 u2(n,u,v); write('u(',n,')=',u);
                                     readln end.
```

12. By a slight modification of **joseph1** we get the program **jos_perm** in Fig. 47.4.

13. a) s:=0; n:=0; **repeat** n:=n+1; s:=s+1/n **until** n=10000;
 b) s:=0; n:=10000; **repeat** s:=s+1/n; n:=n-1 **until** n=0;
 c) When two positive floating-point numbers with different exponents are added, the digits of the smaller number which are in places below the least significant digit of the larger number are mostly neglected. At best, they are taken into account in the rounding of the result. Thus, if we form the sum starting with the smallest terms, the error will be smaller than if we start with the largest terms. It would be even better to add groups of small terms and then add the resulting totals instead of adding the terms individually to the total.

14. The computations suggest

$$(1) \qquad \mathbf{round}\left(\frac{n}{2}\right) + \mathbf{round}\left(\frac{n}{4}\right) + \mathbf{round}\left(\frac{n}{8}\right) + \ldots = n.$$

This can be proved as follows. The formula is true for $n = 0$. If we increase n by 1, the term $\mathbf{round}(n/2^k)$ in the sum (1) is unchanged except when n reaches a value which is an odd multiple of 2^{k-1}, and then it increases by 1. For each value of n there is one such k and therefore the sum of the series is n.

15. a) for x:=1 to 100 do for y:=1 to 100 do
          if abs(x*x-x*y-y*y)=1 then writeln(x,' ',y);
 We get pairs of of successive Fibonacci numbers.
 b) We observe that the solutions are

$$(1) \qquad (\mathrm{fib}(n + 1), \mathrm{fib}(n)) \qquad n = 1, 2, \ldots.$$

It is easy to verify that

$$(2) \qquad \text{if } w = u + v, \text{ then } w^2 - wv - v^2 = -(v^2 - vu - u^2).$$

This equation enables us to start from the solution $(1,1) = (\mathrm{fib}(2), \mathrm{fib}(1))$ and obtain the family of solutions (1).

Next we show that if x and y are > 0 and

(3) $$|x^2 - xy - y^2| = 1$$

then (x, y) is in the family (1).

If $x = 1$ then (3) implies $y = 1$ so we have the first member of the family (1). If $x > 1$, (3) implies $y < x$. By (2), $(y, x - y)$ is also a solution with positive components. We can continue to descend as long as the x-value is > 1, i.e. until we reach $(1,1)$. We can now reverse the process and get back to the (x, y) we started with, and this amounts to forming the Fibonacci sequence.

16. We give 5 different solutions, which are successively more sophisticated and efficient. A sixth program will be given later.

```
program equisum1;          program equisum2;          program equisum3;
var a,b,c,d,e,f:           var a,b,c,p:               var a,b,s,p:integer;
integer; count:longint;    integer;count:longint;        count:longint;
begin count:=0;            begin count:=0;            begin count:=0;
  for a:=0 to 9 do           for s:=0 to 27 do          for s:=0 to 27 do
  for b:=0 to 9 do           begin p:=0;                begin p:=0;
  for c:=0 to 9 do             for a:=0 to 9 do           for a:=0 to 9 do
  for d:=0 to 9 do             for b:=0 to 9 do           for b:=0 to 9 do
  for e:=0 to 9 do             for c:=0 to 9 do           if (a+b<=s) and
  for f:=0 to 9 do             if a+b+c=s                 (a+b>=s-9)
  if a+b+c=d+e+f               then p:=p+1;               then p:=p+1;
  then                         count:=count+p*p           count:=count+p*p
  count:=count+1;            end;                       end;
  writeln(count)             writeln(count)             writeln(count)
end.                       end.                       end.

program equisum4;                    program equisum5;
var a,b,c,s:integer;                 var i:integer; count:longint;
    p,count:longint;                 function p(s:integer):integer;
begin count:=0;                      begin
  for s:=0 to 13 do                    if s=0 then p:=1
  begin p:=0;                          else
    for a:=0 to 9 do                     if s<=9 then p:=p(s-1)+s+1
      for b:=0 to 9 do                   else p:=100-p(s-10)-p(17-s)
        if (a+b<=s)                    end;
        and (a+b>=s-9)               begin count:=0;
        then p:=p+1;                   for i:=0 to 13 do
  count:=count+p*p                       count:=count+sqr(p(i));
  end;                               count:=2*count;
  writeln(2*count:0:0)               writeln(count)
end.                                 end.
```

The programs equisum1 to equisum4 use 1000000, 28000, 2800, 1400 comparisons, respectively. In equisum2 we denote by $p(s)$ the number of triples (a, b, c) with fixed sum s. For each such sum $s = a + b + c$ there are $p(s)$ triples (d, e, f) with the same sum $s = d + e + f$. So $p(s) * p(s)$ is the total number of cases with sum s. This is added up for $s = 0$ to 27. In **equisum3** we save the inner c-loop by means of the test **if** (a+b<=s) **and** (a+b>=s-9), which is satisfied for $0 \leq c \leq 9$. For given s, a, b the variable c is also given by $c = s - a - b$.

In **equisum4** we use the bijection $abcdef \leftrightarrow a'b'c'd'e'f'$ between blocks with equisum property, where $a' = 9 - a$, etc. If $s = a + b + c$ then $s' = a' + b' + c' = 27 - s$. So triples (a, b, c) with sum s and $27 - s$ are equinumerous. Here s runs from 0 to 13 and the total count is doubled at the end.

The recursive program **equisum5** is based on the recurrences

$$p(s) = p(s - 1) + s + 1 \quad \text{for} \ s \leq 9,$$
$$p(s) = 100 - p(s - 10) - p(17 - s) \quad \text{for} \ s > 9.$$

Here is an explanation: A triple abc with sum s is determined by its two digits a, b. Also, any pair of digits a, b with

(1) $s - 9 \leq a + b \leq s$

will form such a triple with $c = 9 - a - b$. Thus $p(s)$ is the number of solutions of (1). To count these, consider the 100 lattice points (a, b) in or on the boundary of the square $S : 0 \leq a, b \leq 9$. For $s \leq 9$, $p(s)$ is the number of lattice points in or on the boundary of the triangle cut off from S by the line $a + b = s$. For $s > 9$ we can think of the lattice points in the region (1) as the lattice points in S from which we remove the lattice point in or on the boundary of the triangle $a + b \leq s - 10$ and also the lattice points in or on the boundary of the triangle $a + b \geq s + 1$. The last triangle is congruent to the triangle $a + b \leq 17 - s$.

17. c) We know that for all nonnegative integers k, q, i, $a(2^i(4k + 1)) = 1 \neq a(2^i(4(k + q) + 3)) = 0$. The difference of the arguments is $2^{i+1}(2q + 1)$. For any period p, $6p$ would certainly have this form so the sequence can not be periodic.

Section 6

5. a) We want to find $x = 7^{9999}$ mod 1000. We first find the smallest power of 7 which is 1 mod 1000.

   `p:=1; i:=0; repeat i:=i+1; p:=7*p mod 1000 until p=1;`

   We get $i = 400$. Since $9999 = 400 * 24 + 399$ we have $7^{9999} \equiv 7^{399}$ (mod 1000).

   `p:=1; for i:=1 to 399 do p:=7*p mod 100;`

   Now $p = 143$, which is the solution x.

   To get $y = 7^{9999}$ mod 10000 we must use **real** or **longint** variables to avoid integer overflow. A completely different solution is based on the program **lincom**. We want to find $x = 7^{9999}$ mod 1000, or $7x = 7^{10000}$ mod 1000. Now $7^{10000} = 7^{400*25} = 1$ mod 1000. Thus $7x \equiv 1$ mod 1000 or $7x + 1000y = 1$. The program **lincom** gives $x = 143$, $y = -1$, and so $x = 143$. b) and c) may require **relincom**.

7. a)   `sum:=0; for i:=1 to n do sum:=sum+i;`

   b)   `function sum(n:integer):integer;`
           `begin if n=1 then sum:=1 else sum:=sum(n-1)+n end;`

8. a)   `function bin(n:integer):byte;`
                   `begin bin(n div 2); write(n mod 2) end;`

   b)   `repeat write(n mod 2); n:=n div 2 until n=0;`

   In b) the digits are printed in reverse order. In a) the digits are printed in correct order. But if we switch the bin- and write-procedures we get the bits in reverse order.

9.   `d:=0; repeat d:=d+n mod 10; n:= n div 10 until n=0;`

10.   `program digdig;`
                `var n:integer;`
                `function digsum(n:integer):integer;`
                `var m,s:integer;`
                `begin m:=n; s:=0;`
                  `repeat s:=s+m mod 10; m:=m div 10 until m=0;`
                  `digsum:=s`
                `end;`
                `begin`
                  `write('n='); readln(n);`
                  `repeat n:=digsum(n) until n<10;`
                  `writeln(n)`
                `end.`

   $f(n)$ is 9 if n is a multiple of 9; otherwise it is n mod 9.

11.   `rev:=0;`
        `repeat rev:=10*rev+n mod 10; n:=n div 10 until n=0;`
        `writeln(rev);`

12.
```
        function c(n,s:integer):integer;
        begin
           if (s=0) or (n=0) then c:=1
           else c:=c(n-1,s-1)+c(n-1,s)
        end;
```

13.
```
        function c(n,s:integer):integer;
        begin
           if s=0 then c:=1 else c:=c(n-1,s-1)*n/s
        end;
```

15. $q(n) = \text{fib}(n+1)/\text{fib}(n)$ approaches $q = 1.6180339887\ldots = (\sqrt{5}+1)/2$. Show by induction that $q(n) - q = (-1)^n/\text{fib}(n)/q^n$.

16.
```
        program div_by_d;
        var  a,b,c,d:longint; n:integer;
        begin
          write('d='); readln(d); a:=0; b:=1; n:=1;
          repeat
            c:=a; a:=b; b:=b+c; n:=n+1;
            if b mod d=0 then write(' fib(',n,')= ',b);
          until b>1200000000;
          readln end.
```

17. a) The sequence completes a cycle if two successive terms repeat. This must happen because of the pigeonhole principle.

  b) Addition mod m is invertible. Hence the sequence can be extended uniquely into the past. But in a loop with a tail the end of the tail has two predecessors.

c)
```
program FibPeriodMODm;           repeat L:=L+1;
var a,b,c,m,L:longint;             c:=a; a:=b; b:=b+c mod m
begin write('m='); readln(m);    until (a=0) and (b=1);
  a:=0; b:=1; L:=0;                writeln(L)
                                 end.
```

18. a) $L(p^n) = L(p) * p^{n-1}$ for every prime p.

  b) $L(p_1^{a_1} p_2^{a_2} \cdots p_r^{a_r}) = \text{lcm}(L(p_1^{a_1}), L(p_2^{a_2}), \cdots, L(p_r^{a_r}))$. In particular,

$$L(10^n) = \text{lcm}(3 * 2^{n-1}, 2^2 * 5^n) = \begin{cases} 12 * 5^n & \text{for} \quad n \leq 3 \\ 15 * 10^{n-1} & \text{for} \quad n < 3 \end{cases}$$

19. a) $L \mid p-1$.   b) $L \mid 2(p+1)$.   c) $L = 20$.

21. $\text{Zeta}(3) = 1 + 1/2^3 + \cdots + 1/n^3 + 1/2(n+1/2)^2 = 1.2020569032\ldots$ from $n = 200$ on.

22. Let A_d be the event $\gcd(x,y,z) = d$. A_d is equivalent to the occurrence of four independent events

$$d \mid x, \quad d \mid y, \quad d \mid z, \quad \gcd(x/d, y/d, z/d) = 1.$$

Thus

$$P(A_e) = \frac{1}{d} * \frac{1}{d} * \frac{1}{d} * p_3 = p_3/d^3, \text{ where } p_3 = P(\gcd(x,y,z) = 1).$$

Since $A_1 \cup A_2 \cup A_3$ will occur with probability 1, we have

$$p_3 \sum_{d \geq 1} 1/d^3 = 1, \text{ or } p_3 = 1/\sum_{d \geq 1} 1/d^3 \ .$$

23. Let A_d be the event $\gcd(x_1, \cdots, x_k) = d$. That is

$$d \mid x_1, \ \cdots, \ d \mid x_n, \ \ \gcd(x_1/d, \cdots, x_n/d) = 1.$$

Then $P(A_d) = p_k/d^k$, where $P(A_1) = 1$. Thus

$$p_k \sum_{d \geq 1} 1/d^k = 1 \to p_k = 1/\sum_{d \geq 1} 1/d^k \ .$$

24. a) will be proved after c).

First we show that the process we gave for generating the Morse-Thue sequence is equivalent to the following:

Denote by S the operation, applicable to a sequence of 0's and 1's, which replaces each 0 by 01 and each 1 by 10. Observe that 10 is the complement of 01 so that if we apply S to a pair of complementary sequences we get a pair of complementary sequences. We claim the initial segment of 2^n digits of the Morse-Thue sequence is $S^n(0)$. Indeed this is clear for $n = 0$ and $n = 1$. Now $S^{n+1}(0) = S^n(01)$, which is $S^n(0)$ followed by $S^n(1)$. Since the S-operator takes complements into complements, what we have now is the rule of formation of the Morse-Thue sequence originally given.

Observe that the operation S can be applied to the entire infinite Morse-Thue sequence and it leaves the sequence unchanged. Suppose now that $x(n) = 0$. In the Morse-Thue sequence $x(n)$ is preceded by n digits. If we apply S to the sequence we replace $x(n) = 0$ by 01, and the n digits preceding it by $2n$ digits. So in this case $x(2n) = x(n)$ and $x(2n + 1) = 1 - x(2n)$. The case $x(n) = 1$ follows similarly. This proves c); the assertion b) is just another way of phrasing the first part of c).

To prove a), we note that the two equations c) together imply that $x(n)$ and $x(2n + 1)$ are always different. If the sequence were ultimately periodic and $x(n)$ were in the periodic part of the sequence, we could conclude that $n + 1$ is not a multiple of the period. The same would be true of $n + 2, n + 3, \ldots$ but this is impossible.

c)
```
      function x(n:integer):byte;
      begin
        if n=0 then x:=0
        else if odd(n) then x:=1-x(n-1) else x:=x(n div 2)
      end;
```

d) is just a way of expressing the definition of the sequence.

```
function x(n:integer):byte;
var p:integer;
begin
  if n<2 then x:=n
  else begin
    p:=1; while n>=p do p:=p+p;
    p:=p div 2; x:=1-x(n-p)
  end
end;
```

e) is true because the sequence of $n+1$ digit binary numbers is obtained from the sequence of all numbers with up to n digits by putting a 1 and possibly some 0's in front of them.

```
function x(n:integer):byte;
var s,d:integer;
begin s:=0;
  repeat d:=n mod 2; n:=n div 2; s:=(s+d) mod 2until n=0;
  x:=s
end;
```

25. b) One can argue the same way as for the Morse-Thue sequence above.

```
function x(n:integer):byte;
begin
  if n<3 then x:=n div 2
  else if n mod 3<2 then x:=x(n div 3) else x:=1-x(n div 3)
end;
```

26. You save about 12% in running time.

27.
```
program pascal_triangle;
var a,b,i,j,n:integer; x:array0..100] of integer;
begin
  write('n=');readln(n);
  for i:=0 to n do x[i]:=0;
  for j:=0 to n do
  begin
    a:=0; b:=1; writeln;
    for i:=0 to j do
    begin
      x[i]:=a+b; write(x[i]:5); a:=b; b:=x[i+1]
    end;
  end
end.
```

Use $n = 15$ for input.

c) Replace the 5th line from below by

```
x[i]:=(a+b) mod 2; write(x[i]); a:=b; b:=x[i+1];
```

28. a)
```
for i:=1 to 800 do
   if (i-1) mod 3=0 then
   if (i-1) mod 4=0 then
   if (i-1) mod 5=0 then
   if i mod 7=0 then
      writeln(i)
```
A better approach is this: i-1 is a multiple of 3, 4, 5, i.e., of 60. So we must check the terms $60i + 1$ for divisibility by 7.
```
for i:=1 to 49 do
   if (60*i+1) mod 7=0 then
      writeln(60*i+1);
```

b) One gets $n = 301 + 420t$. For $t = 0$ we get $n = 301$ eggs of weight ≈ 17 kg. For $t = 1$ we get $n = 721$ eggs of weight ≈ 40 kg, which can be carried by a strong man on his back, but not by an old woman in her basket.

c) One generalization: By taking them a, b, c at a time she got remainders r, s, t, but by taking them d at a time, it came out even:

```
for i:=1 to max do
if (i-r) mod a=0 then
if (i-s) mod b=0 then
if (i-t) mod c=0 then
if i mod d=0 then writeln(i);
```
If the smallest nonnegative solution is s then all solutions are $x = s + t * \mathrm{lcm}(a, b, c)$ (Chinese remainder theorem.)

30. a) Let $m = a + b$. Then $b \equiv -a \pmod{m}$. In every step $a \bmod m$ is replaced by $2a \bmod m$. Hence the process will terminate if for some i, $2^i a$ is a multiple of m. This will be the case if and only if a is divisible by the largest odd divisor of m.

b) If a and b have a common divisor, divide it out. So we can assume without loss of generality that $\gcd(a, m) = 1$. We get pure periodicity if for some n, $2^n a \equiv a \pmod{m}$. This means $(2^n - 1)a$ is divisible by m. This implies $2^n \equiv 1 \pmod{m}$. So m divides $2^n - 1$ and it must be odd. Conversely, if m is odd, $2^n \equiv 1 \pmod{m}$ for some n and the least such n is the length of the cycle.

31. $f(0) = 0$, $f(n) = n \bmod 2 + f(n \text{ div } 2)$ for $n > 0$

32. $n = 5186$

33.
```
program involution;
var i,n: integer;
function g(x:real):integer;
var n:integer;
begin n:=1; while n<=x do n:=n+n; g:=n end;
function f(n:integer):integer;
begin
   if n=1 then f:=1
   else if odd(n) then f:=f((n+1) div 2)+g((n-1)/4)
   else f:=f(n div 2)+g((n-2)/2)
end;
begin
   write('n='); readln(n); writeln(f(f(n))=n)
end.
```

SOLUTIONS

Sections 7-9

3.
```
c:=1;
repeat c:=c+4;
  for a:=1 to c-1 do
  begin b1:=sqrt(c*c-a*a); b:=trunc(b1);
    if a<b then if frac(b1)=0 then if gcd(a,b)=1 then
      writeln(a:7,b:7,c:7)
  end
until c>=max;
```
This algorithm works up to $\mathtt{max} = 180$. Then we get integer overflow. That is, for larger c we need **longint** or **real** variables.

5. See references for Chapter II, sections 7-9.

7. $c_4 = \pi/2$

8. $c_{2k} = \pi^k/k!$, $c_{2k+1} = 2(2\pi)^k/(1 \times 3 \times \cdots \times (2k+1))$. The volume of the unit ball approaches 0 as k goes to infinity. The proportion goes to 0 even faster.

10. $\pi = 3.1415927054$ and $\pi = 3.1415926535$. All digits of the second number are correct. See references for Chapter II, Section 9.

Exercises for Section 10.4

2. We show first that the sequence of numbers whose ternary representations contain no 2's contains no three numbers a, b, c such that $a + b = 2c$. The digits of $2c$ are all 0's and 2's. Since the digits in a and in c are all 0's and 1's, the equation could hold only if the 1's in a were in the same place as the 1's in b, which would mean $a = b$.

Next we show that every number which is not in the sequence forms an arithmetic progression with two smaller numbers which are in the sequence. Let c be a number which has one or more 2's in its ternary representation. Let b be obtained by replacing these 2's by 1's, and let a be obtained by replacing them by 0's. Then a, b, c form an arithmetic progression. Since the only numbers exluded from our sequence are the ones which form an arithmetic progression with two smaller numbers in the sequence, our sequence is the one given by the greedy algorithm.

Exercises for Section 10

2. The "obvious" program for the U-sequence is

```
program unique;
var i,j,k,l,n,can: integer;
  a:array[1..200] of integer;
begin write('n='); readln(n);
  a[1]:=1; a[2]:=2; i:=3; can:=3;
  repeat l:=0;
    for j:=1 to i-2 do
      for k:=j+1 to i-1 do
        if a[j]+a[k]=can then l:=l+1;
      if l=1 then begin a[i]:=can; i:=i+1 end;
      can:=can+1
    until i>n;
    for j:=1 to i-1 do write( a[j],' '); readln
end.
```

Comment: can is candidate currently being tested. If it can be represented in one way (l=1) then it is stored in an array $a[1..n]$ of successful candidates.

3. An "obvious" program is:

```
program unique1;
var i,j,k,l,can,n:integer; a:array[0..200] of integer;
begin write('n='); readln(n);
  a[0]:=0; a[1]:=1; i:=2; can:=2;
  repeat l:=0;
    for j:=0 to i-1 do
      for k:=0 to  i-1 do
        if a[j]+2*a[k]=can then l:=l+1;
      if l=0 then begin a[i]:=can; i:=i+1 end;
      if l>1 then write('There is no such sequence. i=',i);
      can:=can+1
    until i>n;
    for j:=0 to i-1 do write(a[j]:5);
  readln end.
```

Use 50 or 100 for n.

4. a)
```
program first_fib;
  var i,d:integer;a,b:real;x:array[0..9] of integer;
  begin a:=1;b:=1;
    for i:=1 to 9 do x[i]:=0;
    for i:=1 to 10000 do
    begin
      d:=trunc(a); x[d]:=x[d]+1; b:=b+a; a:=b-a;
      if a>=10 then begin b:=b/10;a:=a/10 end
    end;
    writeln('digit':10,'frequency':10); writeln;
    for i:=1 to 9 do writeln(i:10,x[i]:10)
  end.
```

For b) and c) one gets the same frequencies.

5. $q^2 \nmid n$ with probability $1 - 1/q^2$ for any prime q. The probability that a

"random" integer is square free is

$$p = \prod_{\text{all primes}} (1 - 1/q^2) = 6/\pi^2.$$

6. Look up the solution of problem 22 of Sections 11 to 16. There you will find an efficient program for the Frobenius Problem.

7. c)
```pascal
program collatz;
var i, b, count:longint;

procedure next(n:longint);
begin
  count:=count+1;
  if n>1 then
    if  odd(n) then next(3*n+1)
    else next(n div 2)
end;

begin
  write('a,b =');  readln(i,b);
  writeln('i':10,'count':11);  writeln;
  repeat
    count:=0;  next(i);  writeln(i:10, count:10);  i:=i+1;
  until i>b;
readln end.
```

This is one of many versions using a **procedure** to get from n to its successor $f(n)$.

Sections 11-16

1.
```pascal
program rotation;                      (* swaps blocks A, B of *)
var a,b,c,d,i,j,k,n,t:byte;            (* lengths n − k and k. *)
x:array[0..255] of byte;
begin write('k,n=');read(k,n);   else        {case c<d}
a:=0; b:=n;  d:=k; c:=n-k;         begin i:=b-c;
for i:=0 to n-1 do x[i]:=i;          while i<>b do
while (c<>0) and (d<>0) do            begin
begin                                    t:=x[i]; x[i]:=x[i-d];
  if d<=c then                           x[i-d]:=t; i:=i+1
  begin i:=a;                          end;
    while i<>a+d do begin               b:=b-c; d:=d-c
      t:=x[i]; x[i]:=x[i+c];          end
      x[i+c]:=t; i:=i+1 end;        end;
    a:=a+d; c:=c-d                  for i:=0 to n-1 do write(x[i]:4)
  end { end of case d<=c}           end.
```

2. The recursive program is clear. Here is the skeleton of the iterative program:

```
b[0]:=1; for i:=1 to n do b[i]:=0;
for i:=1 to n do
if odd(i) then b[i]:=b[i-1] else b[i]:=b[i-2]+b[i div 2];
```

6. a) 2498   b) 1 229 587

7. 63 992

8. 4562

9. $a(200,8)=104\ 561$ with $D=\{1,2,4,10,20,40,100,200\}$

   Scale by 2, so that coins become integers.

10.
```
c[0]:=1;
for k:=1 to n do c[k]:=0;
for i:=1 to s do
for k:=n downto a[i] do
c[k]:=c[k]+c[k-a[i]];
```
Here c[k] is the number of solutions of $a[1]x_1 + \cdots + a[s]x_s = k$. The numbers c[0],...,c[n] are stored in the array c[0..n].

11.
```
for i:=1 to s do readln(a[i]);
for i:=0 to n do c[i]:=0;
for x₁ :=-1 to 1 do
  for x₂ :=-1 to 1 do
    for x₃ :=-1 to 1 do
      for x₄ :=-1 to 1 do
      begin sum:=
        a[1]*x1+a[2]*x2+
        a[3]*x3+a[4]*x4;
        if sum >=0 then
          c[sum]:=c[sum]+1
      end;
for i:=0 to n do write(c[i],' ');
```
This straightforward program is valid for $s = 4$ and

$$n = a[1] + a[2] + a[3] + a[4].$$

It is easy to see how to generalize and complete the program. Unfortunately its time complexity is $O(3^s)$. So it is only feasible for small s, say $s = 12$. The following program has complexity which is proportional to $n * s$, i.e., $O(n * s)$.

```
program rep3;
var i,k,n,n1,s:integer;
    a:array[0..50] of integer; b:array[0..1000] of integer;
begin write('no. and list of weights '); read(s);
for k:=1 to s do read(a[k]);
n:=0; for k:=1 to s do n:=n+a[k]; n1:=n+n;
b[0]:=1; for i:=1 to n1 do b[i]:=0;
for i:=1 to s do
  for k:=n1 downto a[i] do
    if k>=2*a[i] then b[k]:=b[k]+b[k-a[i]]+b[k-2*a[i]]
    else b[k]:=b[k]+b[k-a[i]];
for i:=n to n1 do write(' w=',i-n,': ',b[i],';');
end.
```

12. c) **program** TwoThreeFive;
  var i,n:integer;
  function f(n:integer):boolean;
  var i:integer;
  begin
   for i:=2 **to** 5 **do**
    while n **mod** i=0 **do** n:=n**div** i;

  f:= n=1;
  end;
  begin
   write('n='); read(n);
   for i:=1 **to** n **do**
    if f(i) **then** write(i:8);
  readln **end**.

17. **program** revit;
 var a,b,m,n:integer;
 begin write('n='); read(n);
 a:=1; b:=0; m:=n;
 while m<>0 **do begin**

  if odd(m) **then** b:=a+b
  else a:=a+b;
  m:=m **div** 2
 end;
 writeln('f(',n,')=',b)
 end.

18. a) **function** f(n:integer):integer;
  begin
   if (n=1) **then** f:=1
   else if (n=3) **then** f:=3
   else if not odd(n) **then** f:=f(n **div** 2)
   else if (n-1) **mod** 4=0 **then** f:=2*f(1+n **div** 2)-f(n **div** 4)
   else {(n-3) **mod** 4=0} f:=3*f(n **div** 2)-2*f(n **div** 4)
  end;

 b) In the interval 1..1988 the relation $i = f(i)$ is satisfied 92 times.

 g) f[1]:=1; f[2]:=1; f[3]:=3;
  for i:=4 **to** n **do**
   if not odd(i) **then** f[i]:=f[i **div** 2]
   else if (i-1) **mod** 4=0 **then** f[i]:=2*f[(i+1) **div** 2]-f[i **div** 4]
   else f[i]:=3*f[i **div** 2]-2*f[i **div** 4];

21. $CBA = (A^R B^R C^R)$. Write the corresponding program. It uses the procedure **reverse** four times.

22. The following algorithm can be used in the exploration of the Frobenius Problem:
    b[0]:=1; **for** k:=1 **to** n **do** b[k]:=0;
    for i:=1 **to** s **do**
     for k:=a[i] **to** n **do** b[k]:=b[k]+b[k-a[i]];

Here is an explanation: Let $b(k, i)$ be the number of ways of representing k as a sum of numbers from among $a(1), \ldots, a(i)$. The same summand may occur several times. Representations which differ only in the order of the terms are considered the same. Let $b(0, i) = 1$ and $b(k, 0) = 0$ for $k > 0$. (This expresses the useful convention that the empty sum has value 0). A representation of k as a sum of numbers from among $a(1), \ldots, a(i)$ is one of two kinds: either it has summand $a(i)$ or it is obtained from a representation of $k - a(i)$ by adding $a(i)$ to it. This gives the recursion formula

(1) $b(k, i) = \begin{cases} b(k, i-1) & \text{if } k < a(i); \\ b(k, i-1) + b(k - a(i), i) & \text{if } k \geq a(i). \end{cases}$

It is easy to write a program which computes the arrays $b(k,1)$, $b(k,2)$, ... by means of (1). Observe that once $b(k,i)$ has been computed, $b(k,i-1)$ is no longer needed so instead of a two-parameter array b[k,i] we can use a 1-parameter array b[0..n] and recompute it for successive values of i. In this way we get the algorithm above.

23. Let $p(k,i)$ be the number of ordered i-tuples of decimal digits with sum k. We have $p(0,i) = 1$ for $i = 0, 1, \ldots$ and $p(k,0) = 0$ for $k = 1, 2, \ldots$. It will simplify our formulas to define $p(k,i) = 0$ for $k = -1, -2, \ldots, -9$ and all i. A recursive formula for $p(k,i)$ $(k, i > 0)$ can be obtained as follows: The i-tuples with sum k and last digit $1, \ldots, 9$ are obtained from the i-tuples with sum $k-1$ and last digit $0, \ldots, 8$ by adding one to the last digit. The number of the latter i-tuples is $p(k-1,i) - p((k-1)-9, i-1)$. The number of i-tuples with sum k and last digit 0 is $p(k, i-1)$. Combining these facts we get

$$(1) \qquad p(k,i) = p(k-1,i) + p(k,i-1) - p(k-10,i-1), \qquad k,i > 0.$$

We know the number of ordered triples with sum k is the same as the number of ordered triples with sum $27 - k$. Hence the number of 6-tuples we want is the same as the number of 6-tuples such that the sum of all 6 digits is 27; we get one kind from the other by replacing each of the last 3 digits d_i by $9 - d_i$. Thus $p(27, 6)$ is an answer to our question. A recursive program based on (1) requires 5 minutes on a 12-MHZ AT. Its iterative version equisum6 takes less than 0.02 seconds.

```
program equisum6;
var i,k :integer;   p:array[-9..27,0..6] of longint;
begin
   for i:=0 to 6 do for k:=-9 to 27 do p[k,i]:=0;
   for i:=0 to 6 do p[0,i]:=1;
   for i:=1 to 6 do
      for k:=1 to 27 do p[k,i]:=p[k-1,i]+p[k,i-1]-p[k-10,i-1];
   writeln(p[27,6]:10)
end.
```

Section 17

8.
```
program chain;
var p,max:real;
{$I realprim}
begin write('max=');
   readln(max); p:=max-1;
   repeat p:=p+2
   until prime(p,3.0)
   and prime(p+1,3.0);
   writeln(p:0:0,' ', p+p+1:0:0)
end.
```

p	2p+1
1013	2027
10061	20123
100043	200087
1000151	2000303
10000079	20000159
run time error	
Output for max=100000000	

By means of the program chain we get the pairs of primes in Table I.

We can speed up the program **chain**. Change the **repeat** ... **until** loop as follows:

```
repeat
  p:=p+2.0; a:=prime(p,3.0); if a then b:=prime(p+p+1.0,3.0)
until a and b;
```

In the program each time both p and $2p + 1$ are tested for primality. In the modified program $2p + 1$ is only tested if p is prime. In the modified program we must also declare the boolean variables a and b.

9. The Euler polynomial $f(x) = x^2 + x + 41$ is especially rich in primes. From 0 to 2398 there are 1199 primes, i.e. 50%. From 0 to 4000 there are 1860 primes, i.e. 46.5%.

```
program Euler_polynomial;
const n=2398;
var i,count:integer; x,f:real;
  {$I realprim}
begin   count:=0;
  for i:=0 to n do
  begin
    x:=i; f:=x*x+x+41.0;
    if prime(f,3.0) then count:=count+1
  end;
  writeln('count=',count)
end.
```

16. $n = 4^k(8q + 7)$.

17. a) $8q + 7$ is not the sum of 3 squares.

   b) If $4^k(8q + 7)$ is a sum of 3 squares then so is $4^{k-1}(8q + 7)$

20. $6578 = 1^4 + 2^4 + 9^4 = 3^4 + 7^4 + 8^4$

21. a) `program` sieveout;

```
var x,y,i,n:integer; a:array[1..10000] of byte;
begin write('n='); readln(n);
  for i:=1 to n do a[i]:=0;
  for x:=1 to n do for y:=1 to n div x do a[x+y+x*y]:=1;
  for i:=1 to n do if a[i]=1 then write(i,' ');
  writeln;
end.
```

3 5 7 8 9 11 13 14 15 17 19 20 21 23 24 25 26 27 29 31 32 33 34 35 37 38 39 41 43 44 45 47 48 49 50 51 53 54 55 56 57 59 61 62 63 64 65 67 68 69 71 73 74 75 76 77 79 80 81 83 84 85 86 87 89 90 91 92 93 94 95 97 98 99

Study this sequence. Transform the term $x + y + xy$ so that you can see which integers are contained in this list. The output is for $n = 100$.

Additional Exercises for Sections 1-18

1.  ```
    program skip;
    var i,n,max:integer;
    begin
        write('max='); readln(max); n:=1;
        for i:=1 to max do
            if i<round(n+sqrt(n)) then write(i:8) else n:=n+1;
        readln end.
    ```

This program lists the numbers skipped by our function. For input 1000 it gives the squares up to 1024. Show that exactly the squares are skipped!

2. The function skips the triangular numbers, (numbers of the form $n(n + 1)/2$.

3. The function skips the numbers $(kn - k + 1)n + \lfloor (k - 1)/4 \rfloor$.

4. a) The function skips the numbers $\lfloor (n + k - 1)n)/k \rfloor$.
    b) Combine the results in 3. and 4 a).

7.  ```
    program sum_3_cubes;
    var x,y,z,n,max:integer;
    begin write('n='); read(n); max:=trunc(exp(ln(n)/3))+1;
        for x:=0 to max do
        for y:=x to max do
        for z:=y to max do
            if x*x*x + y*y*y + z*z*z = n then
                writeln(x,'^',3,'+',y,'^',3,'+',z,'^',3,'=',n)
    end.
    ```

Here x, y, z are nonnegative integers. To see that numbers of the form $9n \pm 4$ are not representable, consider $x^3 + y^3 + z^3$ modulo 9.

10. Gauss has shown that all positive integers can be represented as a sum of at most three triangular numbers. The following program checks this up to n=1000.

    ```
    program triangle;
    {prints nothing if all numbers in 1..max≤32767 are representable}
    var i,x,y,z,d1,d2,d3,d,n,max:integer;
        a:array [0..1000] of 0..1;
    begin write('n='); readln(n); max:=n*(n+1) div 2;
        for i:=1 to max do a[i]:=0;
        for x:=0 to n do for y:=0 to n do for z:=0 to n do
        begin
            d1:=x*(x+1) div 2; d2:=y*(y+1) div 2; d3:=z*(z+1) div 2;
            d:=d1+d2+d3; if d<=max then a[d]:=1
        end;
        for i:=1 to max do if a[i]=0 then write(i,' ');
        readln end.
    ```

11. In the program **mode** we find the mode of the increasing sequence $a[i] = \lfloor \sqrt{i} \rfloor$:

```
program mode;
var i,m,n,f,newf:integer; a:array[1..1200] of integer;
begin write('n='); readln(n);
for  i:=1 to n do a[i]:=trunc(sqrt(i));
 m:=a[1]; f:=1; newf := 1; a[n+1] := a[n] - 1;
for i:=1 to n do
  if a[i]=a[i+1] then newf:= newf + 1
  else if newf>f then begin f:= newf; m:= a[i]; newf:=1 end;
  writeln('m=', m, ' f=', f);
  readln end.
```

12.
```
        for i:=0 to m1 do f[i]:=3*i+1;
        for i:=0 to n1 do  g[i]:=4*i+1;
        k:=0; m:=0; n:=0;
        while (m<m1) and (n<n1) do
        if g[n]>f[m] then m:=m+1
        else
           if g[n]=f[m] then
             begin
               write(f[m],' '); k:=k+1; m:=m+1; n:=n+1
             end
           else n:=n+1;
        writeln; write('k=',k);
```

13. c) A closed formula for g is $g(n) = \lfloor (n+1)t \rfloor$ with $t = (\sqrt{5} - 1)/2$.

14. c) The points $(n, h(n))$ are well approximated by a straight line $y = tx$ through the origin. By substituting $h(n) \approx tn$ into the functional equation for h we get $tn \approx n - t^3(n-1)$ or $t^3 + t \approx 1 + t^3/n$. For $n \to \infty$ we see that t must satisfy $t^3 + t - 1 = 0$. The program **bis** gives $t = 6.8232780381E - 01$. So $h(n) = \text{round}(nt)$ and $h(n) = \text{trunc}((n+1)t)$ are possible candidates. Check them as far out as the capacity of your computer permits. My computer tells me that up to $n = 16000$, $h(n) = \text{round}(n * t)$ is correct in about 85% of all cases and deviates from the correct value of $h(n)$ by $+1$ or -1 in the remaining cases. $\text{trunc}((n+1) * t)$ deviates from $h(n)$ in over 20% of all cases by $+1$ or -1. Can you find an exact formula for $h(n)$?

16.
```
program sextuple;
var n, max: longint;
 {$I realprim}
begin write ('max='); readln(max); n:=3;
repeat if prime(n,3) then if prime(n+4,3)
  then if prime(n+6,3) then if prime(n+10,3)
  then if prime(n+12,3) then if prime(n+16,3)
  then writeln(n:10,n+4:10,n+6:10,n+10:10,n+12:10,n+16:10);
   n:=n+2
until n>max
end.
```

```
17. program euler_real;
    var i,n,max:integer; x,y,pow2:real;
    begin pow2:=4; n:=2;
    writeln('x':6,'y':6,'n':6);
    writeln;
    repeat n:=n+1; pow2:=pow2+pow2;
      max:=
      trunc(sqrt((pow2-1)/7)+1) div 2;
      for i:=1 to max do
      begin x:=i+i-1;
        y:=sqrt(pow2-7*x*x);
        if frac(y)=0 then
          writeln(x:6:0,y:6:0,n:6)
      end
    until n>=20
    end.
```

x	y	n
1	3	4
1	1	3
1	5	5
3	1	6
1	11	7
5	9	8
7	13	9
3	31	10
17	5	11
11	57	12
23	67	13
45	47	14
91	87	16
89	275	17
93	449	18
271	101	19
85	999	20

Table 1

By running the program **euler_real** we get the data in Table 1. How do you get from the current x, y to the x, y in the next line? Study these data very closely. You will observe in Table 1 that any x is one half of the sum or difference of the two preceding x, y. From the current x, y you get the next y by computing $(7x + y)/2$ or $|7x - y|/2$. If the next x is a sum then the next y is a difference, and vice versa. That is we have the two transformations S, T :

$$S : (x,y) \to ((x + y)/2, \ |7x - y|/2), \quad T : (x,y) \to (|x - y|/2, \ (7x + y)/2)$$

Prove this by induction! You will have to prove in addition that one of the two new pairs consists of odd elements.

```
21. program  MinComElem;
    var  i,j,k:integer; f,g,h:array[0..21] of longint;
    begin
    f[0]:=0; f[1]:=1;
    for  i:=1 TO 20 do
      begin f[i+1]:=f[i]+f[i-1]; g[i]:=i*i; h[i]:=9*i end;
    i:=1; j:=1; k:=1;
    repeat
      while  f[i]<g[j] do begin i:=i+1; if i>20 then halt end;
      while  g[j]<h[k] do begin j:=j+1; if j>20 then halt end;
      while  h[k]<f[i] do begin k:=k+1; if k>20 then halt end;
      until (f[i]=g[j]) and (g[j]=h[k]);
    writeln(i:5,j:5,k:5,f[i]:8);
    readln end.
```

29. a) 381654729 b) 3816547290

30. **program** bell_triangle;
```
var i,j,n:integer; oldrow,newrow:array[-1..20] of longint;
begin  write('n='); readln(n); oldrow[-1]:=1;
  for j:=0 to n do
  begin  newrow[0]:=oldrow[j-1];
    for i:=1 to j do newrow[i]:=newrow[i-1]+oldrow[i-1];
    for i:=0 to j do write(newrow[i]:10); writeln;
    for i:=0 to j do oldrow[i]:=newrow[i]
  end
end.
```

The following is an explanation of the triangle algorithm for computing the Bell numbers.

For $k = 0, 1, .. n$, let $B_{n,k}$ be the number of partitions of $\{1, \ldots, n\}$ such that the other elements of the set to which n belongs are all $\leq k$. We have $B_{n,0} = B_{n-1}$ and $B_{n,n-1} = B_n$.

The partitions counted by $B_{n,k+1}$ are of two kinds. First, those in which the element $k + 1$ is not in the same set as n. There are $B_{n,k}$ such partitions. Second, those in which $k + 1$ and n are in the same set. Removing the element $k + 1$ from all of these, we see that the number of these partitions is $B_{n-1,k}$. Thus $B_{n,k+1} = B_{n,k} + B_{n-1,k}$.

32. c) Both are correct.

33.   **program** olympiad;
```
var i,j,k,rows:integer;
begin write('rows='); readln(rows); j:=1;
  for i:=1 to rows do
  begin
    for k:=1 to i do begin write(j:4); j:=j+2 end;
    writeln; j:=j-1
  end
end.
```

35. Each number occurs twice except 1, which occurs four times, and the other powers of 2, which occur three times.

Section 23

23. The following algorithm generates a random permutation:

```
for i:=n downto 2 do
begin k:=1+random(i); copy:=x[i]; x[i]:=x[k]; x[k]:=copy end;
```

Section 24

24. Since the period is $p - 1$, all of $1, 2, \ldots, p - 1$ occur as remainders if we divide a positive integer which is $< p$ by p. The digits we obtain from these remainders are the integer parts of $10/p, 20/p, \ldots, 10(p-1)/p$. The number of times we get the digit d is equal to the number of multiples of 10 in the interval $dp \leq x < (d + 1)p$. That number is at least $\lfloor p/10 \rfloor$ and at most $\lceil p/10 \rceil$. The proof for pairs, triples etc. is similar.

Section 25

2. The pure periodicity follows from the fact that the stream of digits can be uniquely extended into the past.

3. a) Consider the sequence mod 2: 11011 11011 11011 .... The numbers 1,2,3,4 taken mod 2 are 1,0,1,0. This pattern does not occur.
   b) This is 1,0,0,1 mod 2, and so it cannot occur.
   c) Extend the sequence one step into the past and you get 519831138....
   d) This block will occur again since the sequence is purely periodic.

4. $x(n) = (x(n - 3) + x(n - 5)) \bmod 2$ and $y(n) = (4y(n - 4) + y(n - 5)) \bmod 5$ have periods $2^5 - 1$ and $5^5 - 1$. The period of the resulting sequence is the least common multiple of these two numbers. Since they are prime to each other it is their product $31 * 3124 = 96844$.

5. The combined sequence has period $(2^{17} - 1)(5^7 - 1) = 10239790804$.

Section 27

3. We get the infinite binary word as follows: $w_1 = 0$, $w_2 = 001$, $w_3 = w_2 w_2 w_1$. By induction we get $w_{k+1} = w_k w_k w_{k-1}$. Let a_k and b_k be the number of zeros and ones in w_k. Then $a_{k+1} = 2a_k + a_{k-1}$, $b_{k-1} = a_k$, $t_k = a_k/a_{k-1}$, $t_{k+1} = a_{k+1}/a_k = 2 + 1/t_k$. For $k \to \infty$ we get $t = 2 + 1/t$, or $t = \sqrt{2} + 1$. The ratio a_k/b_k tends to an irrational limit. Thus the sequence is not periodic. If it were periodic, t_k would tend to the rational ratio of zeros to ones in one period. For the infinite binary word we have zeros/ones $= (\sqrt{2}+1)/1$, zeros/(zeros+ones) $= (\sqrt{2}+1)/(2+\sqrt{2}) = 1/\sqrt{2}$, ones/(zeros+ones) $= 1/(2+\sqrt{2})$. So every $(2+\sqrt{2})$th digit is a 1. The n-th 1 should have place number $\approx (2 + \sqrt{2})n$. We find empirically that the greatest integer part of the last expression is the place number of the n-th 1. This is tested by the following program. It first constructs n bits of the word. For input u it then finds the place number i of the u-th 1 and compares with $f(u) = \text{trunc}((2 + \sqrt{2}) * u)$.

```
program Bin_Word;
const n=32000;
var i,j,u:integer; x:array[1..n] of 0..1;
begin write('u='); readln(u); x[1]:=0; i:=1; j:=1;
  repeat
    if x[i]=0 then begin
      x[j]:=0; x[j+1]:=0; x[j+2]:=1; j:=j+3 end
    else begin x[j]:=0; j:=j+1 end;
    i:=i+1
  until j>n-3; {the construction of the word is finished}
  j:=0; i:=0;
  repeat i:=i+1; if x[i]=1 then j:=j+1 until j=u;
  write('i=',i,trunc((2+sqrt(2))*u):8)
end.
```

We also give a proof that always $i = \text{trunc}((2 + \sqrt{2})n)$. Let S be the operator which performs the following substitutions on binary sequences: it replaces each 0 by 001 and each 1 by 0.

Lemma 1. *For any sequence $D = d_1 \ldots d_n$ of binary digits the proportions of 0's among the digits of D and $S(D)$ are on opposite sides of $\frac{1}{\sqrt{2}}$.*

Proof. Let the number of 0's in D be z. Let n' be the number of digits in $S(D)$ and let z' be the number of 0's among them. Then

$$(1) \qquad\qquad n' = n + 2z, \qquad z' = n + z.$$

It is easy to check that the ratio $\frac{z}{n}$ would remain unchanged by S if it had the value $\frac{1}{\sqrt{2}}$. Since this is irrational the ratio can not have this value for a finite sequence, but it is instructive to examine how the difference $z - \frac{n}{\sqrt{2}}$ is changed by the substitution S. We find

$$(2) \qquad\qquad z' - \frac{n'}{\sqrt{2}} = -(\sqrt{2} - 1)(z - \frac{n}{\sqrt{2}}).$$

So S not only changes the sign of the difference $z - \frac{n}{\sqrt{2}}$ but also decreases its magnitude by the factor $0.414\ldots$. Interestingly, this is the reciprocal of the limit of the factors by which S *increases* the lengths of sequences as the proportion of 0's approaches the stable value. In the case of the initial segments $0, S(0), S(S(0)), \ldots$ of our sequence, the ratios $\frac{n}{z}$ are the continued fraction approximants of $\sqrt{2}$, as one can deduce from the recursion formulas.

Lemma 2. *Let $I_n = a_1 \ldots a_n$ be an initial segment of our sequence. Let z_n be the number of 0's in the segment. Then $z_n > \frac{n}{\sqrt{2}}$ if $a_n = 0$ and $z_n < \frac{n}{\sqrt{2}}$ if $a_n = 1$.*

Roughly speaking, this says that the ratio of digits hovers so close to the invariant ratio that the last digit can tip it in its own direction.

Proof. One can verify the claim for small values of n. Assume $n > 12$ and that Lemma 2 has been proved for $I_1, \ldots, I_{n-1}$.

If $a_n = 1$ then $I_n = S(I_m)$ for some $m < n$, and the last digit of I_m is 0. Then Lemma 2 follows from the induction hypothesis and Lemma 1. If I_n terminates with a 0, there are two possibilities:

a) $I_n = S(I_m)$ where I_m ends with 1. Here the induction hypothesis and Lemma 1 again give the required result.

b) The final 0 of I_n is the first or the second 0 of a 001 which comes from substituting for a 0 in a precursor initial segment. For definiteness let it be the first 0. Then $I_n = (S(I_m))0 = S(I_m 1)$. Think of the sequence $I_m 1$ as the initial segment up to and including the first 1 beyond a_m, with the 0's beyond a_m deleted. Then we see that the induction hypothesis implies that the proportion of 0's in $I_m 1$ is $< \frac{1}{\sqrt{2}}$ and hence by Lemma 1 the proportion in I_n is $> \frac{1}{\sqrt{2}}$. The case when the final 0 of I_n is the second 0 of a 001 can be dealt with similarly.

We are ready to prove our formula for the number of digits n up to and including the k-th 1. Lemma 2 tells us $k > (1 - \frac{1}{\sqrt{2}})n$. Also, $a_{n+1} = 0$ since every 1 in the sequence is preceded by at least two 0's. Applying Lemma 2 to I_{n+1} we get $k < (1 - \frac{1}{\sqrt{2}})(n + 1)$. We can write these two inequalities as $n < k(2 + \sqrt{2}) < n + 1$, q. e. d.

For a lattice point interpretation of this problem, see the section on continued fractions in Felix Klein's Elementary Mathematics from an Advanced Point of View, Macmillan 1939.

9. $n = 2053$.

Sections 31-32

1. Represent a nonincreasing sequence of at most m numbers, each of which is $\leq n$, by columns of unit squares with bases on the x-axis. If we look at such a figure sideways, it becomes a representation of a nonincreasing sequence of at most n numbers, each of which is $\leq m$.

2.
```
program TwoSamp;
const n=18; k=9; s=107;  d:array[1..18] of integer=
        (0,8,10,10,11,14,14,14,14,14,16,16,18,19,20,21,22,27);
var  i,x,y,z:integer; q:array[0..n,0..k,0..s] of integer;
begin
for z:=0 to s do
  for y:=0 to k do
    for x:=0 to n do q[x,y,z]:=0;
for x:=0 to n do
  for  z:=0 TO s do q[x,0,z]:=1;
for  y:=1 to k do
  for  x:=y to n do
    for  z:=0 to s do
      if z>=d[x] then q[x,y,z]:=q[x-1,y,z]+q[x-1,y-1,z-d[x]]
      else q[x,y,z]:=q[x-1,y,z] ;
writeln('q=',q[n,k,s]);
readln end.
```

```
6. program TwoSamIt ;
   const n=18; k=9; s=107;  d:array[1..18] of integer=
         (0,8,10,10,11,14,14,14,14,14,16,16,18,19,20,21,22,27);
   var x,y,z,min:integer; q:array[0..k,0..s] of longint;
   begin
     for z:=0 to s do q[0,z]:=1;
     for x:=1 to k do
       for z:=0 to s do q[x,z]:=0;
     for x:=1 to n do
     begin
       if k<x then min:=k else min:=x;
       for y:=min downto 1 do
         for z:=s downto d[x] do q[y,z]:=q[y,z]+q[y-1,z-d[x]]
     end;
     writeln('q=',q[k,s]);
   readln end.
```

This program will be used for larger problems. So **q[k,s]** has been declared a longint.

Section 40

```
1. program TwoCompleteSets;
   const n=6;
   var i,spins,count,r:integer; x:array[1..n] of byte;
   begin spins:=0; randomize;
     for i:=1 to n do x[i]:=0;
     for count:=1 to 2*n do
     begin
       repeat r:=1+random(n); spins:=spins+1 until x[r]<=1;
       x[r]:=x[r]+1
     end;
     writeln('spins=',spins)
   end.
4. program CoverMultCase;
   var i,j,m,count:integer; r,s,rcount,scount:byte;
       x,y:array[1..8] of byte;
   begin  randomize; write('m='); readln(m); count:=0;
   for i:=1 to m do
   begin rcount:=0; scount:=0;
     for j:=1 to 8 do begin x[j]:=0; y[j]:=0 end;
     repeat
       r:=1+random(8); s:=1+random(8);
       count:=count+1;
       if x[r]=0 then begin x[r]:=1; rcount:=rcount+1 end;
       if y[s]=0 then begin y[s]:=1; scount:=scount+1 end
     until (rcount=8) or (scount=8)
   end;
   writeln(count/m:8:5)
   end.
```

5. **program** CoverHypgCase is obtained from CoverMultCase by setting up a list of occupied squares and rejecting a random choice if the square is already occupied. See the disk.

Additional Exercises for Sections 21-40

1. **program** craps;
   ```
   var i,n,p,t,count:integer;
   begin write('n='); read(n); randomize; count:=0;
   for i:=1 to n do
   begin t:=random(6)+random(6)+2;
     if (t=7) or (t=11) then count:=count+1
     else if (t>3) and (t<12) then
     begin p:=t;
       repeat t:=random(6)+random(6)+2;
   if t=p then count:=count+1
       until (t=p) or (t=7);
     end;
   end;
   writeln(count/n:7:5)
   end.
   ```

2. **program** CoinGame;
   ```
   var x,y,z,u:byte; i,abby,n:integer;
   begin write('n='); read(n); abby:=0; randomize;
   for i:=1 to n do
   begin
     y:=random(2); z:=random(2); u:=random(2);
     repeat
       x:=y; y:=z; z:=u; u:=random(2)
     until (z*u=1) and (x=y);
     if x=0 then abby:=abby+1
   end;
   writeln(abby/n:8:6)
   end.
   ```

3. $E(D_n^2) = n$. 4. $E(D_n^2) = n$. 5. $E(D_n^2) = n$.

6. About 0.495. 7. $E(D_n^2) = n$.

8. **program** rand_dir_walk;
   ```
   var i,j,m,n:integer; a,h,r,sum,x,y,z:real;
   begin write('m,n='); readln(m,n); randomize; sum:=0;
     for j:=1 to m do
     begin x:=0; y:=0; z:=0;
       for i:=1 to n do
       begin h:=2*random-1; r:=sqrt(1-h*h); a:=2*pi*random;
         x:=x+r*cos(a); y:=y+r*sin(a); z:=z+h
       end;
   ```

```
        s:=s+x*x+y*y+z*z
    end;
    writeln(sum/m:10:10)
end.
```

The program **rand_dir_walk** performs m times a random walk of n steps and finds the mean of the squares of the distances from the origin at the end of each walk. Choose $m = 1000$ and $n = 10, 20, 50, 100$ and you will realize that the expectation of the square of the distance is n. Our interpretation of random choice of a point on a unit sphere S is this: It falls into a subset M of S with probability area$(M)/4\pi$. We use the theorem of Archimedes that a spherical zone of height h on a sphere of radius 1 has area $2\pi h$. We choose h at random from $(-1, 1)$ by means of $h = 2 * \text{random} - 1$. The geographical longitude a of the point is chosen by means of $a = 2\pi * \text{random}$. If we set $r = \sqrt{1 - h * h}$, then $x := x + r * cos(a)$, $y := y + r * sin(a)$, $z := z + h$.
We find that $E(D_n^2) = n$.

10. **program** PokerTest;
```
    var i,j,d,difs,max,dist,pair,twopair,   { difs=different nos. in hand}
        triple, trippair, quad, quint, n:integer;
        x:array[0..9] of integer;
```
begin
```
write('n='); read(n); randomize;
dist:=0; pair:=0; twopair:=0; triple:=0;
trippair:=0; quad:=0; quint:=0;
for j:=1 to n do
begin
    for i:=0 to 9 do x[i]:=0; difs:=0;
    for i:=1 to 5 do begin  d:=random(10); x[d]:=x[d]+1 end;
    for i:=0 to 9 do if x[i]>0 then difs:=difs+1;
    case difs of           { case is an alternative to a chain of }
      1: quint:=quint+1; { if difs=.. then .. else statements }
      2,3: begin max:=2;
              for i:=0 to 9 do if x[i]>max then max:=x[i];
              if difs=3 then
                if max=3 then triple:=triple+1
                else twopair:= twopair+1
              else      (* difs=2 in this case *)
                if max=4 then quad:=quad+1
                else trippair:=trippair+1
           end;
      4: pair:=pair+1;
      5: dist:=dist+1;
    end;    (* of cases *)
end;
write(dist/n:7:4, pair/n:7:4, twopair/n:7:4, triple/n:7:4,
      trippair/n:7:4, quad/n:7:4,quint/n:7:4);
readln; end.
```

For $n = 10000$ we got the printout 3069 5063 1046 679 95 47 1.

15. a) The program **palindrome** generates m n-words from the alphabet $\{0, 1, 2\}$ and counts those which start with a palindrome. The proportion of such words is an estimate for P_n.

```
program palindrome;
label 0;
const n=9; m=10000;
var i,j,k,l,r,count:integer;
    x:array[1..n] of byte;
begin count:=0; randomize;
  for k:=1 to m do
    begin
    for i:=1 to n do
      x[i]:=random(3);
      j:=2;
    repeat l:=1; r:=j;
      while (x[l]=x[r]) and
                     (l<r) do
      begin l:=l+1; r:=r-1 end;
      if l>=r then
        begin count:=count+1;
        goto 0 end;
      j:=j+1
    until j>n;
0:end;
  writeln(count/m)
end.
```

A small table for checking the program:

n	P_n	
2	1/3	
3	5/9	
4	17/27	=0.63
5	55/81	=0.679
6	169/243	=0.6955
7	517/729	=0.70919
8	1561/2187	=0.71376
9	4709/6561	=0.7177
10	14153/19683	=0.71905
11		0.7203
12		0.72072
14		0.72125
16		0.72142
18		0.721481581

16.
```
program total_tour_of_cube;
var v,  {vertex number, runs from 0 to 7}
    r,  {random integer from 1..3}
    i,  {index variable running through the f-array}
    c,  {the no. of vertices visited in the current tour}
    tl, {counts the steps in the current tour}
    tt:integer;  {number of completed total tours}
    sum:longint; {sums the steps of all total tours}
    b:array[1..3] of 0..1; {the binary repr. of v}
    x:array[0..7] of integer;{counts visits of v in one tour}
    f:array[0..200] of integer; {frequency of tour length tl}
begin tt:=0; sum:=0; writeln; randomize;
  for i:=1 to 200 do f[i]:=0;
  repeat
    b[1]:=0; b[2]:=0; b[3]:=0; c:=0; tl:=-1;
    for v:=0 to 7 do x[v]:=0;
    repeat
      v:=4*b[1]+2*b[2]+b[3]; x[v]:=x[v]+1;
```

```
   if x[v]=1 then c:=c+1;
   r:=1+random(3); tl:=tl+1; b[r]:=1-b[r]
 until c=8;
 f[tl]:=f[tl]+1; tt:=tt+1
until tt>=1000;
i:=7;
repeat
  if f[i]<>0 then begin
    write(i:11,f[i]:5); sum:=sum+i*f[i];
    if i mod 5=0 then writeln
  end;
  i:=i+1
until i>200;
writeln; writeln('mean tour length=',sum/1000:20:10)
end.
```

I started and Aoki completed a computation of the exact value, which is $1996/95 = 21.0105$. Later I checked this with *Derive*; the run took 20 seconds.

Exercises for Section 41

2. ```
program triple_birthday;
const m=365; n=10000; p=731;{must be a triple among p birthdays}
var i,j,r:integer; x:array[1..m] of integer;
    {x stores birthdays}
    y:array[1..p] of integer; sum: real; {y stores waiting times}
begin   randomize;
  for i:=1 to p do y[i]:=0;
  for i:=1 to n do
  begin
    for j:=1 to m do x[j]:=0;
    j:=0;
    repeat   {samples the m birthdays until one}
    j:=j+1; r:=1+random(m); x[r]:=x[r]+1 {of them occurs}
    until x[r]=3;   {for the 3rd time.}
    y[j]:=y[j]+1 {Waiting time was j, so y[j] is increased}
  end;
  sum:=0;
  for j:=1 to p do sum:=sum+j*y[j];
  writeln(sum/n:0:4)
end.
```

Ten repetitions of this program gave 88.77 for the expected waiting time.
```

Additional Exercises for Sections 1-41:

2. The formula is $f(a,b,c) = 3abc/(a+b+c)$. Lennart Råde of Sweden put this
problem to a number of people in the last 20 years, but nobody could solve
it. On January 17, 1992 I ran the program ThreeTowerGame and its output
suggested the formula which I then proved. (See the end of the Crash Course
in Probability.)

```
program ThreeTowerGame;
var a,b,c,r,s,t,i,n: integer; sum: real;
    x:array[0..2] of integer;
begin randomize; write('a,b,c='); readln(a,b,c);
  sum:=0.0; n:=10000;
  for i:=1 to n do
  begin x[0]:=a; x[1]:=b; x[2]:=c;
    repeat r:=random(3); sum:=sum+1; x[r]:=x[r]-1;
      s:=(r+1) mod 3; t:=(r+2) mod 3;
      if random(2)=0 then x[s]:=x[s]+1 else x[t]:=x[t]+1
    until x[r]=0
  end;
  writeln('mean time=',sum/n:5:5)
end.
```

Section 45

We assume the random sets are chosen by flipping a fair coin to decide for
each element whether it should be in the set. After the first guess, G is in a set
containing r other numbers which were picked at random from the set of $n-1$
numbers other than G, etc. The guessing ends when $r = 0$.

```
program RandGuess;
const n=5; m=10000;
var i,j,r,s,count:integer; sum:longint;
begin
sum:=0;
for j:=1 TO m DO
begin
  r:=n-1;
  repeat
    s:=0; for i:=1 to r do s:=s+random(2);
    r:=s; sum:=sum+1
  until r=0;
end;
writeln(sum/m:10:4);
readln; end.
```

Section 46

3. Program posa is on the disk in slightly different versions for Turbo Pascal
    for IBM-compatibles, Turbo Pascal for the Macintosh and Think Pascal (for
    the Macintosh). It finds closed knight's tours on $m \times n$ chessboards and draws
    them on the screen. The program is too long to be listed here. It contains
    explanations as comments interspersed with the program on the disk.
       The program is written so that, apart from the graphic part, it should be easy
    to adapt to finding Hamilton circuits in any graph by changing the adjacency
    table.

4. The following program tries to find a knight's path visiting every square of an
    $(m-2) \times (n-2)$ chessboard, using Warnsdorf's method. The array b[i,j] will
    contain the ordinal numbers of the squares in the path. The array is bordered
    with two dummy rows on all 4 sides. This enables us to list the squares accessible
    by a knightjump from a given square (of the actual chessboard) in the same way,
    whereas otherwise we would have to do it differently for squares near or on the
    edge. We initialize b[i,j] to be 0 on the chessboard proper and 1 on the border
    rows. This will cause the algorithm not to put squares of the border rows into
    the knight's path. At each stage the algorithm starts from the current position
    (i,j). It checks all squares reachable by changing the coordinates by at most 2.
    If s, t are the amounts added to the coordinates, the condition for reachability by
    knightjump has the simple form $|st| = 2$. For each square reachable from (i,j),
    we check how many squares reachable from it are still free. We select as the next
    square the one with coordinates imin, jmin, which is a free square reachable
    from (i,j) and has fewer free neighbors than any of the squares previously tried,
    and not more than any tried after it. In the program, c becomes the number
    of free neighbors of a square and min becomes the least value of c among the
    neighbors of (i,j). The construction terminates once we have selected a free
    square with no free neighbors. If this happens before mn squares have been
    covered, the algorithm has failed. Note that if the current square has several
    free neighbors with min free neighbors of their own, choosing the first one our
    algorithm finds is arbitrary; one could make these choices at random. Then a
    second run may succeed if the first one fails. A related program, WrdfClsd can
    be found in its own file on the disk. It tries to find a *closed* knight's tour by
    using Warnsdorf's selection criterion, and backtracking when it gets stuck.

```
program Warnsdroff;
const m=10; n=10; { The size of the board is (m-2)x(n-2) }
var i,j,s,t,p,min,ip,jp,c,ss,tt,imin,jmin: integer;
    b:array[-1..m, -1..n] of integer;
begin
    for s:=-1 to m do for t:=-1 to n do b[s,t]:=1;
    for s:=1 to m-2 do  for t:=1 to n-2 do b[s,t]:=0;
    write('Start coords.='); read(i,j); b[i,j]:=1; p:=1;
```

```
repeat
  min:=8;
  for s:=-2 to 2 do for  t:=-2 to 2 do
    if abs(s*t)=2 then begin
      ip:=i+s; jp:=j+t;
      if b[ip,jp]=0 then begin
        c:=0;
        for ss:=-2 to 2 do for tt:=-2 to 2 do
          if (abs(ss*tt)=2) and (b[ip+ss,jp+tt]=0) then
            c:=c+1;
          if c<min then begin imin:=ip; jmin:=jp; min:=c end
      end
    end;
  i:=imin; j:=jmin; p:=p+1; b[i,j]:=p;
until min=0;
if p<(m-2)*(n-2) then writeln('STUCK AT STEP ',p);
for i:=1 to m-2 do begin
  writeln; writeln; for j:=1 to n-2 do write(b[i,j]:4)
end; readln
end.
```

Section 47

```
3. program savings_boxes;
   var copy,count,first,i,j,k,n,r,rep:integer;
       P:array[1..100] of integer;
   begin
     write('n,k,rep='); readln(n,k,rep); count:=0; randomize;
     for j:=1 to rep do
     begin
       for i:=1 to n do P[i]:=i;
       for i:=n downto 2 do  {Construct random permutation}
       begin
         r:=1+random(i); copy:=P[i]; P[i]:= P[r]; P[r]:=copy
       end;
       i:=0;
       repeat
         i:=i+1; first:=i;
         if P[i]>0 then
           repeat r:=first; first:=P[first]; P[r]:=-P[r]
           until first=i
       until i=k;
       i:=0;
       repeat i:=i+1 until (i=n) or (P[i]>0);
       if P[i]<0 then count:=count+1
     end;
     writeln('P=', count/rep:8:5);
   readln end.
```

By simulation we find $p(n, k) = k/n$. Here is one of the proofs from the report on the competition in the Hungarian high school periodical Matematikai Lapok.

We ignore k for the time being. We consider the following procedure for opening all boxes. Break box 1. Open as many boxes with keys as you can. The last of those has key 1 in it. Now break the lowest numbered box which is still locked, open all the boxes you can, etc.

Let us call the order in which the keys appear in this procedure the *retrieval permutation P'* corresponding to the *hiding permutation P*. Given a retrieval-permutation P', we can reconstruct the hiding permutation P which produces it. Hence every permutation of keys is a possible retrieval-permutation. If all permutations are equally likely to occur as hiding permutations then they are also equally likely to occur as retrieval permutations.

We note that the key in the last place in a retrieval permutation is the key of the last box which was broken. Let the number of that key be l. When we broke box l, all boxes with smaller numbers were already open, so to open all boxes we needed to break one whose number was $\geq l$.

Let us return now to the opening procedure of exercise 3, which consists of breaking the first k boxes at the start and not breaking any more. We can open all boxes if and only if the last entry l of the retrieval permutation is one of 1, 2,..., k. The probability of this is k/n.

Section 48

2. c) Let
$$x(n) = \lfloor nt \rfloor, \quad y(n) = n + \lfloor nt \rfloor = \lfloor n(1+t) \rfloor = \lfloor nt^2 \rfloor.$$
We need to show only that the sequences $x(n)$ and $y(n)$ together contain every positive integer exactly once. This is an instance of S. Beatty's theorem. One way of proving it is as follows. For a given positive integer n, the number of terms $< n$ in the sequence $t, 2t, 3t, \ldots$ is $\lfloor n/t \rfloor$. The number of terms $< n$ in $t^2, 2t^2, 3t^2, \ldots$ is $\lfloor n/t^2 \rfloor$. Since t is irrational, n/t and n/t^2 are both irrational, but $n/t + n/t^2 = n(1+t)/t^2 = n$. Thus
$$\lfloor n/t \rfloor + \lfloor n/t^2 \rfloor = n - 1.$$
This is the number of terms in the two sequences taken together. By taking $n = 1, 2, 3, \ldots$ we see that there is exactly one multiple of either t or t^2 in each of the intervals $(1, 2)$, $(2, 3)$, ... Hence every positive integer is in exactly one of the sequences $\lfloor nt \rfloor$, $\lfloor nt^2 \rfloor$.

4. L consists of all multiples of $k + 1$ including 0.

5. L consists of all multiples of 3 including 0.

6. L consists of all multiples of 4 including 0.

12. L consists of all multiples of 6 including 0.

Section 49

6. The following does the computations for a), b), c) and f).

```
program GoldPerm1;
const n=100;
var i,j,k,w:integer; v,z:real;
    u,x:array[1..n] of integer; y:array [0..n] of real;
begin z:=(sqrt(5)-1)/2; y[0]:=0; (* Sentinel for sorting. *)
for i:=1 to n do begin x[i]:=i; y[i]:=i*z-trunc(i*z) end;
for i:=2 to n do begin
  v:=y[i]; w:=x[i]; j:=i;
  while y[j-1]>v do
    begin y[j]:=y[j-1]; x[j]:=x[j-1]; j:=j-1 end;
  y[j]:=v; x[j]:=w
end;
writeln('The golden permutation:');
for i:=1 to n do write(x[i]:4);
writeln; writeln('Forward differences in the golden permutation:');
for i:=1 to n-1 do write(x[i+1]-x[i]:4);
writeln; writeln;
writeln('Distance from x[i] to x[i]+1 in the golden perm.:');
i:=0;
repeat i:=i+1; j:=0;   { checks c) }
  if x[i]=n then write ('n':4)
  else begin
    repeat j:=j+1 until x[j]-x[i]=1; write(j-i:4); end;
until i=n;
writeln; writeln('Press CR for the inverse permutation.'); readln;
for i:=1 to n do u[x[i]]:=i;
writeln('The inverse permutation:');
for i:=1 to n do write(u[i]:4);
writeln; writeln('Forward differences in the inverse permutation:');
for i:=1 to n-1 do write(u[i+1]-u[i]:4);
writeln; writeln;
writeln('Distance from u[i] to u[i]+1 in the inverse perm.:');
i:=0;
repeat i:=i+1; j:=0;   { checks c) }
  if u[i]=n then write ('n':4)
  else begin
    repeat j:=j+1 until u[j]-u[i]=1; write(j-i:4); end;
until i=n;
readln end.
```

Section 62

1. If a random real x, $0 < x < 1$, is developed into a continued fraction, then the partial denominators 1, 2, 3, 4 will occur with approximate proportions $\log_2(4/3) \approx 0.41504$, $\log_2(9/8) \approx 0.16992$, $\log_2(16/15) \approx 0.09311$, $\log_2(25/24) \approx 0.05890$. The proof is very difficult.

Section 63

1.   Suppose that after the first transformation the largest value occurring among the x_i is M and that a smaller positive value also occurs. If only one such value occurs then after the next step there will be two such values next to each other. Let x_i, $x_{i+1}, \ldots, x_j$ be an interval of maximal length such that it begins and ends with a positive entry $< M$ and none of the entries in between is $= M$. Then if we apply the transformation once more, the interval x'_{i-1}, $x'_i, \ldots, x'_j$ will have positive entries $< M$ at both ends and no entry M inside. Thus the length of the longest such interval increases by at least 1 at each step and after at most $n - 1$ steps no entry M is left. We see that after at most $n(M - 1)$ steps we shall reach a set such that all the positive x_i have the same value.

3.   If the polygonal set x consists of 0's and 1's, then Tx also consists of 0's and 1's, and the number of 1's is even. Also, if x contains an even number of 1's, there are two w's such that $Tw = x$. They are complements of each other. Thus if n is odd, then within the set of x which have an even number of 1's, each element has a unique predecessor under T. It follows that all such x are on cycles.

4.   Applying the transformation T twice to a cycle with an even number of entries is the same as applying it once to the cycle consisting of the even numbered entries and once to the cycle consisting of the odd numbered entries.

8.   a) The following program plays Bulgarian solitaire, starting with any number of stacks of any sizes. It stops when, arranged in increasing order, the i-th stack contains i or $i + 1$ cards, and it prints out the stacks in increasing order.

```
program  bulsolit;
var  i,j,k,v,iter:integer;  flag:boolean;
     stack:array[0..1000] of integer;
begin
  for i:=1 to 1000 do stack[i]:=0; stack[0]:=30000;{sentinel}
  writeln('Stack sizes in nonincreasing order, 0 at end');
  k:=0; repeat k:=k+1;  read(stack[k]) until stack[k]=0;
  k:=k-1; iter:=0;   { k is the # of stacks put in}
  repeat  iter:=iter+1;
    for i:=1 to k do stack[i]:=stack[i]-1;
    v:=k; i:=k+1;                {v is the new stack}
    while stack[i-1]<v do {insert v into correct position}
      begin stack[i]:=stack[i-1]; i:=i-1 end;
    stack[i]:=v;                 {v inserted}
    k:=1; repeat k:=k+1 until stack[k+1]=0; {finds current}
    flag:=true;                         {number of stacks}
    for i:=1 to k do
      flag:=flag and ((stack[k+1-i]=i) or (stack[k+1-i]=i+1))
  until flag;
  for i:=k downto 1 do write(stack[i]:5);{ prints final stack sizes}
  writeln; writeln(iter,' iterations');
readln; end.
```

Section 64

4.   The path programs, **PathFast** and **PathSure,** implement Knuth's method
and show some of the paths on the screen. They stop and display statistics and
the last path when either a record-breaking X-value comes up or the program
has generated as many additional paths as you asked for.
    We can expect that there will be X values which will come up very seldom
but will be so large that they will make a significant and possibly a major
contribution to the expectation of X. The existence of such X-values forces
us to generate paths until a sufficient number of them have come up so that
the one or two largest do not account for too much of the average if we want a
reliable estimate of X.
    In view of the above, we should try to find a way to generate paths which will
not assign extremely small probabilities to large numbers of paths. If we use
equal probabilites whenever we have to choose between two or three possible
ways to extend our path, long winding paths will have small probabilities, i.e.
large X-values. To counteract this, we incorporated a parameter **bias** in our
programs. If **bias** > 0 (it must be < 1) then our algorithm gives some preference
to points which are in a direction away from the target. This increases the
probabilities of at least some of the long winding paths and thus decreases the
magnitude of the largest X-values. The way this is accomplished is arbitrary
and crude. Even so, using a bias of about 0.15 seems to give much less variable
results in a given number of trials than using 0 bias. The reader may wish to
try biasing the choices in ways that take more account of the geometry. For
instance, one may get better results by increasing the bias as one gets closer to
the target point.
    Program **PathFast** uses a random path generating algorithm which is not
guaranteed to reach the target point; about half the paths end in deadends.
If you look at the reasoning on which Knuth's algorithm is based you will see
that algorithms which produce deadends may be used. Paths which end in a
deadend should be assigned X-value 0.
    People generating paths with dice can exclude turns which lead into a deadend
visually; they seem to have done that in the experiment described in Section 64.
Program **PathSure** can simulate such an experiment. It excludes nodes which
lead into deadends by performing procedure **freenodes** before selecting a new
node to be added to the path. It constructs the set **accessibles,** i.e. nodes
accessible from the target corner through nodes not in the existing part of the
path. These are the nodes from which the target corner can still be reached.
The set **accessibles** is constructed by fanning out from the target corner and
adding nodes which are not in the path and connected to the nodes already
in **accessibles,** until no more such nodes can be found. This procedure is
very slow. It could probably be speeded up by using arrays instead of sets. We
wished to present a program using the rudimentary set operations available in
Turbo Pascal.
    The speed disadvantage of **PathSure** is less than the time needed to construct
a path would suggest. **PathSure** is much likelier to produce very long paths than
PathFast since the latter has many chances of getting into a deadend when

the path is long. Therefore the extreme X-values are smaller in the PathSure algorithm (very roughly, by a factor of 50). Hence an estimate of a given accuracy can be obtained from a much smaller sample with PathSure.

One of the statistics displayed is the weighted average of the percentage of free space. The objective of this statistic is to get an idea of the average free space over all paths. It would be misleading to average over the paths we generate because our methods produce short paths, which leave much free space, with greater probability than long ones. To approximate the true average, we have to attach weights to the contributions of the paths of our sample. If we take weights proportional to the X-values, then the expected value of the weighted averages is the unweighted average of the free space, or any other feature, over all paths.

The programs are long and we do not list them here. Explanations of their details can be found as comments in the programs on the disk.

Section 65

4. *Empirical Exploration.* We give two programs, one without and one with a goto.

```
program coin_stack;
var i,j,n,steps: integer;
    t,flag:boolean;
    x:array[1..1000]
       of boolean;
begin
write('n=');  readln(n);
for i:=1 to n do x[i]:=true;
i:=0; steps:=0;
repeat i:=i+1;
  steps:=steps+1;flag:=true;
  for j:=1 to i+1 div 2 do
  begin
    t:=not x[j];
    x[j]:=not x[i+1-j];
    x[i+1-j]:=t;
  end;
  for j:=1 to n do
    flag:=flag and x[j]
  if i=n then i:=0
until flag;
writeln('n=',n,
  ' f(',n,')=',steps)
end.
```

```
program coin1_stack;
label 0;
var i,j,n,steps: integer;
    x:array[1..1000]of boolean;
begin
write('n=');  readln(n);
for i:=1 to n do x[i]:=true;
i:=0; steps:=0;
0:i:=i+1; steps:=steps+1;
  for j:=1 to i+1 div 2 do
  begin
    t:=not x[j];
    x[j]:=not x[i+1-j];
    x[i+1-j]:=t;
  end;
  for j:=1 to n do
    if not x[j] then
    begin
      if i=n then i:=0;
      goto 0
    end;
  writeln('n=',n,
    ' f(',n,')=',steps)
end.
```

Additional Exercises for Sections 1-65

1. Let the number of people who have heard the joke be inf and the number of people who have told it to someone who heard it em. (inf stands for infected, to indicate similarity to models of epidemics, and em stands for embarrassed, for having told a familiar joke.) The next person to be told could be one who heard it already, other than the person telling it, or someone new. We model this by taking an integer at random from $2 \ldots n$. If the number is $\leq inf$ we add 1 to em, otherwise we add 1 to inf. When $em = inf$ the process stops and we record how many people have not heard the joke. The experiment is repeated m times. 11 repetitions of the program Joke gave the median output 0.2035 and mean output 0.2032. That is, about 20% of the population does not hear the joke.

```
program Joke;
const n=1000; m=100;
var j,r,inf,em,count:integer; sum:longint;
begin sum:=0; randomize;
for j:=1 to m do begin
  inf:=1; em:=0;
  repeat
    r:=2+random(n-2); if r>inf then inf:=inf+1 else em:=em+1
  until em=inf;
  sum:=sum+n-inf
end;
write(sum/m/n:0:8);
readln end.
```

If n is large and we set $inf/n = x$ and $em/n = y$ then the two quantities grow approximately in accordance with the differential equation $dy/dx = x/(1-x)$, $y(0) = 0$, until y becomes equal to x, which happens when $2x = \ln(1-x)$. Then $1 - x$, the proportion of people who did not hear the joke, is $0.2031878699799799\ldots$

2. You will get all numbers of the form $2^m 3^n - 1$.

5. The following program checks Watson's claim with interpretation a):

```
program checkit;
var i,j,k:integer; count,max:longint;
    a,b,c, x,y,z,xx,yy,zz,xxx,yyy,zzz:real;
begin
randomize; write('no. of trials='); read(max); count:=0;
for k:=1 to max do
begin
  x:=random; y:=random; z:=random;      (* This is faster than *)
  xx:=random; yy:=random; zz:=random;      (* using arrays *)
  xxx:=random; yyy:=random; zzz:=random;
  a:=sqr(x-xxx)+sqr(y-yyy)+sqr(z-zzz);
  b:=sqr(xx-x)+sqr(yy-y)+sqr(zz-z);
  c:=sqr(xxx-xx)+sqr(yyy-yy)+sqr(zzz-zz);
  if a<b+c then if b<c+a then if c<a+b then count:=count+1
end;
writeln(chr(7),'70*freq. of ac. tri.=', 70*count/max:12:4);
readln; end.
```

We have $33/70 = 0.47142857\ldots$ On the other hand, five runs of the program **checkit** with max$=10000$ resulted in 0.4564, 0.4549, 0.4637, 0.4600, 0.4564, which is close to, but consistently below 0.47. So this does not seem to be the right interpretation. By trying interpretation b) we get 0.472. This seems to be the correct interpretation. One can get the exact value by algebraic computation. One can show that the required probability is the same as for the following simpler problem: Let S be a sphere with north pole N. Pick two points A, B inside S at random. Find the probability that $\triangle NAB$ is acute. If one takes the north pole as the origin of polar coordinates, evaluating the triple integrals representing the probability of getting an obtuse angle at the north pole or at another vertex becomes a laborious exercise in third-term calculus.

Restricting the random points to a cube or a sphere is arbitrary. The following argument seems to reduce the completely unrestricted problem to the simplified problem above.

Let A be that one of the three points chosen which is farthest from the origin O. Then all we know about B and C is that they randomly chosen points inside the sphere about O with radiusOA. Thus the probability that $\triangle ABC$ is acute is the same as the probability that a triangle formed by the North Pole of a sphere and two points chosen at random inside it is acute.

The weakness of this argument is that it tacitly uses the notion of selecting a point at random from an infinite space, without giving preference to any part of the space. Assuming that this can be done leads to contradictions. For example, take the two dimensional case of our problem. Let A be the first point we pick, B the second, C the third. Consider the probability P that the angle at C is $> 45°$. This will happen if C is in the union of the interiors of the two circles through A and B with radii $AB/\sqrt{2}$. We can place infinitely many disjoint copies of this region in the plane. The probability that C will be in any one of these ought to be P also. The sum of these probabilities must be < 1, hence P can not be positive, it must be 0. Thus the probability that the angle at C is $> 45°$ is 0. But there is nothing special about the last point C in our problem. The three points could be selected by three people simultaneously, and they could tell us in random order. So the probability that a random triangle has any angle $> 45°$ is 0, which is absurd.

6. The following program first computes the array $f[1..n]$. Then for input q, it prints out $f(q)$. To exit, enter 1.

```
program self_descr_seq;
var i,j,k,m,n,q:integer; f:array[1..32000] of integer;
begin write('n='); readln(n); for i:=1 to n do f[i]:=0;
  i:=0; j:=0; f[1]:=1; f[2]:=2;
  repeat i:=i+1; k:=f[i];
    for m:=j+1 to j+k do f[m]:=i;
    j:=j+k
  until m>n;
  repeat write('q='); readln(q); writeln('f(',q,')=', f[q])
  until q=1
end.
```

We get $c * 10000^d \approx 356$, $c * 5000^d \approx 232$. Hence $2^d \approx 356/232 = 89/58$, $d \approx \ln(89/58)/\ln(2) = 0.61775\ldots$, $c \approx 356/10000^d = 1.23035\ldots$. So $f(n) \sim 1.23035n^{0.61775}$. Check $f(n) = \lfloor c\, n^d + 0.5 \rfloor$ with the correct value of $f(n)$. Can you do better? In Turbo Pascal we compute powers using $n^d = \exp(d \ln(n))$.

8. a)
```
    program merge_sort_comp;
    var n:integer;
    function com(n:integer):integer;
    begin
      if n=1 then com:=0
      else com:=com(n div 2)+com((n+1) div 2)+n-1
    end;
    begin
      write('n='); readln(n);
      writeln('n=',n,'   com(',n,')=',com(n))
    end.
```

b) $\mathrm{com}(n) = n\lceil \log_2 n \rceil - 2^{\lceil \log_2 n \rceil} + 1$

11.
```
program ThreePlayers;
var a,b,c,i,m,r,x,y,z:integer; j:longint;
begin write('a,b,c='); readln(a,b,c); m:=10000; j:=0;
  for i:=1 to m do
  begin x:=a; y:=b; z:=c;
    repeat r:=random(3)+1; x:=x-1; y:=y-1; z:=z-1; j:=j+1;
      if r=1 then x:=x+3 else if r=2 then y:=y+3 else z:=z+3
    until (x=0) or (y=0) or (z=0)
  end;
  writeln(j/m:0:4);
end.
```
Simulation did not help me with four players.

12.
```
program intertwined_functions;
var i,k,n:integer; t:real;
function M(n:integer):integer;forward;
function F(n:integer):integer;
  begin if n=0 then F:=1 else F:=n-M(F(n-1)) end;
function M;
  begin if n=0 then M:=0 else M:=n-F(M(n-1)) end;
begin
  write('k,n='); readln(k,n); t:=(sqrt(5)-1)/2;
  for i:=k to n do
    writeln(i:5, F(i):5, round(i*t):5, trunc((i+1)*t):5)
end.
```

If you plot the functions $F(n)$ and $M(n)$ you will observe that both are strongly clustering about the same straight line through the origin: $F(n) \sim tn$, $M(n) \sim tn$. For t we get $tn = n - t^2(n-1)$, or, for large n, $t^2 + t = 1$ with the solution $t = (\sqrt{5} - 1)/2$. The program finds $F(i), M(i), \lfloor it + 0.5 \rfloor, \lfloor (i+1)t \rfloor$ from k to n. The agreement is exact with a few exceptions, when the difference between any pair is at most 1.

14. a) $g(a, b, c) = ab + bc + ca$;

   b) $g(a, b, c, d) = ab + ac + ad + bc + bd + cd$;

   c) $g(x_1, \ldots, x_n) = \sum_{i < k} x_i x_k$

20. The point (x, y) can be reached from $(1,1)$ if and only if $\gcd(x, y) = 2^n$, $n = 0, 1, 2, \ldots,$.

26.
```
program twinperm;
var d,i,j,k,m,n,copy,count:integer;
     x:array[1..200] of integer;
begin randomize; write('n='); readln(n); m:=n+n; count:=0;
for j:=1 to 10000 do
begin
   for i:=1 to m do x[i]:=i;
   for i:=m downto 2 do {generates random permutation}
   begin
     k:=1+random(i); copy:=x[i]; x[i]:=x[k]; x[k]:=copy
   end; {end of random permutation}
   i:=0;
   repeat i:=i+1; d:=abs(x[i]-x[i+1]) until (d=n) or (i>=m-1);
   if d=n then count:=count+1
end;
writeln(count/10000:0:4); readln
end.
```

26. b) $p[10000] = 0.36787024156 \approx e^{-1} = 0.36787944117\ldots$. This is also the probability of getting a random permutation of $\{1, \ldots, n\}$ without a fixed point. See Ex. 3 of this section. But the convergence to the asymptotic value e^{-1} is slower.

27. For $x = 26861$ we have $\pi_1(x) = 1473$, $\pi_3(x) = 1472$

28. c) (1) is exact for $n \geq 1 + s(b-1)$, where $2s$ is the word length and b is the base.

32. The program **pentagon** picks the negative terms to be inverted in random order.

```
program pentagon;
label  0, 1, 2;
var  n,count,i,j,p:integer; x:array[0..4] of integer;
     flag: boolean; ans:char;
begin
  write('n= '); readln(n);
2:for i:=0 to 4 do x[i]:=n; x[2]:=1-4*n;
  writeln; for i:=0 to 4 do write(x[i]:4); writeln;
  flag:=true; count:=0; j:=0;
```

```
0:flag:=true; for i:=0 to 4 do if x[i]<0 then flag:=false;
  if  flag then goto 1;
  if  x[j]<0 then
  begin count:=count+1; p:=x[j];
    if  j=0 then x[4]:=x[4]+p else x[j-1]:=x[j-1]+p;
    if  j=4 then x[0]:=x[0]+p else x[j+1]:=x[j+1]+p;
    x[j]:=-p;
    for i:=0 to 4 do write(x[i]:4); write('    ');
    if (count mod 3)=0 then writeln;
  end;
  j:=(j+1+random(4)) mod 5; goto 0;
1:writeln;writeln('count= ',count);
  write('Another run? ( y or n )');
  readln(ans); IF ans='y' then goto 2;
end.
```

We find that the algorithm always stops and the number of steps is $20n - 10$ no matter how we pick the next operation, although the sets of numbers which come up do depend on that.

The key to the proof is to find an integer-valued, non-negative function $f(x_1, x_2, x_3, x_4, x_5)$ of the vertex labels whose value decreases when the given operation is performed. All but one of the students who solved the problem found the same function,

$$f(x_1, x_2, x_3, x_4, x_5) = \sum_{i=1}^{5}(x_i - x_{i+2})^2, \quad \text{where} \quad x_6 = x_1, \ x_7 = x_2.$$

If, say, x_4 is flipped, $f_{new} - f_{old} = 2sx_4 < 0$. A decreasing sequence of nonnegative integers must be finite, so the algorithm must stop.

Bernhard Chazelle (Princeton) gave an argument which shows why the number of steps is determined by the initial values. He considers the infinite multiset S of all sums $s(i,j) = x_i + \cdots + x_j$ with $1 \le i \le 5$ and $i \le j$, where $x_6 = x_1$ etc. A multiset is a set which can have equal elements. In this multiset one element changes from negative to positive and all the others are either unchanged or switched with others when we perform an operation. For instance, if x_4 is changed from negative to positive, and the new values are denoted by primes, $s'(4,4) = -s(4,4)$ changes from negative to positive whereas $s'(4,5) = s(5,5)$, $s'(5,5) = s(4,5)$, $s'(4,6) = s(5,6)$, $s'(5,6) = s(4,6)$ etc. There are only finitely many negative elements in S, since $s > 0$. The number of steps until stop is the number of negative elements of S. The x_i need not be integers for this argument to be valid.

Summary of Notation and Formulas

$a \mid b$: a divides b, i.e., b is a multiple of a for integers a, b.

$a \mathrel{..} b$: the set $\{a, a+1, \ldots, b-1, b\}$ of integers.

$a[1..n]$: the array $a[1]$ to $a[n]$, or the vector with components $a[1]$ to $a[n]$.

$a[1..n, 1..m]$: the two dimensional array $a[i, k]$ for $1 \le i \le n$, $1 \le k \le m$, that is, an $n \times m$ matrix.

$\lfloor x \rfloor$: floor of x, i.e. the largest integer $\le x$. In number theory usually denoted by $[x]$.

$\lceil x \rceil$: ceiling of x, i.e. the smallest integer $\ge x$.

$\gcd(a, b)$: greatest common divisor of the integers a, b. In number theory usually denoted (a, b).

$\operatorname{lcm}(a, b)$: least common multiple of the integers a, b.

$p \Rightarrow q$: The statement p implies the statement q.

$p \Leftrightarrow q$: p is true if and only if q is true.

$\mapsto$: mapping or function symbol. $a \mapsto b$: a is mapped into b.

$a \leftarrow b$: the same as $a:=b$ in Pascal. a is replaced by b. $\leftarrow$ is the assignment operator.

$a \leftarrow b \leftarrow c \leftarrow 1$: a, b, c are replaced by 1.

while P **do** Q; R; S **od**: while the condition P is true, repeat the operations Q, R, S. We sometimes use do...od in informal descriptions of algorithms, instead of Pascal's do begin...end.

$(a_1, \ldots, a_n)$: an n-tuple, a sequence, a vector with components a_i. The same as $a[1..n]$.

$f(n) \sim g(n)$: $f(n)$ is asymptotically equal to $g(n)$, i.e. $f(n)/g(n) \to 1$ for $n \to \infty$.

$f(n) = O(g(n))$: There is a constant C such that $|f(n)| \le C\, g(n)$.

Harmonic series: $1 + \frac{1}{2} + \frac{1}{3} + \cdots$. The n-th harmonic number $H_n = 1 + \frac{1}{2} + \cdots \frac{1}{n}$ is the n-th partial sum of the series.

$\phi(n) = n \prod_p (1 - 1/p)$, where the product ranges over all distinct prime factors of n. This is the number of integers in $1..n$ which are relatively prime to n. $\phi(n)$ is called Euler's phi function.

$\sum_{n \ge 1} 1/n^2 = \pi^2/6$, a classic result of Euler.

$\binom{n}{s}$: the number of s-subsets of an n-set. $\binom{n}{s} = \frac{n}{s}\binom{n-1}{s-1} = \frac{n(n-1)\ldots(n-s+1)}{s(s-1)\ldots 1} = \frac{n!}{s!(n-s)!}$, is called n choose s, or the binomial coefficient n over s. Here $n!$, pronounced n factorial, is the product $1 \cdot 2 \cdot 3 \cdots n$. By definition $0! = 1$.

$\gg$: much larger than. The precise meaning of this phrase should be clear from the context.

$\approx$: approximately equal. The precise meaning of this should be clear from the context.

A Short Summary of Turbo Pascal

General Remarks

The author used the compiler Turbo Pascal for IBM-compatibles and we think that is what most of our readers will have. In the next section we discuss the slight changes you may have to make in some programs if you are using another Pascal compiler.

The first draft of this book was written when Version 3 of Turbo Pascal was current. Version 4 brought several improvements, such as the integer type longint. Changes in Turbo Pascal for IBM-compatibles since version 4 (version 6.0 is current as of this writing) have been directed mainly to help people write applications with graphical interfaces resembling the Macintosh. The limitations of the compiler from the point of view of a mathematical user, e.g. that no more than 64K of memory can be used for all variables combined, irrespective of how much memory the computer has, remain uncorrected.

To help make Pascal programs portable, the American National Standards Institute and the Institute of Electrical and Electronics Engineers issued a definition of the Pascal computer language, standard 770X3.97-1983. In response to a need for additional features, they later issued a standard for Extended Pascal, 770X3.160-1989. Turbo Pascal does not fully conform to the 1983 standard. It also has features not prescribed by that standard which are convenient but make the programs less portable. Pascal compilers found on larger computers in colleges are likely to conform at least to the 1983 standard in full.

Our book explains as much of Pascal as we need and this summary should be of added help as a reference. For a complete introduction to Pascal we recommend Programming with Pascal, by Byron S. Gottfried. (Schaum's Outline Series, McGraw-Hill Publishing Co.)

One can produce graphics with microcomputer Pascal compilers but the instructions are not the same on different compilers. The standards do not cover graphics. Of the programs on our disk, Posa and Paths produce graphics. They should run on IBM-compatibles with any of the graphic cards, and we give the modifications needed to make them run with TPMac and Think Pascal (for the Macintosh).

Compiler directives

We need to mention two things about Turbo Pascal and numeric coprocessor chips. If you write the compiler directive {$N+} under the line with the program name, Turbo Pascal will write code to utilize the numeric coprocessor if your computer has one, or the coprocessor emulation software included with Turbo Pascal if the computer has no coprocessor. This makes available the real type *extended,* which has a 63-bit mantissa. If a program is compiled in the $N+$ state, the values of the predefined *real*-valued functions and π are available in *extended* precision.

When the coprocessor is used, Turbo Pascal tries to use the stack in the coprocessor for recursive programs, which is too small for that purpose. Because of this, compile the recursive programs in the default $N-$ state, in which the coprocessor is ignored.

Two other compiler directives should be mentioned here. The range check directive, {$R+} makes TP send an error message if any array index gets out of range, which is a great help in debugging. The *include* directive, {$I name of a file}, which can be anywhere except between a **begin** and an **end,** causes the compiler to include the file in the compilation. This avoids the need to paste in definitions of functions which are used in several programs.

Pascal punctuation marks and symbols

Pascal compilers treat capital letters as interchangeable with their lowercase equivalents. They ignore spaces and line breaks, except those occurring in character strings.

;   separates two statements, end of declaration, end of function body.

,   separates list items.

:   separates variable:type, function:type, label:type,

     expression:field length, field length:# of decimal places.

..   first element..last element.

=   used after the identifier of a newly declared constant or type. (Also used as an operator, see below.)

.   end of program.

Brackets (pairs of symbols which pull together as a whole what they enclose)

( )   Used in algebraic expressions and to enclose parameter lists.

[ ] or (. .)   array bound or index brackets

' '   character string delimiters. write(a) writes the value of the variable a, while write($'a'$) writes the letter a.

begin...end are brackets used after **then, else** and **do,** which would apply only to the immediately following statement without these.

     The statement part (executable part) of a program or a function or procedure definition must start with **begin** and stop with **end.**

{ } or (* *) comment delimiters. We may insert explanations of any length between these. The Pascal compiler will skip them. Exception: compiler directives, which start with {$.

Numerical constants and functions

constants **pi** $= \pi$ **maxint** $= 2^{15} - 1 = 32\,767$, **maxlongint** $= 2^{31} - 1 = 2\,147\,483\,647$, **true, false**.

abs(x)   absolute value of x. It is of the same type (*real, integer, longint* etc.) as x.

arctan(x)   The angle in radians in $(-\pi/2, \pi/2)$ with tangent x.

chr(x)   the character whose ASCII number is x, assuming the compiler was intended for the English language.

cos(x)   cosine of x radians.

exp(x) e^x with $e \approx 2.7182818284$.

frac(x)   fractional part of x. For both positive and negative numbers it shaves off the part before the decimal point. This function is not required by the Pascal standard.

int(x)   shaves off the part behind the decimal point. The result is of type *real* if the program is compiled in {$N-} mode and *extended* if compiled in {$N+} mode. int($x$) is not a standard Pascal function.

ln(x)   natural logarithm of x; x must be > 0.

odd(x)   has the value *true* if the integer or longint x is odd, *false* if it is even.

ord(x)   can be applied to variables of *ordinal type*, i.e. variables whose possible values can be numbered with integers in a natural way. ord(x) returns the number of x. If x is a character, ord(x) is its ASCII number. The boolean values *false, true* have the ordinals 0, 1.

pred(x)   is $x - 1$ for an integer x; it can be applied to any "ordinal" variable type, e.g. **pred**('c')=b.

random generates a random number of type *real*, uniformly distributed in [0,1).

random(n) picks an integer at random from the set $0 .. n - 1$.

randomize initializes the RNG with a number from the system clock. Without **randomize** Turbo 4 and later always give the same sequence of random numbers.

RNG random number generator.

round(x) is a function with value of type *longint*. It is the integer nearest to x; if x is halfway between two integers it is the one farther from 0.

sin(x) The sine of an angle of x radians.

sqr(x) square of x. Used instead of $x * x$ if x is complicated. Has same data type as x.

sqrt(x) square root of x. Has data type *real* even if x is the square of an integer.

succ(x) successor of x if the type of x is an ordinal type. For a variable x of integer type, **succ**(x) $= x + 1$. Applied to a character, **succ** gives the next character: **succ**('x') = y.

trunc(x) shaves off the part of x following the decimal point. The result has type *longint*.

Arithmetic operators (functions of 2 variables)

+, -, *, /   Note that sums and products of integers are computed mod 2^{16} (longints, mod 2^{32}) with no message of overflow.

div a div b is a/b with the digits after the decimal point omitted; a, b must be integer types (integer, longint, byte or shortint).

mod a mod b is the remainder when $|a|$ is divided by $|b|$. The sign is that of a/b. (In standard Pascal b must be positive and $a \bmod b$ is that value of $a - kb$, k integer, which is in $[0 .. b - 1]$.)

shl, shr a shl b, where a and b must be integers, shifts the binary representation of a to the left by b places. Spaces left empty are filled with 0's. shr shifts to the right. shl, shr are not in standard Pascal.

and, or, not, xor   (the last one, exclusive or, is not in standard Pascal). Turbo Pascal allows integer type arguments in these operators. They are performed pairwise on corresponding binary digits and yield an integer type result. 0 stands for false and 1 for true. In standard Pascal logical operators act only on boolean arguments.

Logical and relational operators

and, or, not, xor   (the last one, exclusive or, is not in standard Pascal). Use plenty of parentheses in expressions involving both arithmetic and logical operators. Since Turbo Pascal logical operators accept integer variables, unexpected parsings or failures to parse can occur.

=, <, >, <> ($\neq$), <= ($\leq$), >= ($\geq$) are also operators in Pascal; they return boolean values. They can be applied to variables of any ordinal type. Thus $'a'<'b'$ is true; if a, b are boolean variables, a<>b has the truth table of the *exclusive or*, and a<=b has the truth table of the conditional, a$\Rightarrow$b.

Some predefined simple variable types

boolean (a variable with values true or false), char (the values are characters), integer, real are standard Pascal types. Turbo Pascal has additional number types, byte (range of values $0..255$), longint, extended (a real type discussed above) and others, which we do not use.

Read and write procedures

read($x, y, \ldots$) reads x, $y, \ldots$ which can be typed in separated by spaces or CR's.

readln($x, y, \ldots$) (readline) is, for keyboard inputs, similar to read except that read does not swallow the last CR. (A readln without an argument makes the computer wait for a CR; readln(a,b,c) is equivalent to read(a,b,c); readln).

write($x, y, z, \ldots$) and writeln($x, y, z, \ldots$) (writeline) write on the screen. writeln(x) adds a CR at the end. writeln without an argument sends a CR to the screen. write(Lst,($x, y, \ldots$)) writes on the printer instead of the screen. It may be necessary to put uses Printer; under the program name line to make write(Lst,$\ldots$) work.

Flow control

```
for...to...do...
for...downto...do...
while...do
repeat...until...
if...then
if...then...else
case ... of ...:
goto....
```

Reserved and predefined words

The following words are predefined in Turbo Pascal or in Extended Pascal. Redefining them will not always cause an error but it is better to use other words. You can also use the list as a starting point to find out about additional features of Turbo Pascal or Extended Pascal.

abs, absolute, and, arctan, arg, assembler, asm, array, begin, bind, binding, BindingType, Boolean, bound, capacity, card, case, char, chr, cmplx, complex, const, constructor, cos, date, DateValid, day, destructor, dispose, div, do, downto, empty, else, end, eof, eoln, epsreal. EQ, exp, extend, external, false, file, far, for, forward, function, GE, get, GetTimeStamp, goto, GT, if, im, implementation, in, index, inline, input, integer, interface, interrupt, label, LastPosition, LE, length, ln, LT, maxchar, maxint, maxreal, minreal, minute, mod, month, name, NE, near, new, nil, not, object, odd, of, or, ord, output, pack, packed, page, polar, pred, private, procedure, program, put, re, read, readln, readstr, real, record, repeat, reset, rewrite, round, second, SeekRead, SeekUpDate, SeekWrite, set, shl, shr, sin, sqr, sqrt, StandardInput, StandardOutput, string, substr, succ, text, then, time, TimeStamp, TimeValid, to, trim, true, trunc, type, unbind, unit, unpack, until, update, uses, var, while, virtual, with, write, writeln, writestr, xor, year.

For Users of Other Pascal Compilers

There are good Pascal compilers for probably every type of computer introduced since the Commodore 64 and the Apple II. If you are using a Pascal compiler other than Turbo Pascal v. 4 or later, some of our programs may not compile or run properly. Here are suggestions of what may be causing the difficulty and how to remedy it.

The result of the computation flashes onto the screen but immediately disappears. Put a **readln** just before the final **end.** This should keep the output on the screen until you press Carriage Return. (If one **readln** does not work, two may.)

The program gets into an endless loop or the last few digits of your result differ from the ones in the book. The cause of either of these occurrences could be that numbers of type **real** are represented in your implementation of Pascal by fewer bits than in Turbo.

In Turbo Pascal v. 6, a variable of type **real** has a 39-bit mantissa. You can use program **mantissa** to find out how many bits your compiler allocates to the mantissa of a variable of type **real**.

```
program mantissa;
var x,y:real; n:integer;
begin n:=0; x:=1;
repeat
  x:=2*x; n:=n+1; y:=x+1;
until x=y;
write(n,' bit mantissa.');
readln end.
```

If we try to solve an equation with an accuracy which we can not attain because of roundoff errors, we get into an endless loop. For instance, the program **bis** in Section 3 throws Turbo Pascal for the Macintosh into an endless loop because variables of type **real** have 24-bit mantissas in TPMac. The remedy is to replace the condition **f(a)-f(b) < 1E-09** with a less stringent one or to use variables of type **extended** instead of **real**.

Your compiler does not recognize the variable type **byte.** Replace **byte** by [0..255].

Your compiler objects to applying Boolean operators to integers. It may have operators with other names which produce the same results with integers as the logical operators do in Turbo Pascal. If this is not the case, but the integers to which the program applies the logical operations are restricted to 0 and 1, use mod 2 arithmetic. "False" is numbered 0, and the arithmetic equivalent of, say, a **or** b applied to 1-bit integers is $(a + b - ab)$ mod 2.

Your compiler does not recognize **xor.** This is the exclusive **or,** i.e. a **xor** b means one of a, b is true and one is false. If a, b are 1-bit integers, a **xor** b is $(a + b)$ mod 2.

The Turbo Pascal functions int(x) *and* frac(x) *are not predefined in your compiler.* These are not standard Pascal functions. You can paste or *include* the definitions given here before the other function definitions in the program.

```
function frac(x:real):real;        function int(x:real):real;
var m,y:real;                      begin
begin m:=maxint+1;                     int:=x-frac(x)
   y:= x-m*trunc(x/m);             end;
   frac:=y-trunc(y)
end;
```

Your Pascal implementation does not have the integer type **longint**. *The* largest value an integer variable can have is usually $2^{15} - 1$. You will have to use variables of type **real** to represent larger integers. The range of integers which can be represented exactly as numbers of type **real** can be determined by means of the program **mantissa** given above. The functions **divr**, **modr** can be used to replace **div** and **mod** when we use variables of type **real**.

```
function divr(a,b:real):real;      function modr(a,m:real):real;
begin                              begin
   divr:=int(a/b)                      modr:=a-m*int(a/m)
end;                               end;
```

Note that **modr** computes the Turbo Pascal **mod** function which gives values in the range $[-m+1 .. 0]$ if a is negative and m is positive. In standard Pascal the range is always $[0 .. m - 1]$ and m must be positive.

Your compiler does not recognize the shift operators a **shl** *b or a* **shr** *b.* These operators have integer type arguments. The first is equivalent to b multiplications by 2 and the second to b operations a **div** 2.

Printing results using **write(Lst,...)**, as explained at the end of Section 2, is not part of standard Pascal. The ways to send results or programs to printer or disk are different for different compilers.

Arrays of constants are not available in standard Pascal. This Turbo feature is used for putting a list of data into a program, as in **"rematch"** in the section on matched pairs in Chapter IV. Standard Pascal has no convenient way of incorporating a list of data into a program.

In case the implementation you are using does not have constant arrays, you can get data into programs the way we shall do it now for **"rematch"** (Fig. 28.1).

Define **d** as an array of variables instead of constants and declare an additional integer variable **i**. Insert the following list of assignment statements for the array just after the beginning of the statement-part (where the program execution starts).

```
i:=1;     d[i]:=6;
i:=i+1;   d[i]:=8;
i:=i+1;   d[i]:=14;
   :         :
i:=i+1;   d[i]:=75;
```

The characters in the text of the assignment statements between consecutive numbers of the number list are always the same. Therefore we can generate the assignment statements as follows: Take the list of 15 integers with commas separating them, as in "rematch" in Sect. 28 and on the disk, and put it in your program as a comment line, i.e., enclosed by $(* ... *)$. Use the "find and replace" command of a text editor to replace the commas between the entries by the sequence of symbols   ;CRi:=i+1;d[i]:=   Here CR stand for carriage return. Then enter the characters before 6 in the first line, and the

";"at the end of the last line. If your editor has no provision for inserting a CR into the replacement string, just insert the rest of the symbols and then break the lines into pieces which are not too long for your compiler.

A random number generator is not part of Standard Pascal, since in some applications speed is the main consideration while quality is essential in others. In case your Pascal compiler lacks a random number generator, you can program the one predefined in Turbo Pascal v. 4, 5 and 6 (see Section 39) or use a similar one, given below, which has one of the sets of constants recommended in Numerical Recipes by William H. Press et al., Cambridge U. Press, 1st ed., p. 198. The variable **myseed** has to be declared in the variable list of the main program since we need to preserve it for the next call to **random**. It should be a **longint** if that is available (see below); if not, it should be **real**.

You have to assign an initial value to **myseed** in the main program. Turbo Pascal's **randomize** does that with an input from the system clock. You can write your own procedure **randomize**. If you can't find out how to access the system clock, make **randomize** ask for an input from the keyboard.

```
function random: real;
const m=139968.0; a=205.0; c=29573.0;
begin
  myseed:=myseed*a+c; myseed:=myseed-myseed*trunc(myseed/m);
    random:= myseed/m;
end;

function random(n:integer):integer;
const m=139968.0; a=205.0; c=29573.0;
begin
  myseed:=myseed*a+c; myseed:=myseed-m*trunc(myseed/m);
  random:=trunc(n*myseed/m);
end;
```

If the type **longint** is available, declare **myseed** to be **longint**, write the constants without decimal points and replace the line after **begin** in both definitions by

$$\text{myseed} := (\text{myseed} * a + c) \ \textbf{mod} \ m;$$

Standard Pascal and even Turbo Pascal for the Macintosh regard IBM Turbo's function names ''**random**'' and ''**random(n)**'' as duplicate uses of the same identifier and reject attempts to introduce both in the same program. None of our programs uses both functions.

References

Abbreviations:
AMM: American Mathematical Monthly;
MM: Mathematics Magazine.

Chapter 1

Two excellent sources on bold gamble, suitable for novices, are P. Billingsley: The Singular Function of Bold Play, Am. Scientist 71(1983), 392-397 and Rényi [1985].
The most extensive account on the Josephus Problem is given by Ahrens [1901]. The program josephl is from Knuth [1968], vol.1, 2nd ed., problem 1.3.3-33. The surprising depth of the Josephus permutation is treated in Herstein and Kaplansky [1974].
The algorithm in Exercise 7 is from Domoryad [1964].

Chapter 2

Section 6. Most of the section is classical and goes back at least to the last century. 6.6 can be found in Dijkstra [1976]. See also his 1985 article in the Mathematical Intelligencer 8.1 , and Gries [1981], 19.2, Exercise 1. Exercise 18 is treated in D. D. Wall, Fibonacci series modulo m, AMM 67 (1960), 525-532. Exercises 24-25 are inspired by Jacobs [1969].

Sections 7-9. Formula (8) in Exercise 5 is proved in Hilbert/Cohn-Vossen [1952] and Shanks [1962]. References for Exercise 10 are C.T. Rajagopal and M.S. Rangachari, On an Untapped Source of Medieval Keralese Mathematics, Archive for History of Exact Sciences 18 (1978), 81-101 and On Medieval Kerala Mathematics, 35(1986), 91-99.

Section 10. The sequence U in Exercise 2 is due to S. Ulam. See Guy [1981], C4. The sophisticated algorithm for U is due to M.C. Wunderlich, The Use of Bit and Byte Manipulation in Computing Summation Sequences, BIT, 11 (1971), 217-224. The sequence in Exercise 3 is due to L. Moser, MM 35 (1962), 37-38.

Exercise 6: The most extensive account on the Frobenius Problem is by E.S. Selmer, On the linear diophantine problem of Frobenius. J. reine angew. Math. 293/294(1977), 1-17. The case $k = 3$ is (sort of) solved by E.S. Selmer and Ö. Beyer in the same journal, v. 301, 161-170. It is followed by a simplification due to Ö. J. Rödseth, v. 301, 171-178. See also Harold Greenberg, Journal of Algorithms, 1988, 343–353.

Exercise 8: The $3n + 1$ problem is treated thoroughly in J.C. Lagarias AMM 92(1985), 3-23.

Exercise 10: 0, 1, 144 are the only Fibonacci squares. The proof can be found in J.H.E. Cohn, Square Fibonacci Numbers, Fibonacci Quart. 2 (1964), 109-113. Exercise 11 is due to J.H. Conway. See Guy [1981], E17.

Section 11-16. Switching adjacent blocks by means of reversals is part of the folklore. See Gries [1981] and Bentley [1986], Column 2. An efficient algorithm for switching nonadjacent blocks can be found in J.L Mohammed and C.S. Subi, An Improved Block-Interchange Algorithm, Journal of Algorithms 8 (1987), 113-121.

The sequence in section 16 goes back to the last century. See G. de Rham, Un peu de mathématique à propos d'une courbe plane. Elemente d. Math. v. II, No 4, 5 (1947), 73-104.

Sections 17-18. 17.2 is treated in M.L. Stein/P.R. Stein, New Experimental Results on the Goldbach Conjecture, MM 38(March 1965), 72-80, and F.Mosteller, A data-analytic look at Goldbach counts, Statistica Neerlandica, 26(1972), No. 3, 227-242.

For 17.3 see also F. Mosteller, An Empirical Study of the Distribution of Primes and Litters of Primes, in T.A. Bancroft, ed.: Statistical Papers in Honour of C.W. Snedecor (1972), 245-257.

Additional Exercises for Sections 1-18: Exercises 1 and 2 can be found in Honsberger [1970], Essay 12.

Exercise 10: Gauss has shown that any positive integer is the sum of 3 triangular numbers.

Exercises 13 to 15 are from Hofstadter [1979].

Chapter 3

Section 23. All these algorithms are inefficient for large n and $s \ll n$. E.g., from a telephone directory of New York with n entries s entries are to be chosen at random for a phone interview. Problems of this kind are treated efficiently by D.E. Knuth in Communications of the Association for Computing Machines, May 1986. He develops an algorithm with space and time complexity $O(s)$.

Section 25. See Golomb [1967]. Knuth's generator is from Knuth [1981], v.2, 2nd ed.

Section 26. See S. Wolfram, Random Sequence Generation, Advances in Appl. Math. 7(1986), 123-169.

Section 27. The Period Finding Problem is treated very efficiently in many places, but in ways not suitable for beginners. Our treatment is from Knuth [1981], v.2, 2nd ed., Exercise 3.1.-6.

Chapter 4

Section 37. The probability that nobody remains lonesome is from Engel [1987], page 239.

Section 38. The expected length of the longest success run was found by A. Rényi. See Mark F. Schilling, The Longest Run of Heads, The College Mathematics Journal, May 1990.

Chapter 5

Section 42. Sorting is treated extensively elsewhere, especially in Knuth [1973], v. 3. So we just give examples of a few important methods.

Section 43. R.W. Floyd and R. Rivest present (see Communications of the Association for Computing Machines, March 1975) a selection algorithm that uses $n + k + O(\sqrt{n})$ comparisons. Ours uses about $2n$ comparisons and in the worst case (computation of the median) between $3n$ and $4n$ comparisons. See Bentley [1988], column 15.

Section 46. These and other kinds of games are treated in R.A. Epstein [1967].

Section 49. The version of the subset sum problem $\sqrt{1}, \ldots, \sqrt{n}$ is due to R.W. Floyd. See Aho, Hopcroft and Ullman [1983], Exercise 1.4. The program gray_stack is from Reingold, Nievergelt and Deo [1977], p. 178. A popular account on Gray Codes is M. Gardner, Gray Codes, Scientific American v. 227, August 1972.

Chapter 7

Section 62. See C.D. Olds, Continued Fractions, [1963].

Section 64. See D.E. Knuth, Mathematics and Computer Science: Coping with The Finite, Science, vol. 194, 1235-1242, 17 Dec. 1976.

Section 65. I was able to guess the general formula for problem 11 in less than ten minutes and then verify it. In spite of many hours of hard work I was not able to guess the formula for the four players game. If we could guess it, we could verify it, maybe by using Derive. It is easy to write down a functional equation for the formula.

Bibliography

1. A. V. Aho, J. E. Hopcroft and J. D. Ullman, Data Structures and Algorithms, AW 1983.
2. W. Ahrens, Mathematische Unterhaltungen und Spiele, Teubner 1901.
3. J. Bentley, Programming Pearls, AW 1986.
4. J. Bentley, More Programming Pearls, AW 1988.
5. E. W. Dijkstra, A Discipline of Programming, Prentice-Hall 1976.
6. A. P. Domoryad, Mathematical Games and Pastimes, Macmillan 1964.
7. A. Engel, Stochastik, Ernst Klett Verlag, Stuttgart 1987.
8. R. A. Epstein, Theory of Gambling and Statistical Logic, AP 1967.
9. S. W. Golomb, Shift Register Sequences, Holden-Day 1967.
10. B. S. Gottfried, Programming with Pascal, MacGraw Hill 1985. Schaum Outline.
11. R. L. Graham, D. E. Knuth and O. Patashnik, Concrete Mathematics, AW 1988.
12. D. Gries, The Science of Programming, Springer-Verlag 1981.
13. R. K. Guy, Unsolved Problems in Number Theory, Springer-Verlag 1981.
14. I. N. Herstein and I. Kaplansky, Matters Mathematical, Harper & Row 1974.
15. D. Hilbert and S. Cohn-Vossen, Geometry and the Imagination, Chelsea 1952.
16. D. Hofstadter, Gödel, Escher, Bach, Basic Books 1979.
17. R. Honsberger, Ingenuity in Mathematics, MAA, NML 23, 1970.
18. K. Jacobs, Selecta Mathematica I, Springer-Verlag 1969.
19. D. A. Klarner (ed.), The Mathematical Gardner, Wordsworth International 1981.
20. D. E.Knuth, The Art of Computer Programming, vol. 1, 2nd ed. 1973, vol. 2, 2nd ed. 1981, vol. 3 1973, AW.
21. C. D. Olds, Continued Fractions, MAA, NML 9, 1963.
22. E. M. Reingold, J. Nievergelt and N. Deo, Combinatorial Algorithms PH 1977.
23. A. Rényi: A Diary on Information Theory, Wiley 1985.
24. L. A. Santalo, Integral Geometry and Geometric Probability, AW 1976.
25. D. Shanks, Solved and Unsolved Problems in Number Theory, Spartan Books 1962. 2nd (1978) and 3rd (1985) editions by Chelsea.
26. W. Sierpinski, Elementary Theory of Numbers, Hafner 1964.
27. H. Solomon, Geometric Probability, CBMS.
28. N. N. Vorob'ev, The Fibonacci Numbers, New Classics Library 1983.

Index

Look first in Contents

Look first in Contents

Look first in Contents

6